The Role of Vascular Dysfunction in Neuronal Degeneration and Cognitive Impairment

The Role of Vascular Dysfunction in Neuronal Degeneration and Cognitive Impairment

Guest Editor

David Lominadze

MDPI

Basel • Beijing • Wuhan • Barcelona • Belgrade • Novi Sad • Cluj • Manchester

Guest Editor
David Lominadze
Department of Surgery
University of South Florida
Tampa
USA

Editorial Office
MDPI AG
Grosspeteranlage 5
4052 Basel, Switzerland

This is a reprint of the Special Issue, published open access by the journal *Biomolecules* (ISSN 2218-273X), freely accessible at: https://www.mdpi.com/journal/biomolecules/special_issues/Vascular_Neuronal.

For citation purposes, cite each article independently as indicated on the article page online and as indicated below:

Lastname, A.A.; Lastname, B.B. Article Title. *Journal Name* **Year**, *Volume Number*, Page Range.

ISBN 978-3-7258-6105-7 (Hbk)
ISBN 978-3-7258-6106-4 (PDF)
https://doi.org/10.3390/books978-3-7258-6106-4

Contents

About the Editor

David Lominadze

David Lominadze, Department of Surgery, University of South Florida (USF), Tampa, Florida, USA. The main research interests include microrheology, microcirculation, and cardiovascular and neurovascular pathologies. His work addressed the effects of blood rheological properties on blood flow, role of copper deficiency on platelet thrombogenesis and microcirculation, effects of erythrocyte aggregability during hypertension, and the role of hyperfibrinogenemia during traumatic brain injury (TBI).

After graduating from the Tbilisi State University (TSU), Tbilisi, Georgia, Dr. Lominadze worked in the Laboratory of Physiology and Pathology of Cerebral Circulation, Iv. Beritashvili Institute of Physiology (IvBIP), Georgian Academy of Sciences (GAS), Tbilisi, Georgia. In 1990, he received a Ph.D. degree in Human and Animal Physiology and became the Head of the Microrheology Section, Microcirculation Research Center (MRC), IvBIP, GAS. From February 1992, he was the Assistant Head of the MRC. In September 1992, Dr. Lominadze joined the Center for Applied Microcirculatory Research (CAMR) and the Department of Physiology and Biophysics (DPB), University of Louisville (UofL), Louisville, KY, USA, as an Adjunct Assistant Professor. He also worked as a Leading Research Scientist at TSU (Jan.-Sept. 1999). He emigrated to the USA in 1999 and worked as an Assistant Research Scientist (till June 2000) in the CAMR, and then in the DPB, UofL. In 2002, he became an Assistant Professor in the DPB. Since 2002, his work involving the effects of microcirculation during various inflammatory diseases has been continuously funded first by the AHA and then by the NIH. Since 2010, Dr. Lominadze has been working predominantly on testing mechanisms involving cerebrovascular and neuronal dysfunctions resulting in cognitive changes during TBI. In 2019, he joined the Department of Surgery at USF, where he continues to work, addressing vascular effects on neuroinflammation during various neurodegenerative diseases.

Preface

Neurodegeneration inevitably leads to cognitive impairment, significantly altering the quality of life of affected people. Neuroinflammatory diseases such as Alzheimer's disease, multiple sclerosis, traumatic brain injury, and stroke are associated with memory reduction caused by the impairment of neuronal function. Over the past decade, greater attention has been given to the vascular effects involved in cognitive decline. The topic of vascular cognitive impairment and dementia has been identified as one of the research priorities by the National Institutes of Health (USA). Many of the neuroinflammatory responses are associated with changes in systemic circulation and brain vasculature.

The scope of this Special Issue is to identify some mechanisms involved in neurodegeneration and cognitive decline that originate in the vasculature. A wide variety of these processes can be explained by the functional complexity of the neuro-vascular unit and the number of cells and proteins that can affect vasculo-neuronal interactions during various neuroinflammatory and neurodegenerative diseases. The present Special Issue is appealing to a wide range of scientists and health specialists working in the field of neuroinflammation.

David Lominadze
Guest Editor

biomolecules

MDPI

Article

The Role of Nuclear Factor-Kappa B in Fibrinogen-Induced Inflammatory Responses in Cultured Primary Neurons

Nurul Sulimai [1], Jason Brown [1] and David Lominadze [1,2,*]

1 Department of Surgery, University of South Florida Morsani College of Medicine, Tampa, FL 33612, USA
2 Department of Molecular Pharmacology and Physiology, University of South Florida Morsani College of Medicine, Tampa, FL 33612, USA
* Correspondence: dlominadze@usf.edu

Abstract: Traumatic brain injury (TBI) is an inflammatory disease associated with a compromised blood–brain barrier (BBB) and neurodegeneration. One of the consequences of inflammation is an elevated blood level of fibrinogen (Fg), a protein that is mainly produced in the liver. The inflammation-induced changes in the BBB result in Fg extravasation into the brain parenchyma, creating the possibility of its contact with neurons. We have previously shown that interactions of Fg with the neuronal intercellular adhesion molecule-1 and cellular prion protein induced the upregulation of pro-inflammatory cytokines, oxidative damage, increased apoptosis, and cell death. However, the transcription pathway involved in this process was not defined. The association of Fg with the activation of the nuclear factor-κB (NF-κB) and the resultant expression of interleukin-6 (*IL-6*) and C–C chemokine ligand-2 (*CCL2*) were studied in cultured primary mouse brain cortex neurons. Fg-induced gene expression of *CCL2* and *IL-6* and the expression of NF-κB protein were increased in response to a specific interaction of Fg with neurons. These data suggest that TBI-induced neurodegeneration can involve the direct interaction of extravasated Fg with neurons, resulting in the overexpression of pro-inflammatory cytokines through the activation of transcription factor NF-κB. This may be a mechanism involved in vascular cognitive impairment during neuroinflammatory diseases.

Keywords: neuroinflammation; transcription factor; C–C chemokine ligand-2; interleukin-6; intercellular adhesion molecule-1 and cellular prion protein

1. Introduction

Traumatic brain injury (TBI) is an inflammatory neurodegenerative disease that is a leading cause of morbidity and mortality in the United States. Unintentional falls and motor vehicle accidents are the most common cause of injury contributing to a TBI-related hospitalization [1]. The impact of a single fall on these patient's life could be detrimental, life-altering, or even fatal. Immediate physical damage and inflammation during TBI could ignite a secondary injury and result in neurodegeneration with cognitive deterioration that could show up weeks, months, or even years later. In some cases, TBI patients can have long-term sequelae including cognitive dysfunction, pain, sleep disorders, and physical disability, collectively known as post-concussion syndrome [2]. Although it is known that the blood–brain barrier (BBB) disruption is an early event during TBI, it may also persist for many years after the initial injury [3].

Fibrinogen (Fg) is an acute-phase reactant protein that is increased during the inflammatory condition [4] that accompanies TBI [5]. We have previously found that during an elevated level of Fg, called hyperfibrinogenemia (HFg), there is enhanced cerebrovascular permeability via caveolar protein-mediated transcytosis and the enhanced formation of Fg-amyloid β and Fg-cellular prion protein (PrP^C) complexes [6]. Extravascular deposits of Fg have been found in a post-mortem brain sample of a TBI patient who survived 18 years after a fall [3]. During chronic inflammatory diseases, such as Alzheimer's disease (AD) or TBI, the finding of Fg, a protein that is mainly produced in the liver [7], in the extravascular space of the brain parenchyma implies that Fg extravasated from the blood vessels. We have shown that at elevated levels, Fg crosses the vascular wall via, mainly, caveolar transcytosis [8–10]. The extravasated Fg that was found in TBI patients [3] certainly can come in contact with neurons and trigger a subsequent signaling responses. We have previously shown that Fg caused the upregulation of pro-inflammatory cytokines in astrocytes [11,12] and neurons [13]. Furthermore, Fg caused an increased generation of reactive oxygen species (ROS) and nitric oxide (NO) in astrocytes [11], in addition to increased ROS, NO, and mitochondrial superoxide in neurons [13]. These pro-inflammatory effects were in part due to the strong association of Fg with its receptors intercellular adhesion molecule-1 (ICAM-1) and PrP^C in astrocytes [11] and neurons [13], which ultimately resulted in increased cell death in astrocytes [11] and neurons [13]. These effects could be a mechanism of neurodegeneration and the resultant memory reduction seen during TBI [6,12]. However, the possible transcription pathway that can be activated in neurons because of an interaction with Fg has never been described. Therefore, in the present work, we investigated whether there is a possible role of nuclear factor-κB (NF-κB) in Fg/neuron interactions. To eliminate any confounding effects from other cells, such as endothelial cells, astrocytes, and other glial cells, we conducted the experiments in vitro, testing the effects of intact Fg and mouse brain neurons.

NF-κB is a ubiquitous transcription factor that is one of the most important modulators of stress and inflammatory gene expression in the nervous system [14]. The activation of the NF-κB is associated with many neurodegenerative diseases including TBI [14–16]. A prolonged activation of NF-κB has also been found up to 1 year after a head injury [15]. However, the specific source of the persistent post-traumatic activation of NF-κB in this study was not clear. The possible roles of transcription factors and the particular interconnection of Fg and NF-κB signaling in neurodegenerative pathology are not known. The objective of the present study was to investigate the involvement of NF-κB in the pro-inflammatory effects of Fg initiated in the microvasculature that result in the short-term memory reduction that we have shown earlier [5,6,12,17].

2. Materials and Methods

2.1. Cells, Antibodies, Reagents, and Materials

Primary mouse brain cortical neurons from C57BL/6 embryonic-day-17 mice (cat. #A15586) were purchased cryopreserved from Thermo Fisher Scientific (Waltham, MA, USA). Nunc™ 12-well (cat. #150628) and 24-well plates (cat. #142475), German #1 glass coverslips (cat. #50-121-5159), phosphate-buffered saline (PBS) (composition: 1.05 mM KH2PO4, 155.17 mM NaCl, and 2.97 mM Na_2HPO_4. $7H_2O$, without Ca^{2+} and Mg^{2+}), ACROS Organics™ Triton™ X-100 (cat. #AC422355000), normal goat serum (cat. #NC9660079), PowerUp™ SYBR™ Green Master Mix (cat. #A25776), Pierce® Radioimmunoprecipitation (RIPA) lysis and extraction buffer (cat. #89900), Laemmli sodium dodecyl-sulfate (SDS) buffer (6X; cat. #J61337), Electron Microscopy Sciences 32% Paraformaldehyde (formaldehyde) aqueous solution (cat. #50-980-494), and Molecular Probes™ ProLong™ Diamond Antifade Mountant with 4′,6-diamidino 2 phenylindole (DAPI) (cat. #P36971) were also from Thermo Fisher Scientific. Primary neuron basal medium (PNBM) (cat. #CC-3256) with a specific additive SingleQuots™ (cat. #CC-4462) was purchased from Lonza (Basel, Switzerland). Poly-d-Lysine hydrobromide (cat. #P0899) and Laminin from Engelbreth-Holm-Swarm murine sarcoma basement membrane (cat. #L2020-1MG), Hirudin from leeches (cat. #H7016-10UN), bovine serum albumin (BSA) (cat. #A7906-10G), the primary antibody against NF-κB p65 (cat. #06-418) that was used in immunofluorescence, and the primary antibody against β-actin (cat. #A1978-200 μL) were purchased from Millipore-Sigma (Burlington, MA, USA). The primary antibody against NF-κB p65 (cat. #06-418) that was used in Western blot was from the Proteintech Group (Rosemont, IL, USA). The inhibitor of NF-κB p65 activity, caffeic acid phenethyl ester

(CAPE) (cat. #2743/10), was purchased from R&D systems (Minneapolis, MN, USA). The rat purified function-blocking antibody (clone: YN1/1.7.4) against mouse ICAM-1 (cat. #116133) was purchased from BioLegend (San Diego, CA, USA), and the prion protein-blocking peptide (cat. #GTX89339-PEP) was purchased from GeneTex (Irvine, CA, USA). Recombinant murine tumor necrosis factor alpha (TNFα) (cat. #315-01A-20UG) and interferon gamma (IFNγ) (cat. # 315-05-20UG) were purchased from Peprotech (Rocky Hill, NJ, USA).

The polyclonal rabbit antibody against human Fg (cross-reacts with mouse) (cat. #A008002-2) was purchased from Dako Cytomation (Carpentaria, CA, USA). TRIzol™ reagent (cat. #15596026), ProLong™ Diamond antifade mountant with 4′,6-diamidino-2-phenylindole (cat. # P36962), the primary antibody against neuronal marker microtubule associated protein 2 (MAP2) (cat. #PA1-10005), goat anti-rabbit IgG (H+L) cross-adsorbed secondary antibody, horseradish peroxidase-conjugated (HRP) (cat. #A16104), goat anti-mouse IgG (H+L) cross-adsorbed secondary antibody, HRP (cat. #A16072), goat anti-chicken IgY (H+L) secondary antibody, Alexa Fluor™ 488 (cat. #A-11039), and goat anti-rabbit IgG (H+L) cross-adsorbed secondary antibody, Alexa Fluor™ 647 (cat. #A-21244) were from Invitrogen (Carlsbad, CA, USA). iScript™ cDNA synthesis kit (cat. #1708891), 10% Mini-PROTEAN TGX Stain-Free™ Gels (cat. #4568033), Immun-Blot® polyvinylidene difluoride membranes (cat. #162-0175), and the Quick Start™ Bradford protein assay kit (cat. #5000201) were from BioRad (Hercules, CA, USA). The TransAM® NF-κB p65 activation assay kit (cat. #40096) was purchased from Active Motif (Carlsbad, CA, USA).

2.2. Experimental Procedures with Cell Cultures

Primary mouse brain cortex neurons were plated at 7×10^5 cells/mL, grown in PNBM containing SingleQuots™, and used on day 7 or 10. The neurons were grown on 24-well plates for gene analysis and cell immunocytochemistry and on 12-well plates for protein level detection. The cell culture plates with integrated #1 German glass coverslips were coated with poly-d-Lysine (30 μg/mL) and laminin (200 μg/mL) for 1 h, at room temperature, before being rinsed twice with PBS prior to cell seeding for optimal neuronal attachment and growth.

The neurons were grown and treated as we reported previously [13]. Prior to treatment with Fg, the neurons were pre-treated for 90 min with a potent and specific inhibitor of nuclear transcription factor NF-κB activation, caffeic acid phenethyl ester (CAPE) [18]. A function-blocking antibody against ICAM-1 or a function-blocking peptide against PrP^C (7 μg/mL, for both) was added to the cells 30 min prior to the treatment with Fg. The neurons were treated with Fg for 1 h. In the control group, an equal volume of PBS was added in place of Fg. As a positive control group, cells stimulated with 200 ng/mL of TNFα along with 200 ng/mL of a murine IFNγ were used. Each experimental group contained hirudin (0.5 U/mL) in order to block the possible thrombin-induced conversion of Fg into fibrin. The cells were kept in an incubator at 37 °C with 5% CO_2.

2.3. Quantitative Real-Time PCR (qRT-PCR)

Total RNA was extracted from mouse astrocytes using the TRIzol reagent (Invitrogen), and reverse transcription was conducted using an iScript cDNA synthesis kit from Bio-Rad (Hercules, CA, USA), following the manufacturer's instructions. The qRT-PCR analysis was carried out using PowerUp™ SYBR™ Green Master Mix (Applied Biosystems, Austin, TX, USA). The PCR cycle parameters were 50 °C for 2 min, followed by 95 °C for 10 min, then 40 cycles at 95 °C for 15 s; the annealing temperature was 56 °C for 1 min. The gene expression levels were determined by QuantStudio 3 from Life Technologies (Carlsbad, CA, USA). The mRNA expression of the target genes was analyzed and normalized to that of 18S, which was used as the housekeeping gene. Data analysis of fold changes in gene expression was performed using the ΔΔCt method and is presented as 2- (average ΔΔCt). The primers used were: *CCL2*—Fwd 5′-GTTGGCTCAGCCAGATGCA-3′, Rev 5′-AGCCTACTCATTGGGATCATCTTG-3′; *IL-6*—

Fwd 5′-GACTTCCATCGAGTTGCCTTCT-3′, Rev 5′-TTGGGAGTGGTATCCTCTGTGA-3′; and *18S*—Fwd 5′-CGGCGACGACCCATTCGAAC-3′, Rev 5′-GAATCGAACCCTGATTCC CCGTC-3′ as a housekeeping gene.

2.4. NF-κB DNA-Binding Activity

2.4.1. Preparation of Nuclear Proteins

Nuclear extracts were collected from treated neurons following the manufacturer's protocol (Active motif). The neurons were washed with cold PBS/protease inhibitor (PI) (dilution—1:100), scraped, and centrifuged (300× *g* for 5 min at 4 °C). The pelleted neurons were resuspended in 1 mL of hypotonic buffer (HB). HB was prepared with 20 mM Hepes, 5 mM NaF, 10 μM Na_2MoO_4, and 0.1 mM EDTA in distilled water adjusted to a pH of 7.5 with NaOH. HB was sterilized by filtering through a 0.2 μm filter and stored at 4 °C until used. The neurons in HB were allowed to swell on ice for 15 min before adding 50 μL of 10% Nonidet P-40 (0.5% final concentration) and vortexing vigorously for 10 s. At this point, the cell membrane was completely lysed, while the nuclear membrane remained intact. The homogenate was centrifuged for 30 s at 4 °C in a microcentrifuge. The supernatant containing the cytoplasmic fraction was removed. The remaining pellet was resuspended in 50 μL of complete lysis buffer and agitated on ice for 30 min on a shaking platform. After being centrifuged for 10 min at 14,000× *g* at 4 °C, the supernatant containing the nuclear extract was collected and stored at −80 °C.

2.4.2. NF-κB p65 Transcription Factor Detection and Quantification

To test for DNA binding, the TransAM p65 activation assay kit (Active Motif) was used. The kit contains a 96-well plate coated with DNA oligonucleotides containing the NF-κB consensus site (5′-GGGACTTTCC-3′) that has been immobilized. We added 5 μg of nuclear extract to each well, and the assay was performed as recommended by the manufacturer (Active Motif). The active form of NF-κB in the nuclear extract that was added specifically bound to the oligonucleotide. After the washing steps, we added a primary antibody used to detect NF-κB p65 that recognizes an epitope on p65 that is accessible only when NF-κB is activated and bound to its target DNA. A secondary HRP-conjugated antibody was used that provided a sensitive colorimetric readout quantified by spectrophotometry. The optical density was measured at 450 nm with an absorbance plate reader and is reported as fold increase with respect to the control group.

2.5. Western Blot Analysis

The neurons were rinsed with cold PBS/PI (1:100 dilution) twice before lysis with the lysing buffer RIPA/PI (1:100). The content of proteins was assessed with a Bradford assay according to the manufacturer's protocol. The lysate of neurons was mixed with an equal volume of 6X Laemmli SDS sample buffer and boiled at 105 °C for 10 min. Equal amounts of protein samples of neurons were loaded onto 10% SDS-polyacrylamide gels and electrophoresed under reducing conditions. After electrophoresis, the sample proteins were transferred onto polyvinylidene difluoride membranes. The membranes were blocked with 5% BSA in TBS-T and then were incubated with antibodies against NF-κB p65 (1:1000, host species rabbit) and β-actin (1:5000, host species mouse) for 2 h at room temperature. After probing with goat anti-rabbit IgG (H+L) cross-adsorbed secondary antibody, HRP (1:10,000) and goat anti-mouse IgG (H+L) cross-adsorbed secondary antibody, HRP (1:20,000) for 1 h at room temperature, the blots were developed using a Bio-Rad Molecular Imager (ChemiDoc XRS+, Hercules, CA, USA). Image analysis was performed using Image Lab TM Version 6.0.1 build 34, Standard edition, 2017, Bio-Rad Laboratories, Inc. The results are expressed as a ratio of the integrated optical density (IOD) of the target protein band to the IOD of the β-actin band in the same lane.

2.6. Immunofluorescence Staining

The neurons were treated with 1 mg/mL of Fg or with PBS (control) for 60 min. After the incubation, the cells were washed with PBS and fixed with 4% paraformaldehyde in PBS for 15 min. The fixed neurons were treated with a permeating solution containing 0.1% Triton™ X-100 for 10 min before blocking with 3% bovine serum albumin in PBS for 1 h. The neurons were incubated with primary antibodies against MAP2, used at 1:500 dilution, and NFκB p65, used at 1:150 dilution, and kept in the dark at 4 °C overnight. The cells were incubated for 2 h at a 1:200 dilution of an appropriate secondary antibody that was conjugated with Alexa Fluor, Goat anti Rabbit 594 and Goat anti chicken 488. The cells were washed with PBS between the incubations. A mounting medium with DAPI was used to label the cell nuclei.

2.7. Image Analysis

The immunofluorescent-stained neurons were observed using an Olympus FV1000 (Olympus Corporation, Tokyo, Japan) laser-scanning confocal microscope, and the obtained images were deconvoluted using the 2D deconvolution algorithm of the CellSens Dimension 1.11 software (Olympus). NF-κB p65 nuclear translocation was revealed by its co-localization with DAPI.

2.8. Statistical Analysis

The data were analyzed using Graph Pad Prism software (San Diego, CA, USA). All data are expressed as mean ± SE. The experimental groups were compared using one-way ANOVA with a pairwise comparison using Tukey's multiple comparison. Differences were considered significant when $p < 0.05$.

3. Results

3.1. Fg-Induced Expression of NF-κB p65 in Neurons

Fg dose-dependently increased the expression of the NF-κB p65 protein in neurons, as shown by Western blot analysis (Figure 1A–C,E). NF-κB p65 protein expression was practically undetectable in neurons treated with vehicle (PBS, control) (Figure 1, inset). Pre-treatment of the cells with CAPE reduced Fg-induced neuronal NF-κB p65 protein expression (Figure 1C,E). Similarly, the use of a function-blocking antibody against ICAM-1 and a function-blocking peptide against PrP^C ameliorated Fg-induced NF-κB p65 expression (Figure 1D,F).

3.2. Effect of Fg on NF-κB p65 DNA-Binding Activity in Neurons

The results of an ELISA-based assay showed that NF-κB p65 DNA-binding activity was dose-dependently increased in Fg-treated neurons compared to the control group (Figure 1G). The use of CAPE (an inhibitor of NF-κB activation), a function-blocking antibody against ICAM-1, and a function-blocking peptide against PrP^C resulted in a reduction of Fg-induced p65 DNA-binding activity (Figure 1G).

3.3. Fg-Induced Translocation of NF-κB p65 into the Nucleus

The expression of NF-κB p65 in the nuclei of neurons treated with 1 mg/mL of Fg was greater than that in the nuclei of neurons treated with vehicle (control) (Figure 2). At the same time, the colocalization of NF-κB p65 with the nuclear marker DAPI in neurons treated with 1 mg/mL of Fg was greater than that in neurons treated with vehicle (control) (Figure 2). Both the expression of NF-κB p65 and the co-localization of NF-κB p65 with the nuclear marker DAPI were significantly reduced by caffeic acid phenethyl ester (CAPE), a function-blocking antibody against ICAM-1, or a function-blocking peptide against PrP^C in neurons treated with Fg (Figures 1 and S1).

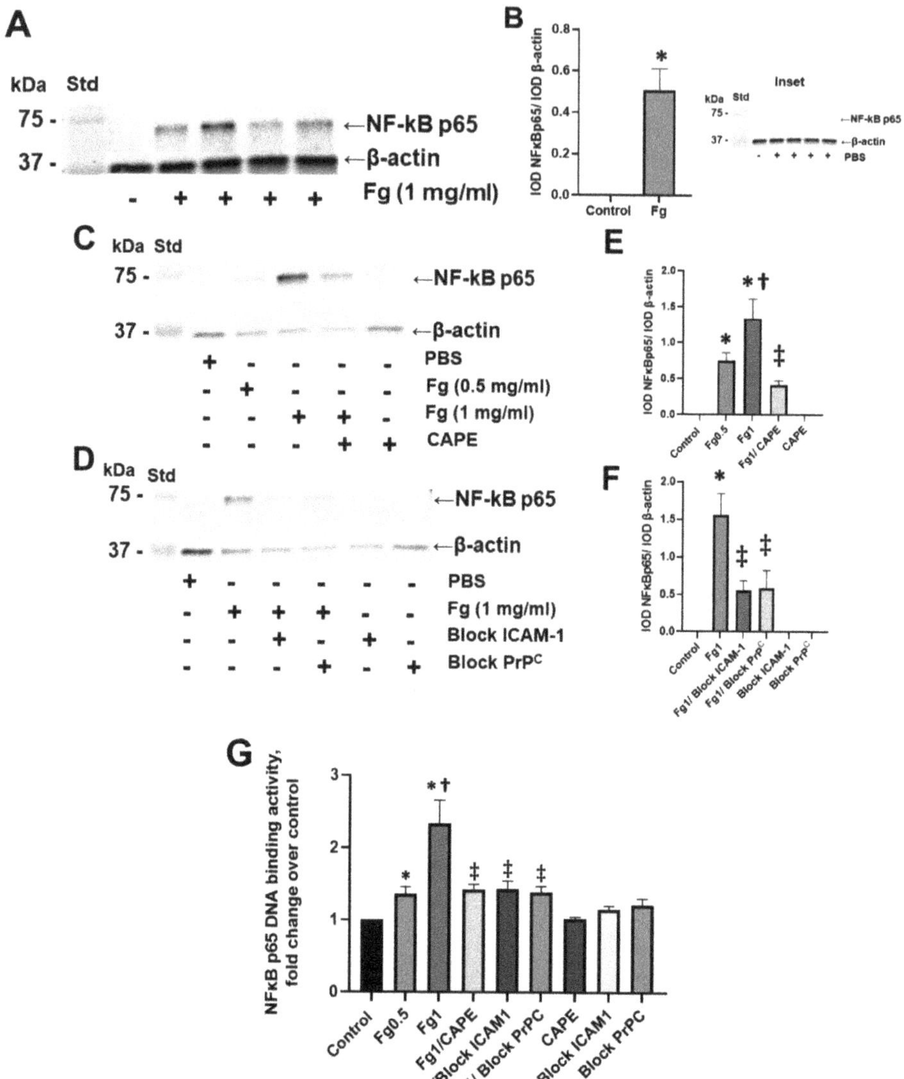

Figure 1. Fibrinogen (Fg)-induced expression of nuclear factor-κB (NF-κB) p65 protein in neurons. Primary mouse brain cortex neurons were treated with 1 mg/mL of Fg or phosphate-buffered saline (PBS) for 1 h. (**A**) The level of the NF-κB p65 protein in neurons was assessed by western blot analysis in Fg-treated and PBS-treated (inset) neurons. (**B**) Summary of the ratios of the integrated optical density (IOD) of NF-κB p65 bands to IOD of β-actin (used as a loading control) bands in their respective lanes. (**C**) Expression of NF-κB p65 proteins in neurons treated with PBS, 0.5 mg/mL of Fg, or 1 mg/mL of Fg in the presence or absence of caffeic acid phenethyl ester (CAPE). (**D**) Expression of NF-κB p65 protein in neurons treated with PBS or 1 mg/mL of Fg in the presence or absence of a function-blocking antibody against ICAM-1 or a function-blocking peptide against PrP^C. (**E**,**F**) Summary of the ratios of IOD of NF-κB p65 bands to IODs of β-actin bands in their respective lanes. (**G**) Expression of the NF-κB p65 protein in Fg-treated neurons analyzed by the TransAM p65 activation assay (an ELISA based method). The optical density was measured at 450 nm. Molecular weight standard (Std). $p < 0.05$ for all; * vs. control; † vs. Fg0.5; ‡ vs. Fg1; n = 3.

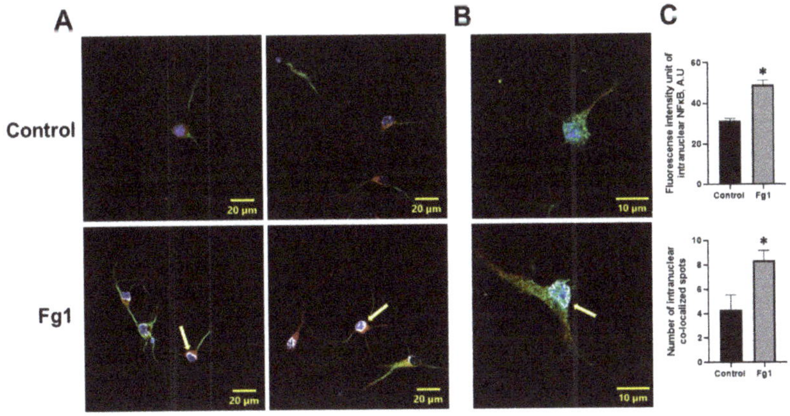

Figure 2. Fibrinogen (Fg)-induced expression of nuclear factor-κB (NF-κB) p65 and its translocation to the nuclei of neurons. Primary mouse brain cortex neurons were treated with 1 mg/mL of Fg (Fg1) or with phosphate-buffered saline (control) for 1 h. (**A**) Representative images show the translocation of NF-κB p65 (red) to the nuclei of Fg-treated neurons defined with a neuronal marker for microtubule-associated protein 2 (MAP2, green). Colocalization of NF-κB with the cell nuclei marked with 4′,6-diamidino-2-phenylindole (DAPI, blue) shown as white (yellow arrows). (**B**) Representative images of the same process at a higher magnification. (**C**) Summary of the fluorescence intensity changes of intranuclear NF-κB staining (upper graph) and the number of spots of co-localized NF-κB and DAPI (lower graph). * $p < 0.05$ vs. control; n = 3.

3.4. *Fg-Induced Expressions of CCL2 and IL-6 mRNAs in Neurons*

A real-time PCR analysis demonstrated that Fg induced a dose-dependent increase in the expression of *CCL2* mRNA (Figure 3A). Similarly, Fg induced a dose-dependent increase in the expression of *IL-6* mRNA (Figure 3B). The use of CAPE reduced Fg-induced upregulation of *CCL2* and *IL-6* genes. In parallel, Fg-induced upregulation of *CCL2* and *IL-6* genes was reduced with a function-blocking antibody against ICAM-1 and a peptide against PrP^C (Figure 3A,B, respectively).

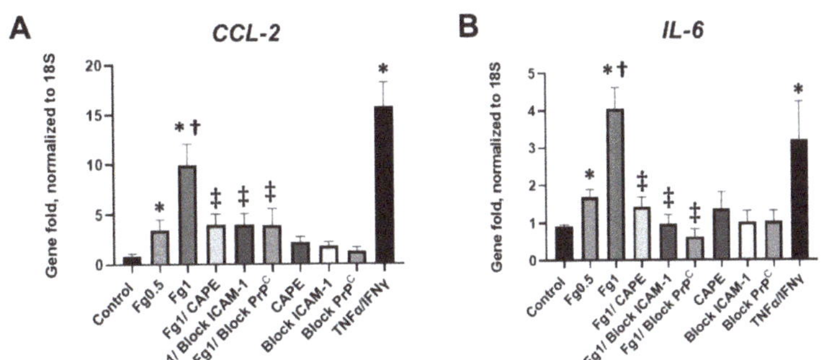

Figure 3. Expression of the mRNA of (**A**) C-C motif chemokine ligand 2 (*CCL2*) and (**B**) interleukin-6 (*IL-6*) in fibrinogen (Fg)-treated neurons, measured by real-time PCR. Primary mouse brain cortex neurons were treated with 0.5 mg/mL of Fg or 1 mg/mL Fg in the presence or absence of caffeic acid phenethyl ester (CAPE), a function-blocking antibody against intercellular adhesion molecule-1 (ICAM-1), or a function blocking peptide against cellular prion protein (PrP^C) for 1 h. Stimulation of neurons with tumor necrosis factor alpha (TNFα) and interferon gamma (IFNγ) was used as a positive control. $p < 0.05$ for all; * vs. control; † vs. Fg0.5; ‡ vs. Fg1; n = 4.

4. Discussion

For the first time, we present evidence that the direct interaction of plasma soluble Fg with neurons in the brain resulted in the activation of the NF-κB transcription factor in these cells. Fg induced an increased expression of NF-κB protein in neurons (Figure 1). The finding that untreated neurons or neurons treated with PBS expressed undetectable NF-κB p65 protein (Figure 1, inset) indicates the specificity of Fg-induced NF-κB activation in neurons in the present study. These results are in agreement with other studies showing that unstimulated cells present a very low level of NF-κB compared to its substantially up-regulated level in Fg-treated endothelial cells [19] or in mononuclear phagocytes [20]. In endothelial cells, 3 mg/mL of Fg caused the activation of NF-κB, and the effect lasted up to 24 h [19], whereas in mononuclear phagocytes, Fg (50 μg/mL) in the presence of Mn^{2+} induced a significant activation of NF-κB [20]. Thus, an interaction of Fg with endothelial cells, mononuclear phagocytes, or neurons positively resulted in the activation of NF-κB in these cells. These results suggest that an interaction of the blood plasma component Fg not only with vascular cells, but also with neurons, resulted in the activation of these cells, which may serve as the mechanism of known deleterious consequences to these neurons.

Although it is known that NF-κB is activated during TBI [14–16], the cause of NF-κB activation after injury has never been associated with Fg. During TBI, NF-κB DNA-binding activity in the injured cerebral cortex increased, with peak binding activity at 3 days after injury, and subsided from 10 to 14 days after the injury [16]. Immediately after injury (from 4 to 24 h), NF-κB activity was mainly observed in the neuronal cells of the affected cortex as well as in astrocytes located in the corpus callosum adjacent to the injury [14,15]. A pulse-like pattern of NF-κB activity in microglial cells was also found [14]. In vascular endothelial cells, NF-κB was detected early (1 h after the injury), and its expression persisted up to 1 year [15]. Our findings that Fg extravasates, even after acute inflammation has subsided during TBI [17], may indicate that there is a possibility of a direct interaction of Fg with neurons. As a result, there is a strong possibility that Fg activates NF-κB in neurons during TBI. It is not surprising that, during severe TBI, the activation of NF-κB was detected soon after a head injury and that the extent of that activity was reduced in a relatively short period of about 2 weeks [14,16]. On the other hand, although it is not specifically stated, the presented characteristics of lesions suggest that injury severity was in the mild-to-moderate range, yet the activation of NF-κB persisted for about one year [15], suggesting that chronic inflammation that lasts that long during mild-to-moderate TBI can still have pathological effects on neurons. Others have shown that BBB disruptions may persist for many years after TBI, and an extensive extravascular Fg disposition has been found in TBI patients who survived 18 years after a fall [3]. Therefore, the long-lasting NF-κB activation contributing to the prolonged (chronic) inflammation and neurodegeneration seen during TBI might be connected, in part, to NF-κB activation in neurons by extravasated Fg and/or its product fibrin. This does not exclude effects of Fg on endothelial cells [19] and mononuclear phagocytes [20] during the neuroinflammatory injury.

Interestingly, blocking the function of ICAM-1 and PrP^C dampened NF-κB DNA-binding activity and NF-κB protein expression (Figure 1D,F,G). The specificity of these effects was confirmed by the reduction of intranuclear of NF-κB and its translocation to the nuclei of neurons when the functions of both ICAM-1 and PrP^C were blocked (Figure S1).

The nuclear translocation of the NF-κB subunit p65 is a crucial step in NF-κB pathway activation. Thus, we determined the effects of Fg on NF-κB p65 translocation from the cytoplasm to the nucleus in neurons using immunocytochemistry. NF-κB p65 immunolabelling confirmed that there was an enhanced intracellular translocation of NF-κB p65 in Fg-treated neurons (Figure 2).

In this study, we also found that Fg-induced upregulation of the proinflammatory cytokines *CCL-2* and *IL-6* was ameliorated in the presence of a function-blocking antibody against ICAM-1 or a function blocking peptide against PrP^C (Figure 3). Collectively, these results suggest that the interaction of Fg with both its receptors ICAM-1 and PrP^C on neurons and the resultant upregulation of the pro-inflammatory cytokines *CCL2* and *IL 6* were associated with NF-κB signaling in these cells.

CCL2 is produced by neurons and glial cells in the brain and has been shown to be upregulated as a result of injury and inflammation [21,22]. Upregulation of CCL2 has also been shown in the cerebrospinal fluid (CSF) from patients with neurodegenerative diseases including mild cognitive impairment and Alzheimer's disease [23]. At the same time, high Fg levels were associated with an increased risk of both AD and vascular dementia [24]. In the current study, we found that Fg caused the upregulation of neuronal CCL2. We have shown that Fg resulted in the overexpression of *CCL2* in astrocytes [11], and others have shown that Fg increased the expression of CCL2 in endothelial cells [19]. Although these effects did not occur in neurons, they still confirm the pro-inflammatory potential of Fg in the brain during neurodegenerative diseases. Extravascularly, Fg, and most likely its product fibrin, have been shown to trigger the recruitment of inflammatory monocytes into the CNS and to induce demyelination [25]. In patients with AD, there was a positive correlation between cognitive impairment and the CCL2 concentration in the CSF [23]. We previously showed that Fg had a significant role in inflammatory responses leading to impairment in cognition [12]. Fg significantly increased cerebrovascular permeability [6], the formation of Fg–PrP^C complexes [26], and the generation of ROS [13] and reduced short-term memory during TBI [5,6,12]. The Fg-induced *CCL2* upregulation in neurons that we found in the current study might be the bridge that connects a pro-inflammatory state to cognitive decline, as we [27] and others have found [23]. Furthermore, CCL2 has been shown to induce an increase in BBB permeability by causing alterations in tight junction proteins in endothelial cells [28,29], which can further contribute to neuroinflammatory processes. Thus, Fg-induced activation of NF-κB could play a role in the signaling mechanism that results in the upregulation of the pro-inflammatory cytokines CCL2 and IL-6, causing the neuropathology seen during neuroinflammatory diseases [23] that is associated with elevated levels of Fg [25].

NF-κB is a pleiotropic transcription factor that is widely expressed in all cell types in the brain. In basal conditions, NF-κB functions to maintain healthy neuronal conditions, synapse growth, and plasticity-related functions [30]. In our study, we found that blocking NF-κB activation in Fg-induced neuronal pathology via a specific inhibitor for NF-κB, such as CAPE, or blocking ICAM-1 and PrP^C could be beneficial in preventing the activation of neurons caused by the neuroinflammation-induced detrimental effects of an elevated blood level of Fg. It is known that CAPE is also an active component of propolis, a resin-like material made by bees that is known to possess anti-inflammatory and antioxidative properties and is widely used as a therapeutic agent [18]. A comprehensive meta-analysis study confirmed a significant reduction in IL-6 and C-reactive protein levels as a result of propolis consumption [31]. Perhaps, propolis can be a useful remedy to alleviate some Fg-induced pathologies associated with NF-κB activation. This is the first report that connects a neuropathology that involves NF-κB signaling in neurons to the Fg protein that is originated from the vasculature and can extravasate only as the result of abnormal vascular permeability. Moreover, we found that at an elevated level of Fg can itself cause an increase in cerebrovascular permeability mainly via caveolar transcytosis [5,6,9,32,33]. A study of the role of Fg-induced NF-κB activation during TBI may provide opportunities for therapeutic interventions or new insights into the mechanisms of action for agents with therapeutic potential in the prevention of neurodegeneration and the resultant memory reduction as a result of neuroinflammatory diseases.

Many neuroinflammatory diseases such as TBI [5,34], AD [24], and stroke [35] are accompanied by an increased blood level of Fg. We, and others, have shown that TBI is associated with an enhanced deposition of Fg in the brain tissue [3,17]. There are reports of perivascular deposition of Fg during AD [36] and stroke-associated cerebral infarction [37,38]. We have shown that during HFg, which is a marker of inflammation [4], Fg deposition in the extravascular space is increased [6]. Combined, these results and the results of the present study, indicate that Fg can interact with neurons during any neuroinflammatory disease that is accompanied an increased deposition of Fg in the brain interstitium and results in neurodegeneration through the activation of NF-κB pathway.

5. Conclusions

This study demonstrated that Fg dose-dependently induced NF-κB activation in neurons. Furthermore, our results showed that the Fg-induced NF-κB activation may be inhibited by blocking known receptors of Fg (ICAM-1 and PrP^C), indicating that this is a specific effect of Fg. The present study, for the first time, clarified that blocking the neuronal receptors ICAM-1 and PrP^C reduced Fg-induced NF-κB activation and the resultant pro-inflammatory cytokine production. Therefore, this report suggests that using NF-κB -specific anti-inflammatory agents or blocking the function of the Fg receptors ICAM-1 and PrP^C should be considered as candidates for the treatment of the Fg-induced neuroinflammatory effects that accompany neurodegenerative diseases and in particular TBI, which is associated with chronic inflammation, deposition of Fg in the brain parenchyma, and increased neurodegeneration resulting in memory reduction. These findings identify possible mechanisms that are involved in the observed effects that originate in the vasculature (vascular permeability, Fg extravasation) and result in neuroinflammation (manifested as the expression of *CCL-2* and *IL-6*), leading to the memory reduction previously found during TBI [6] and collectively called vascular cognitive impairment and dementia.

Supplementary Materials: The following supporting information can be downloaded at: https://www.mdpi.com/article/10.3390/biom12121741/s1. Figure S1: Reduction of fibrinogen (Fg)-induced expression of nuclear factor-κB (NF-κB) p65, through the application of a NF-κB antagonist and the prevention of Fg/neuron interactions, and its translocation to neuronal nuclei. Figures S2–S6: Original images of Western blots shown in Figure 2.

Author Contributions: Conceptualization, N.S. and D.L.; methodology, N.S., J.B. and D.L.; software, N.S.; validation, N.S., J.B. and D.L.; formal analysis, N.S. and D.L.; investigation, N.S. and J.B.; data curation, N.S. and J.B.; writing—original draft preparation, N.S., J.B. and D.L.; writing—review and editing, N.S., J.B. and D.L.; visualization, N.S.; supervision, D.L.; funding acquisition, D.L. All authors have read and agreed to the published version of the manuscript.

Funding: This work was supported by the NIH grant # HL146832 and USF COM HSC-18330 Funds.

Institutional Review Board Statement: Not applicable.

Informed Consent Statement: Not applicable.

Data Availability Statement: Not applicable.

Conflicts of Interest: The authors declare no conflict of interest. The funders had no role in the design of the study; in the collection, analyses, or interpretation of data; in the writing of the manuscript, or in the decision to publish the results.

References

1. Peterson, A.B.; Zhou, H.; Thomas, K.E.; Daugherty, J. Traumatic Brain Injury-Related Hospitalizations and Deaths by Age Group, Sex, and Mechanism of Injury: United States 2016/2017. 2021. Available online: https://www.cdc.gov/traumaticbraininjury/pdf/TBI-surveillance-report-2016-2017-508.pdf (accessed on 1 July 2022).
2. McKee, A.C.; Robinson, M.E. Military-related traumatic brain injury and neurodegeneration. *Alzheimers Dement. J. Alzheimers Assoc.* **2014**, *10*, S242–S253. [CrossRef] [PubMed]
3. Hay, J.R.; Johnson, V.E.; Young, A.M.; Smith, D.H.; Stewart, W. Blood-brain barrier disruption is an early event that may persist for many years after traumatic brain injury in humans. *J. Neuropathol. Exp. Neurol.* **2015**, *74*, 1147–1157. [PubMed]
4. Gabay, C.; Kushner, I. Acute-phase proteins and other systemic responses to inflammation. *N. Engl. J. Med.* **1999**, *340*, 448–454. [CrossRef] [PubMed]
5. Muradashvili, N.; Benton, R.L.; Saatman, K.E.; Tyagi, S.C.; Lominadze, D. Ablation of matrix metalloproteinase-9 gene decreases cerebrovascular permeability and fibrinogen deposition post traumatic brain injury in mice. *Metab. Brain Dis.* **2015**, *30*, 411–426. [CrossRef]
6. Muradashvili, N.; Tyagi, R.; Tyagi, N.; Tyagi, S.C.; Lominadze, D. Cerebrovascular disorders caused by hyperfibrinogenemia. *J. Physiol.* **2016**, *594*, 5941–5957. [CrossRef]
7. Redman, C.M.; Xia, H.U.I. Fibrinogen biosynthesis. *Ann. N. Y. Acad. Sci.* **2001**, *936*, 480–495. [CrossRef]

8. Muradashvili, N.; Qipshidze, N.; Munjal, C.; Givvimani, S.; Benton, R.L.; Roberts, A.M.; Tyagi, S.C.; Lominadze, D. Fibrinogen-induced increased pial venular permeability in mice. *J. Cereb. Blood Flow Metab.* **2012**, *32*, 150–163. [CrossRef]
9. Muradashvili, N.; Tyagi, R.; Lominadze, D. A dual-tracer method for differentiating transendothelial transport from paracellular leakage in vivo and in vitro. *Front. Physiol.* **2012**, *3*, 166–172. [CrossRef]
10. Muradashvili, N.; Benton, R.; Tyagi, R.; Tyagi, S.; Lominadze, D. Elevated level of fibrinogen increases caveolae formation; Role of matrix metalloproteinase-9. *Cell Biochem. Biophys.* **2014**, *69*, 283–294. [CrossRef]
11. Sulimai, N.; Brown, J.; Lominadze, D. Fibrinogen Interaction with Astrocyte ICAM-1 and PrP^C Results in the Generation of ROS and Neuronal Death. *Int. J. Mol. Sci.* **2021**, *22*, 2391. [CrossRef]
12. Muradashvili, N.; Charkviani, M.; Sulimai, N.; Tyagi, N.; Crosby, J.; Lominadze, D. Effects of fibrinogen synthesis inhibition on vascular cognitive impairment during traumatic brain injury in mice. *Brain Res.* **2020**, *1751*, 147208. [CrossRef]
13. Sulimai, N.; Brown, J.; Lominadze, D. The Effects of Fibrinogen's Interactions with Its Neuronal Receptors, Intercellular Adhesion Molecule-1 and Cellular Prion Protein. *Biomolecules* **2021**, *11*, 1381. [CrossRef]
14. Sanz, O.; Acarin, L.; González, B.; Castellano, B. NF-κB and IκBα expression following traumatic brain injury to the immature rat brain. *J. Neurosci. Res.* **2002**, *67*, 772–780. [CrossRef]
15. Nonaka, M.; Chen, X.-H.; Pierce, J.E.S.; Leoni, M.J.; McIntosh, T.K.; Wolf, J.A.; Smith, D.H. Prolonged Activation of NF-κB Following Traumatic Brain Injury in Rats. *J. Neurotrauma* **1999**, *16*, 1023–1034. [CrossRef]
16. Yang, K.; Mu, X.S.; Hayes, R.L. Increased cortical nuclear factor-κB (NF-κB) DNA binding activity after traumatic brain injury in rats. *Neurosci. Lett.* **1995**, *197*, 101–104. [CrossRef]
17. Muradashvili, N.; Tyagi, S.C.; Lominadze, D. Localization of fibrinogen in the vasculo-astrocyte interface after cortical contusion injury in mice. *Brain Sci.* **2017**, *7*, 77. [CrossRef]
18. Natarajan, K.; Singh, S.; Burke, T.R.; Grunberger, D.; Aggarwal, B.B. Caffeic acid phenethyl ester is a potent and specific inhibitor of activation of nuclear transcription factor NF-kappa B. *Proc. Natl. Acad. Sci. USA* **1996**, *93*, 9090–9095. [CrossRef]
19. Guo, M.; Sahni, S.K.; Sahni, A.; Francis, C.W. Fibrinogen regulates the expression of inflammatory chemokines through NF-kappaB activation of endothelial cells. *Thromb. Haemost.* **2004**, *94*, 672–895.
20. Sitrin, R.G.; Pan, P.M.; Srikanth, S.; Todd, R.F., 3rd. Fibrinogen activates NF-kappa B transcription factors in mononuclear phagocytes. *J. Immunol.* **1998**, *161*, 1462–1470.
21. Howe, C.L.; LaFrance-Corey, R.G.; Goddery, E.N.; Johnson, R.K.; Mirchia, K. Neuronal CCL2 expression drives inflammatory monocyte infiltration into the brain during acute virus infection. *J. Neuroinflamm.* **2017**, *14*, 238. [CrossRef]
22. Glabinski, A.R.; Balasingam, V.; Tani, M.; Kunkel, S.L.; Strieter, R.M.; Yong, V.W.; Ransohoff, R.M. Chemokine monocyte chemoattractant protein-1 is expressed by astrocytes after mechanical injury to the brain. *J. Immunol.* **1996**, *156*, 4363–4368. [PubMed]
23. Galimberti, D.; Schoonenboom, N.; Scheltens, P.; Fenoglio, C.; Bouwman, F.; Venturelli, E.; Guidi, I.; Blankenstein, M.A.; Bresolin, N.; Scarpini, E. Intrathecal chemokine synthesis in mild cognitive impairment and Alzheimer disease. *Arch. Neurol.* **2006**, *63*, 538–543. [CrossRef] [PubMed]
24. Van Oijen, M.; Witteman, J.C.; Hofman, A.; Koudstaal, P.J.; Breteler, M.M.B. Fibrinogen is associated with an increased risk of Alzheimer disease and vascular dementia. *Stroke* **2005**, *36*, 2637–2641. [CrossRef] [PubMed]
25. Ryu, J.K.; Petersen, M.A.; Murray, S.G.; Baeten, K.M.; Meyer-Franke, A.; Chan, J.P.; Vagena, E.; Bedard, C.; Machado, M.R.; Rios Coronado, P.E.; et al. Blood coagulation protein fibrinogen promotes autoimmunity and demyelination via chemokine release and antigen presentation. *Nat. Commun.* **2015**, *6*, 8164. [CrossRef] [PubMed]
26. Charkviani, M.; Muradashvili, N.; Sulimai, N.H.; Lominadze, D. Fibrinogen—Cellular Prion Protein Complex Formation on Astrocytes. *J. Neurophysiol.* **2020**, *124*, 536–543. [CrossRef]
27. Muradashvili, N.; Tyagi, R.; Metreveli, N.; Tyagi, S.C.; Lominadze, D. Ablation of MMP9 gene ameliorates paracellular permeability and fibrinogen-amyloid beta complex formation during hyperhomocysteinemia. *J. Cereb. Blood Flow Metab.* **2014**, *34*, 1472–1482. [CrossRef]
28. Stamatovic, S.M.; Shakui, P.; Keep, R.F.; Moore, B.B.; Kunkel, S.L.; Van Rooijen, N.; Andjelkovic, A.V. Monocyte Chemoattractant Protein-1 Regulation of Blood–Brain Barrier Permeability. *J. Cereb. Blood Flow Metab.* **2005**, *25*, 593–606. [CrossRef]
29. Stamatovic, S.M.; Keep, R.F.; Kunkel, S.L.; Andjelkovic, A.V. Potential role of MCP-1 in endothelial cell tight junction 'opening': Signaling via Rho and Rho kinase. *J. Cell Sci.* **2003**, *116*, 4615–4628. [CrossRef]
30. Dresselhaus, E.C.; Meffert, M.K. Cellular Specificity of NF-κB Function in the Nervous System. *Front. Immunol.* **2019**, *10*, 1043. [CrossRef]
31. Shang, H.; Srikanth Bhagavathula, A.; Ali Aldhaleei, W.; Rahmani, J.; Karam, G.; Rinaldi, G.; Clark, C.; Salehisahlabadi, A.; Yuan, Q. Effect of propolis supplementation on C-reactive protein levels and other inflammatory factors: A systematic review and meta-analysis of randomized controlled trials. *J. King Saud Univ. Sci.* **2020**, *32*, 1694–1701. [CrossRef]
32. Tyagi, N.; Roberts, A.M.; Dean, W.L.; Tyagi, S.C.; Lominadze, D. Fibrinogen induces endothelial cell permeability. *Mol. Cell. Biochem.* **2008**, *307*, 13–22. [CrossRef]
33. Muradashvili, N.; Khundmiri, S.J.; Tyagi, R.; Gartung, A.; Dean, W.L.; Lee, M.-J.; Lominadze, D. Sphingolipids affect fibrinogen-induced caveolar transcytosis and cerebrovascular permeability. *Am. J. Physiol. Cell Physiol.* **2014**, *307*, C169–C179. [CrossRef]
34. Kossmann, T.; Hans, V.H.; Imhof, H.G.; Stocker, R.; Grob, P.; Trentz, O.; Morganti-Kossmann, C. Intrathecal and serum interleukin-6 and the acute-phase response in patients with severe traumatic brain injuries. *Shock* **1995**, *4*, 311–317. [CrossRef]

35. Del Zoppo, G.J.; Levy, D.E.; Wasiewski, W.W.; Pancioli, A.M.; Demchuk, A.M.; Trammel, J.; Demaerschalk, B.M.; Kaste, M.; Albers, G.W.; Ringelstein, E.B. Hyperfibrinogenemia and functional outcome from acute ischemic stroke. *Stroke* **2009**, *40*, 1687–1691. [CrossRef]
36. Fiala, M.; Liu, Q.N.; Sayre, J.; Pop, V.; Brahmandam, V.; Graves, M.C.; Vinters, H.V. Cyclooxygenase-2-positive macrophages infiltrate the Alzheimer's disease brain and damage the blood–brain barrier. *Eur. J. Clin. Investig.* **2002**, *32*, 360–371. [CrossRef]
37. Massberg, S.; Enders, G.; Matos, F.C.M.; Tomic, L.I.D.; Leiderer, R.; Eisenmenger, S.; Messmer, K.; Krombach, F. Fibrinogen Deposition at the Postischemic Vessel Wall Promotes Platelet Adhesion During Ischemia-Reperfusion In Vivo. *Blood* **1999**, *94*, 3829–3838. [CrossRef]
38. Shimamura, N.; Matchett, G.; Yatsushige, H.; Calvert, J.W.; Ohkuma, H.; Zhang, J. Inhibition of Integrin αvβ3 Ameliorates Focal Cerebral Ischemic Damage in the Rat Middle Cerebral Artery Occlusion Model. *Stroke* **2006**, *37*, 1902–1909. [CrossRef]

biomolecules

MDPI

Article

A Soluble Platelet-Derived Growth Factor Receptor-β Originates via Pre-mRNA Splicing in the Healthy Brain and Is Upregulated during Hypoxia and Aging

Laura Beth Payne [1,2], Hanaa Abdelazim [1,2], Maruf Hoque [1,2], Audra Barnes [1,2,3], Zuzana Mironovova [1,2], Caroline E. Willi [1,2], Jordan Darden [1,2], Clifton Houk [1,4], Meghan W. Sedovy [1,2], Scott R. Johnstone [1,2,5] and John C. Chappell [1,2,3,4,*]

1 Fralin Biomedical Research Institute (FBRI) at Virginia Tech-Carilion (VTC), Roanoke, VA 24016, USA
2 FBRI Center for Vascular and Heart Research, Roanoke, VA 24016, USA
3 Department of Biomedical Engineering and Mechanics, School of Biomedical Engineering and Sciences, Virginia Tech, Blacksburg, VA 24061, USA
4 Virginia Tech Carilion School of Medicine, Roanoke, VA 24016, USA
5 Department of Biological Sciences, Virginia Tech, Blacksburg, VA 24061, USA
* Correspondence: JChappell@vtc.vt.edu; Tel.: +1-540-526-2219; Fax: +1-540-982-3373

Abstract: The platelet-derived growth factor-BB (PDGF-BB) pathway provides critical regulation of cerebrovascular pericytes, orchestrating their investment and retention within the brain microcirculation. Dysregulated PDGF Receptor-beta (PDGFRβ) signaling can lead to pericyte defects that compromise blood-brain barrier (BBB) integrity and cerebral perfusion, impairing neuronal activity and viability, which fuels cognitive and memory deficits. Receptor tyrosine kinases such as PDGF-BB and vascular endothelial growth factor-A (VEGF-A) are often modulated by soluble isoforms of cognate receptors that establish signaling activity within a physiological range. Soluble PDGFRβ (sPDGFRβ) isoforms have been reported to form by enzymatic cleavage from cerebrovascular mural cells, and pericytes in particular, largely under pathological conditions. However, pre-mRNA alternative splicing has not been widely explored as a possible mechanism for generating sPDGFRβ variants, and specifically during tissue homeostasis. Here, we found sPDGFRβ protein in the murine brain and other tissues under normal, physiological conditions. Utilizing brain samples for follow-on analysis, we identified mRNA sequences corresponding to sPDGFRβ isoforms, which facilitated construction of predicted protein structures and related amino acid sequences. Human cell lines yielded comparable sequences and protein model predictions. Retention of ligand binding capacity was confirmed for sPDGFRβ by co-immunoprecipitation. Visualizing fluorescently labeled sPDGFRβ transcripts revealed a spatial distribution corresponding to murine brain pericytes alongside cerebrovascular endothelium. Soluble PDGFRβ protein was detected throughout the brain parenchyma in distinct regions, such as along the lateral ventricles, with signals also found more broadly adjacent to cerebral microvessels consistent with pericyte labeling. To better understand how sPDGFRβ variants might be regulated, we found elevated transcript and protein levels in the murine brain with age, and acute hypoxia increased sPDGFRβ variant transcripts in a cell-based model of intact vessels. Our findings indicate that soluble isoforms of PDGFRβ likely arise from pre-mRNA alternative splicing, in addition to enzymatic cleavage mechanisms, and these variants exist under normal physiological conditions. Follow-on studies will be needed to establish potential roles for sPDGFRβ in regulating PDGF-BB signaling to maintain pericyte quiescence, BBB integrity, and cerebral perfusion—critical processes underlying neuronal health and function, and in turn, memory and cognition.

Keywords: platelet-derived growth factor receptor-β; pericytes; brain; hypoxia; aging; pre-mRNA splicing

1. Introduction

The cerebrovasculature is essential for sustaining all of the unique cell types within the brain [1]. Neurons in particular depend on adequate perfusion to replenish oxygen and nutrients during synaptic communication, as well as to regulate cerebrospinal fluid composition [2]. Therefore, disruption of brain vasculature compromises neuronal health and function, and exacerbates neurodegeneration that fuels cognitive and behavioral deficits, including memory loss and dementia [3,4]. Among the vascular cell types that can become defective within the brain are pericytes [5]. Pericytes are found deep within cerebral capillary networks, extending along the endothelium and encased in the extracellular matrix that forms the vascular basement membrane or basal lamina [6,7]. Recent studies have suggested vessel-associated pericytes contribute to the blood-brain barrier (BBB), where they "tune" vessel permeability to regulate exchange between the circulation and central nervous system (CNS) [8,9]. Pericyte investment within the brain microcirculation depends in part on tight regulation of platelet-derived growth factor-BB (PDGF-BB) signaling through PDGF receptor-beta (PDGFRβ) on the pericyte cell surface [10,11]. Perturbation of the PDGF-BB pathway can undermine pericyte coverage of cerebral microvessels and disrupt vascular function, especially the regulation of BBB permeability and trafficking [12]. These defects can thereby impair neural activity and longevity [13], damaging cognition and memory.

Pericytes are critical cellular constituents of cerebrovascular networks, as discussed above, playing a variety of roles in sustaining brain health and function [14]. Maintaining BBB stability has emerged as a prominent function for cerebral pericytes, as synaptic communication and neuronal health heavily depend on the tight regulation of the brain microenvironment [15]. Vessel leakage and disrupted vesicular transport across the brain endothelium have been attributed in part to pericyte dysfunction and loss [8,16]. Furthermore, pericytes have been implicated in regulating blood flow within brain capillaries, though not without controversy [17,18]. This particular function would significantly impact the oxygen and nutrient delivery to active neurons. Disruption of these and other presumptive pericyte functions would thus compromise brain homeostasis, leading to neuronal dysfunction and degeneration [19]. These injuries in turn fuel the onset and progression of cognitive and behavioral deficits, including memory loss, among other manifestations [20]. It is therefore critical to better understand the molecular determinants that support pericyte function within the cerebral microcirculation, especially regulators within the PDGF-BB pathway [13], given its pleiotropic and potent effects on pericyte biology.

Akin to vascular endothelial growth factor-A (VEGF-A) for endothelial cells, PDGF-BB is considered a "master regulator" of pericyte activity, including proliferation, survival, chemotaxis/migration, and retention within the capillary wall, among other behaviors [21]. Both VEGF-A and PDGF-BB ligands signal via corresponding transmembrane receptor tyrosine kinases (RTKs)—a class of signaling receptors commonly regulated by naturally occurring soluble counterparts [22]. VEGF Receptor-1 (VEGFR1/Flt-1), for example, largely acts as a "decoy" receptor for VEGF-A [23,24]. A soluble isoform of this receptor (sVEGFR1/sFlt-1) generated by pre-mRNA alternative splicing [25] modulates VEGF-A gradients [26], among other functions [27]. Soluble RTKs may also be released from the membrane by proteolytic cleavage [28], which has been described for sVEGFR-1/sFlt-1 [29]. Release of a soluble isoform of PDGFRβ (sPDGFRβ) by proteolytic cleavage has also been recently reported in scenarios of cerebral mural cell hypoxia and damage [5,30–33]; however, sPDGFRβ production by pre-mRNA alternative splicing remains relatively underexplored.

Soluble isoforms of RTKs can be generated by a variety of mechanisms and in a broad range of neuropathological scenarios. Soluble VEGFR-1/Flt-1, for instance, has recently been implicated in cognitive dysfunction and perturbations in brain structure [34], though the underlying mechanism remains ill-defined. As mentioned above, many RTKs including sVEGFR-1/sFlt-1 can shed extracellular domains via ligand-induced proteolytic cleavage. However, many angiogenic RTKs such as VEGFR1 and VEGFR2/Flk-1 produce high levels of soluble counterparts via pre-mRNA alternative splicing as a mechanism to regulate intracellular signaling [22,26,35–38]. Both soluble VEGF-A receptors are produced by a skipped splice event, followed by an intronic read-through and in-frame termination. This mechanism frequently occurs with RTKs to create soluble "decoy" receptors that (i) lack a transmembrane domain but retain ligand-binding capacity and (ii) can tether to the extracellular matrix (ECM) through heparin-binding domains [22,39]. These soluble negative regulators can be essential to maintain ligand signaling within a critical physiological range, as seen in the VEGF-A pathway, where loss of sFlt-1 leads to endothelial disorganization, vascular dysmorphogenesis, and embryonic lethality through a variety of defects [23,26,40]. Corresponding mechanisms within the PDGF-BB pathway are still emerging [41] alongside a recent surge in appreciation for PDGF-BB-mediated pericyte dysfunction, particularly within cerebral blood vessels.

Here, we explore one potentially essential regulatory feedback element within the PDGF-BB pathway that likely impacts microvascular pericytes and their function within the brain microvasculature. Soluble PDGFRβ has been identified as a cleavage product from vascular mural cells, and cerebral pericytes in particular [31,32]; yet, to our knowledge, sPDGFRβ production by pre-mRNA alternative splicing has not been widely reported. In the current study, we detected sPDGFRβ protein by Western blot in the brain and in a variety of other murine tissues. We selected the mouse brain for follow-on identification of mRNA sequences that correspond to soluble isoforms of PDGFRβ, generating predicted protein structures and related amino acid sequences. Comparable sequences and protein model predictions were also identified using a human mural cell line. Applying co-immunoprecipitation (co-IP) approaches, we also found that sPDGFRβ retains its capacity for ligand binding, an important characteristic for its potential functional relevance. Custom RNA-Scope® probes allowed for fluorescent in situ hybridization and visualization of sPDGFRβ isoform transcript distribution in mouse brain pericytes adjacent to cerebrovascular ECs. In certain brain regions, we were also able to image sPDGFRβ protein distribution throughout the brain parenchyma and adjacent to certain cerebral microvascular networks, such as near the lateral ventricles. In complementary models, we also found that sPDGFRβ transcript and protein levels increased in mice with age, and that hypoxia can also induce elevated production of sPDGFRβ isoform transcripts and protein.

2. Materials and Methods

2.1. Animal Care and Use Approval

All experiments involving animal use were performed following review and approval from the Institutional Animal Care and Use Committee (IACUC) at Virginia Tech. All experimental protocols were reviewed and approved by Virginia Tech Veterinary Staff and the IACUC. The Virginia Tech NIH/PHS Animal Welfare Assurance Number is A-3208-01 (Expires: 31 July 2025).

2.2. Western Blotting

Wild-type (WT, c57BL/6) mice were euthanized at select time points (P7, n = 3, both sexes; P21, n = 3, both sexes; P90, n = 3, both sexes) by asphyxiation to isoflurane (Cat. No. NDC 13985-030-60, MWI Veterinary Supply Co., Boise, ID, USA) and thoracotomy. Liver incisions allowed for immediate collection of whole blood, and for improved cardiac perfusion with Dulbecco's phosphate buffer solution (PBS). Brain removal was performed following cardiac perfusion. Specifically, the calvarium was exposed by sharp dissection and incised along the sagittal and lambdoid sutures using angled scissors. The resulting four cranial flaps were reflected away from the brain using forceps. Brains were severed from the spinal cord and cranial nerves using a spatula, and isolated brains were removed. Internal organs (heart, kidney, liver, intestines) were then excised, followed by collection of skeletal muscles from the hindlimb (gracilis and adductor).

Serum was fractionated from whole blood and treated to remove high-abundance proteins such as albumin and immunoglobulins (BioTechne, MIDR002, Minneapolis, MN, USA). All tissues were harvested or transferred into standard Radio Immuno Precipitation Assay (RIPA) buffer and homogenized. Lysates were clarified and quantified using a Bradford Protein Assay (BioRad, Hercules, CA, USA). Samples were prepared for SDS-PAGE in LDS sample buffer (ThermoFisher Scientific, Waltham, MA, USA) and separated on a 4–12% gradient BOLT gel (ThermoFisher Scientific, Waltham, MA, USA). Proteins were transferred to a PVDF membrane using the Trans-Blot Turbo Transfer System (BioRad, Hercules, CA, USA). EveryBlot Blocking Buffer (BioRad, Hercules, CA, USA) was used to block the membrane and dilute antibodies. Primary antibodies were sequentially incubated or together following the application of a stripping buffer. Primary antibody, polyclonal goat anti-mouse PDGFRβ (R&D, AF1042), was used at 1:500 overnight at 4 °C. Primary housekeeping antibodies—mouse anti-mouse α-tubulin (Sigma, T6199-25UL, St. Louis, MO, USA) or rabbit anti-mouse β-actin (Cell Signaling, Danvers, MA, USA, 4967) were used at 1:5000 or 1:500, respectively. Following washes in TBS + 0.05% Tween-20 (TBS-T), secondary antibodies were incubated at room temperature for 1 h, specifically: donkey anti-goat AlexaFluor647 (1:2000, Jackson ImmunoResearch, West Grove, PA, USA, 705-605-147), donkey anti-goat AlexaFluor488 (1:1000, Jackson ImmunoResearch, West Grove, PA, USA, 705-545-003), donkey anti-rabbit AlexaFluor647 (1:1000, Jackson ImmunoResearch, West Grove, PA, USA, 711-605-152), or donkey anti-mouse AlexaFluor647 (1:5000, Abcam, Cambridge, UK, ab150107). Following washes, the membrane was imaged on a Chemidoc® gel imaging system. For sequential target probing, the imaged membrane was stripped using ReBlot Plus Strong Antibody Stripping Solution (Sigma, St. Louis, MO, USA), according to the manufacturer's instructions. The stripped membrane was re-blotted for a "housekeeping" protein using 1:5000 monoclonal mouse anti-mouse α-tubulin (Sigma, St. Louis, MO, USA, T6199-25UL), overnight at 4 °C, followed by secondary staining with 1:5000 anti-mouse AlexaFluor647 (Abcam, Cambridge, UK, ab150107) and imaging as above. Bands were quantified using volumetric analysis in BioRad ImageLab software (v6.0.1) or integrated density quantification in ImageJ/FIJI.

2.3. RNA Ligase-Mediated Rapid Amplification of 3′ and 5′ cDNA Ends (RLM-RACE)

3′ Ends: Unique *Pdgfrβ* transcript variants were elucidated using the GeneRacer™ Kit (Invitrogen). A whole brain was harvested into Trizol from a postnatal day 90 (P90) female WT (c57BL/6) mouse (n = 1, female), following a thoracotomy and PBS perfusion as described above. Following homogenization, the tissue was processed for RNA isolation and DNase treatment (Zymo Quick-RNA Miniprep). RNA integrity was confirmed via 28S and 18S levels from agarose gel electrophoresis. Reverse transcription was performed with SuperScript III RT and Oligo dT primers. cDNA ends were amplified via touchdown PCR using Platinum Taq High Fidelity Polymerase (Invitrogen) with the GeneRacer 3′ Primer and a forward (FW) gene-specific primers (GSPs), alongside 3 separate negative controls: no template, no GSP, no GeneRacer 3′ primer. Forward GSPs were designed within Exons 4, 5, and 6:

FW GSP-Exon 4	CCCTACGACCACCAGCGAGGTTTC
FW GSP-Exon 5	CGAGAGCATCACCATCCGGTGCATTG
FW GSP-Exon 6	TGCCCTCCCGCATTGGCTCCATCCT

Following confirmation via agarose gel electrophoresis, Nested PCR was performed on each PCR amplified product using the GeneRacer Nested 3′ primer and the following FW GSP primers:

FW GSP-Exon 5 (*nested on GSP Exon 4 product*)	CGAGAGCATCACCATCCGGTGCATTG
FW GSP-Exon 6 (*nested on GSP Exon 5 product*)	TGCCCTCCCGCATTGGCTCCATCCT
FW GSP-Exon 7 (*nested on GSP Exon 6 product*)	GGACGCTGCGGGTGGTGTTCGAGGCTTAT

Products were separated via agarose gel electrophoresis. Individual gel-purified bands (S.N.A.P. columns, Invitrogen) were cloned into the pCR 4-TOPO vector and transformed into One Shot Top10 chemically competent cells and plated overnight. Six colonies were picked from each plate and grown for 16 h in LB broth culture. Plasmids were purified (Monarch Plasmid Miniprep Kit, New England Biolabs, Ipswich, MA, USA), and Sanger sequenced (Virginia Tech Genomics Sequencing Center). Reads were analyzed (DNAstar) and referenced to the mouse *Pdgfrβ* genomic sequence, located within chromosome 18, NC_000084.6. One unique variant was detected, identified in each amplification product, truncating in intron 10 (i10 splice variant). To determine if unique *Pdgfrβ* transcript variants were present in human sources, the described 3′RACE procedure was performed using RNA from human aorta smooth muscle cell lysates. RNA was isolated as described above. Two unique variants were detected, truncating in intron 4 (i4 splice variant) and in intron 10 (i10). The following GSP primers were used:

FW GSP-Exon 3	CCTCACTGGGCTAGACACGGGAGAA
FW GSP-Exon 4 (*also nested onGSP Exon 3 product*)	CTCACTGGGCTAGACACGGGAGAA
FW GSP-Exon 6 (*also nested on GSP Exon 4 product*)	CCTCACTGGGCTAGACACGGGAGAA
FW GSP-Exon 7 (*nested on GSP Exon 6 product*)	CCTGGGAGAGGTGGGCACACTACAA

5′ Ends: The 5′ end of the mouse *Pdgfrβ* i10 splice variant was elucidated using the GeneRacer™ Kit (Invitrogen, Carlsbad, CA, USA). The P90 WT (c57BL/6) female whole brain RNA (described above) was dephosphorylated and decapped, followed by the protocol described above for 3′ end. To elucidate the 5′ end specific to the i10 variant, the following reverse (Rev) gene-specific primers (GSP) were designed within the unique 3′UTR of the i10 variant, with upstream nested primers:

Rev GSP–i10_a	GGCTCAGTACGGCGGGATCAAGGAA
Rev GSP–i10_b	TCAGGCTCAGTACGGCGGGATCAA
Rev GSP–Exon 7 (*nested on GSP i10_a product*)	CCGGAGTCACCCAAGGTACGGTTGT
Rev GSP–Exon 6 (*nested on GSP i10_b product*)	CGGGAGGGCACTCCAAAGAGGTAGT

Two 5′ ends were detected–a short Exon 1 (truncated upstream) and a unique Exon 1 derived from intron 1 sequence.

2.4. Protein Modeling and Functional Domain Mapping

Following elucidation of transcript sequences via RLM-RACE and Sanger sequencing, sequence translations (minus signal sequence) were submitted to the I-TASSER server1–3 for protein structure modeling predictions. The resulting PDB files were submitted to the Cluspro 2.0 server4–7 to predict protein-protein docking with PDGF-BB propeptides (I-TASSER PDB files derived from NM_002608 and NM_011057 for human and mouse, respectively) and heparin binding (selection within Cluspro 2.0). Models were processed using PyMOL2.

The functional PDGFRβ protein domains were mapped to the *Pdgfrβ* exon structure using the UniProt database (for protein) and DNASTAR Lasergene software (for transcripts). Specifically, human domains from accession P09619 were mapped to NM_002609, and mouse domains from P05622 were mapped to NM_001146268. The functional domains tightly corresponded between human and mouse. The crystal structure of complexed human PDGF-BB:PDGFRβ Ig domains 1–3 (D1–D3), PDB-3MJG8, was examined (PyMOL2) to confirm the Uniprot functional IgG 1–3 domain regions, and to decipher ligand binding regions.

2.5. Co-Immunoprecipitation (Co-IP) Assay

Co-IP was performed using the Dynabeads Co-Immunoprecipitation Kit (ThermoFisher Scientific, Waltham, MA, USA), using polyclonal goat anti-mouse PDGF-BB antibody (Novus NBP1-52533) for immunoprecipitation. Briefly, a whole brain from a P21 WT (c57BL/6, n = 1) male mouse was harvested and homogenized in immunoprecipitation buffer, following a thoracotomy and PBS perfusion. Dynabeads were coupled with 10 ug of antibody per mg of beads. Whole-brain lysate (1.5 g protein) was processed with 1.5 g coupled beads according to the manufacturer's instruction, and using the following extraction buffer: 100 mM NaCl, EDTA-free HALT Protease\Phosphatase Inhibitor (ThermoFisher Scientific, Waltham, MA, USA), 2 mM $MgCl_2$, and 1 mM DTT. Aliquots were collected at each step for downstream analysis. Eluted protein was immunoblotted under denaturing conditions (see Western blotting protocol above), using 1:1000 polyclonal goat anti-mouse PDGFRβ (R&D, AF1042), and 1:2000 donkey anti-goat AlexaFluor488 (Jackson ImmunoResearch, West Grove, PA, USA, 705-545-147).

2.6. Fluorescent RNA-Scope® mRNA Labeling and Confocal Microscopy

Full-length *Pdgfrβ*, i10 *Pdgfrβ* splice variant, and *Pecam1* transcripts were detected in situ using the RNA-Scope® Multiplex Fluorescent v2 Assay (ACD, BioTechne, Minneapolis, MN, USA), according to recommended instructions. Briefly, whole brains were harvested from P21 WT (c57BL/6) mice (n = 4, both sexes), following thoracotomy and PBS perfusion, and were immediately snap frozen in optimal cutting temperature (OCT) compound. Slides were prepared with single 20-micron-thick cryo-slices. Samples were fixed (by 4% paraformaldehyde in PBS) and dehydrated. Following slide preparation, probes were hybridized to target RNA, specifically targeting: (a) i10 *Pdgfrβ* unique sequence in intron 10 (custom designed by ACD, 1148911-C3); (b) full-length *Pdgfrβ* sequence from Exon 13–21, corresponding to intracellular kinase domains (custom designed by ACD, 1148921-C2); and (c) *Pecam1* (pre-designed ACD, 316721). Positive control probes were run for each channel (*Polr2a*-C1, *Ppib*-C2, and *Ubc*-C3), as well as a negative control in each channel, *DapB*, targeting a bacterial gene. Signals were developed and amplified, using 1:1500 Opal Dye 520 for C1-*Pecam1*, 1:1500 Opal Dye 570 for C3-i10_*Pdgfrβ*, and 1:2000 TSA Plus Cy5 for C2-*Pdgfrβ* (Akoya, FP1487001, FP1488001, NEL745001, respectively). High-resolution volumetric images were acquired on a Zeiss LSM 880 confocal microscope. Quantification of RNA-Scope® signal colocalization was performed using ImageJ/FIJI software on single z-plane images, following the application of the "IsoData" threshold filter. Pearson's colocalization coefficient was obtained using the "Colocalization" plugin in ImageJ/FIJI.

2.7. Immunohistochemistry and Confocal Microscopy

Normal WT brains (C57BL/6–P1, n = 4–5, both sexes; P21, n = 3, both sexes) collected for immunohistochemistry and confocal microscopy were collected in a similar manner as described above for protein collection with the exception of intracardiac perfusion of 4% paraformaldehyde (in PBS) following PBS perfusion. Prior to sectioning, the olfactory bulbs and cerebellum were removed from the cerebrum using a single-edged razor blade (Garvey, Camberwell, Australia, 40475). Brains were then caudally affixed by superglue (Loctite, Düsseldorf, Germany, 1364076) to the vibratome dish with their ventral surface supported by a block of 4% agarose. They were then submerged in PBS and sectioned into 100 µm slices using half a double-edged razor blade (Electron Microscopy Sciences, Hatfield, PA, USA, 72000) in a vibratome (1000-Plus, Pelco 102, Ted Pella Inc., Redding, CA, USA). Slices were transferred to a 24-well plate and then stored in PBS at 4 °C until processing for immunohistochemistry.

Brain sections were blocked for 1 h at room temperature in PBS and 0.1% Triton 100x (PBS-T) with 1.5% Normal Donkey Serum (Jackson ImmunoResearch, West Grove, PA, USA, 017-000-121). Slices were then incubated at 4 °C overnight in 1.5% Normal Donkey Serum in PBS-T with rat anti-PDGFR-β antibody (Affymetrix, Santa Clara, CA, USA, 14-1402, 1:500), goat anti-CD31/platelet-endothelial cell adhesion molecule-1 (PECAM-1) antibody

(R&D Systems, Minneapolis, MN, USA, AF3628, 1:500), and rabbit anti-glial-fibrillary acidic protein (GFAP) (Millipore Sigma, Burlington, MA, USA, AB5804, 1:500) on a shaker. After incubation with primary antibodies, the slices were washed four times for 15 min at room temperature in PBS-T on a shaker.

Brain slices were then incubated for either 4 h at room temperature or overnight at 4 °C in 1.5% Normal Donkey Serum in PBS-T with donkey anti-rat DyLight 550 (ThermoFisher, Waltham, MA, USA, SA5-10029, 1:1000), donkey anti-goat AlexaFluor488 (Jackson ImmunoResearch, West Grove, PA, USA, 705-545-147, 1:500), donkey anti-rabbit AlexaFluor647 (Jackson ImmunoResearch, West Grove, PA, USA, 711-605-152, 1:1000), and 4′,6-Diamidino-2-phenylindole (DAPI, Sigma, Burlington, MA, USA, D9542, 1:1000). After incubation with secondary antibodies, the slices were washed four times for 15 min at room temperature in PBS-T, and then washed again for 15 min at room temperature in PBS after final incubations of staining combinations. Stained brain slices were mounted in 50% glycerol in PBS, a coverslip applied (22 mm × 22 mm–1.5 thickness, ThermoFisher, Waltham, MA, USA, 12-541-B), and sealed with clear nail polish (Electron Microscopy Sciences, Hatfield, PA, USA, 72180). Images of stained brain sections were acquired with a Zeiss LSM 880 confocal microscope using a 20× objective lens and Zen Black software.

Subcortical regions were identified as being above ventricles and included cortex and subcortical white matter. Images were collected in 2 × 2 tile scans with 10% overlap in 50-100 *z*-axis sections as determined by optimal *z*-axis distance. Using Zen Black image processing capability, the original tile scans were stitched together. Selected brain regions of interest were outlined and saved as new image files for analysis.

2.8. Quantitative Real-Time Polymerase Chain Reaction (qRT-PCR)

Following thoracotomy and PBS perfusion, tissues were harvested into Trizol from WT (c57BL/6) mice (equally distributed by sex, when possible) at each of three time-points: P5 (n = 4, both sexes), P21 (n = 5, both sexes), P90 (3-months-old, n = 4, both sexes). To capture a time-point associated with aging, tissues were collected from 20-month-old mice (n = 4, both sexes) which, while not of the same WT background, was considered as a littermate control for parallel experiments. Specifically, this mouse was negative for Cre-recombinase, but harbored floxed STOP codons within the *Vhl* gene (i.g. *UBC-CreER negative; Vhl$^{lox/lox}$* on the CD1 background) and exhibited no phenotypic changes by various metrics [42]. RNA was isolated using the Zymo Direct-zol RNA Miniprep kit, with DNase treatment. Isolated RNA was further treated for genomic material using a Turbo DNA-Free kit (ThermoFisher Scientific, Waltham, MA, USA). Samples were reverse transcribed using a High-Capacity cDNA RT Kit (Applied Biosystems, Waltham, MA, USA). Each sample was run in triplicate for qRT-PCR (QuantStudio Flex 6). To quantify full-length *Pdgfrβ* transcripts, Applied Biosystem Taqman assay mix Mm01262481_g1 (Fisher Scientific, Waltham, MA, USA) was used, which targets the exon 13-14 boundary. Custom probes were designed and validated to quantify the *Pdgfrβ* i10 splice variant:

Forward Primer	AACTCCATGGGTGGAGATTC
Reverse Primer	GTGAGAGTCATCAGAGCCATC
Probe Sequence	TCACCGTGGTCCCACATTGTGAG

2.9. Cell Culture and Hypoxia Exposure

Primary human coronary artery smooth muscle cells (cASMC, #PCS-100-021, ATCC) were cultured in human vascular smooth muscle cell basal media (#M231500, ThermoFisher, Waltham, MA, USA), supplemented with smooth muscle cell growth supplement (SMGS, #S00725, ThermoFisher, Waltham, MA, USA). All cells were used within 16 population doublings, as recommended by the manufacturer. CASMC were grown in six-well plates to 70–80% confluence, washed with PBS twice, and media changed for smooth muscle cell basal media (M231500, ThermoFisher, Waltham, MA, USA) containing 2% FBS for 3 days to induce cell stall. Stalled cells at 100% confluence were treated with 50 ng/mL recombinant human platelet-derived growth factor-BB (PDGF-BB, 50ng/mL, #PHG0045, ThermoFisher, Waltham, MA, USA) for 24 h prior to harvest.

Mouse embryonic stem cells (ESCs) harboring reporters for cells within endothelial and pericyte lineages were generated as previously described [43]. These "double reporter" (DR)-ESCs were maintained as undifferentiated cells through exposure to leukemia inhibitory factor (LIF). Upon LIF removal, DR-ESC differentiation was initiated, and cells were cultured as floating embryoid bodies (EBs) for 3 days. They were then allowed to adhere to tissue-culture-treated plates and differentiate for 7 additional days into primitive vascular structures, along with other cell types. Hypoxia-exposed cultures were incubated in a gas-tight chamber (STEMCELL Technologies) filled with 3% O_2, 5% CO_2, and 92% N_2 for the last 12 or 24 h of their differentiation (n = 4–6). Control DR-ESC cultures were maintained at atmospheric O_2 (~21%) and 5% CO_2 for the duration of their differentiation. At the conclusion of the experiment, RNA was collected in Trizol and processed as described above to measure relative levels of full-length and i10 variant *Pdgfrβ* transcripts by qRT-PCR.

2.10. Statistical Analysis

Statistical analysis was performed using GraphPad Prism software, version 9.1.2 (226). For the cell-based hypoxia experiment, the data were analyzed by an ordinary one-way ANOVA followed by unpaired Tukey's *t*-test. Outlier data points were removed by applying the Grubbs' test, or the extreme studentized deviate (ESD) method, with $\alpha = 0.05$. Significant differences were determined as P values less than or equal to 0.05.

3. Results

3.1. Soluble PDGFRβ Is Present across Various Murine Tissues with a Distinct Enrichment in the Blood, Kidney, and Brain

Truncated isoforms of PDGFRβ have been detected across a wide range of biological contexts, from in vitro settings, such as differentiating mouse embryonic stem cells [44], to in vivo scenarios, such as the healthy and diseased mouse and human brain [32,33]. To better characterize the abundance of sPDGFRβ across a variety of tissue compartments, we collected protein from normal, healthy WT (c57BL/6) postnatal day 21 (P21) and adult (P90) murine brain, heart, kidney, intestine, liver, skeletal muscle, and blood serum (Immunoglobulin G depleted). Proteins from each of these organ systems, along with recombinant full-length PDGFRβ protein as a positive control (data not shown), were processed for Western blot and immunolabeling for PDGFRβ (Figure 1). We observed bands corresponding to larger protein weights (100–160 kDa) for the recombinant protein control and for most tissues corresponding to full-length PDGFRβ, with varying levels for each tissue. As expected, serum proteins did not appear to contain much, if any, full-length PDGFRβ at higher molecular weights. In contrast, the serum lanes for both time-points displayed prominent bands at a lower molecular weight corresponding to approximately 32 kDa (Figure 1). Notably, ~32 kDa bands were also found with kidney and brain protein lysates, while the other tissues appeared to contain relatively lower amounts of the truncated PDGFRβ isoform within this size range.

3.2. PDGFRβ Is Alternatively Spliced to Generate a Soluble Isoform That Retains PDGF-BB Ligand Binding Capacity

While proteolytic cleavage has been proposed as the primary mechanism for generating soluble PDGFRβ isoforms [31,32], we hypothesized that alternative splicing on the mRNA level might also yield sPDGFRβ isoforms. To test this hypothesis, we collected and isolated high-quality mRNA from a postnatal day 90 (P90) murine brain. We used 3′ rapid amplification of cDNA ends (RACE) to elucidate and Sanger sequence unique 3′ ends of PDGFRβ mRNAs (see Methods for full details). Sequence reads were referenced to the mouse *Pdgfrβ* genomic sequence. One unique variant was detected, identified in each amplification product, truncating in intron 10 (Figure 2, *see Supplemental Table S1 for nucleotide sequence information*). To determine if a similar *Pdgfrβ* variant exists in human cells, the same RNA ligase-mediated rapid amplification of 3′ cDNA ends (RLM-RACE) was applied to RNA from human smooth muscle cell lysates. From this analysis, two unique variants were detected, truncating in intron 10, as in the mouse, but also in intron 4 (Figure 2). Follow-on work is underway to determine if truncation in intron 4 also occurs in the mouse. We also elucidated the 5′ end of the murine intron 10 (i10) *Pdgfrβ* splice variant to generate a complete sequence for this isoform, facilitating protein modeling and development of additional tools and approaches to detect this variant.

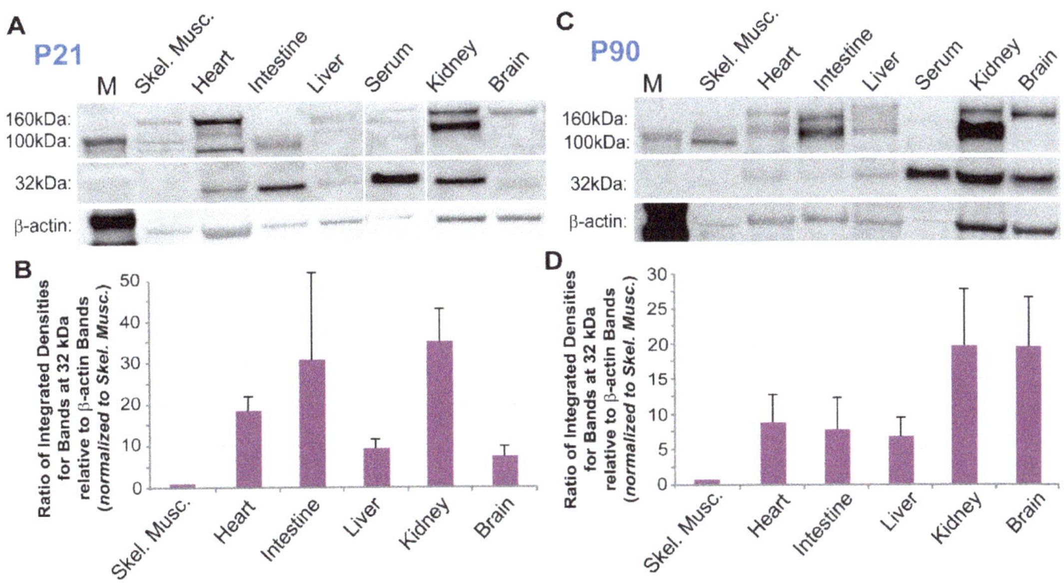

Figure 1. A Truncated Isoform of PDGFRβ was Detected alongside Full-Length Receptors across a Range of Murine Tissues at Different Ages. (**A**) Representative images of a Western blot for proteins from select murine tissues at postnatal day 21 (P21) (from n = 3 biological replicates, both sexes) detected by immunolabeling for PDGFRβ and β-actin–lane 1: molecular size marker (M), lane 2: skeletal muscle (Skel. Musc.), lane 3: heart, lane 4: intestine, lane 5: liver, lane 6: blood serum, lane 7: kidney, lane 8: brain. (**B**) Graph of the ratio of integrated densities for bands at the 32 kDa size range relative to β-actin bands for each tissue indicated, all of which were normalized to the skeletal muscle samples. Bars are averages with standard deviations. (**C**) Representative images of a Western blot for proteins from select murine tissues at postnatal day 90 (P90) (from n = 3 biological replicate, both sexes) detected by immunolabeling for PDGFRβ and β-actin–lane 1: molecular size marker (M), lane 2: skeletal muscle (Skel. Musc.), lane 3: heart, lane 4: intestine, lane 5: liver, lane 6: blood serum, lane 7: kidney, lane 8: brain. (**D**) Graph of the ratio of integrated densities for bands at the 32 kDa size range relative to β-actin bands for each tissue indicated, all of which were normalized to the skeletal muscle samples. Bars are averages with standard deviations.

Drawing from the sequence information established for these variants, we sought to generate protein structure modeling predictions to map functional domains. These models allowed us to (i) predict important domains such as PDGF-BB and heparin binding regions, and (ii) compare these domains across species. Sequence translations (without signal sequence) from both putative mouse and human sequences were submitted for protein model creation (Figure 3, *see Supplemental Table S2 for detailed amino acid predictions*). Functional PDGFRβ protein domains were also mapped to the *Pdgfrβ* exon structure. Functional domains for the murine and human sequences displayed a high degree of homology. Furthermore, protein docking predictions and analysis of known crystal structures of complexed human PDGF-BB with PDGFRβ supported the idea that these isoforms generated by pre-mRNA alternative splicing likely retain their ligand binding capacity. To directly address the potential of sPDGFRβ isoforms to bind PDGF-BB, we performed a co-immunoprecipitation (co-IP) assay. Specifically, we selected a PDGF-BB antibody for protein capture by adsorbing these antibodies to beads, and incubating antibody-coated beads with whole brain protein lysate from a normal, healthy P21 WT (c57BL/6) mouse. Proteins captured by the anti-PDGF-BB antibodies were eluted and immunoblotted under denaturing conditions using PDGFRβ antibody labeling and detection (Figure 3). We detected truncated PDGFRβ variants in the whole brain lysate and in the eluted fraction following co-IP, with minimal to no detection in the depleted flow-through fraction of supernatant following immunoprecipitation. These data are consistent with the notion that soluble isoforms of PDGFRβ retain their capacity to bind PDGF-BB ligands, including sPDGFRβ proteins generated by pre-mRNA alternative splicing.

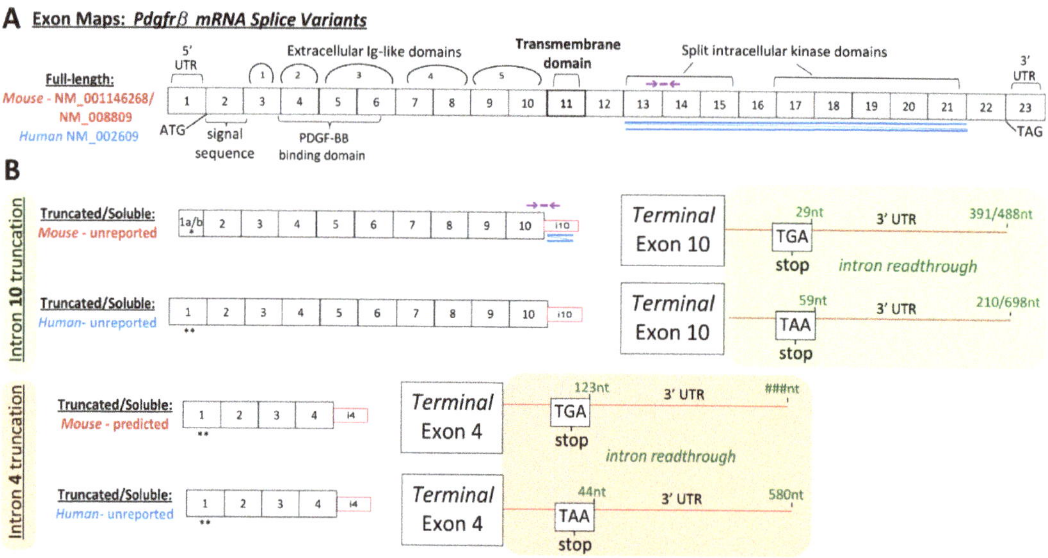

Figure 2. Exon Maps of Full-Length and Truncated *Pdgfrβ* mRNA Transcripts were Constructed Based on Known and RLM-RACE-Generated Sequences. (**A**) Full-length *Pdgfrβ* mRNA exon map based on murine (red accession number) and human (blue accession number) sequences with various regions labeled: signal sequence, PDGF-BB binding domain, five extracellular Ig-like domains, the transmembrane domain, and split intracellular kinase domains. Double blue lines indicate the region used to design RNA-Scope® probes for in situ transcript detection. Purple line and arrows indicate the approximate region corresponding to qRT-PCR probes for relative quantitation of mouse full-length *Pdgfrβ* transcripts. (**B**) *Pdgfrβ* mRNA exon maps for soluble isoforms truncated at intron 10 (green background) or intron 4 (orange background) from both mouse and human. Double blue lines indicate the region used to design RNA-Scope® probes for in situ transcript detection, 3′ UTR intron "read-through regions are shown corresponding to each truncation location and species. Purple line and arrows indicate the region used to design custom qRT-PCR probes for relative quantitation of i10 mouse *sPdgfrβ* transcripts. Single asterisk (*) indicates an established 5′ sequence, while double asterisk (**) indicates 5′ sequences that are currently unresolved for these variants.

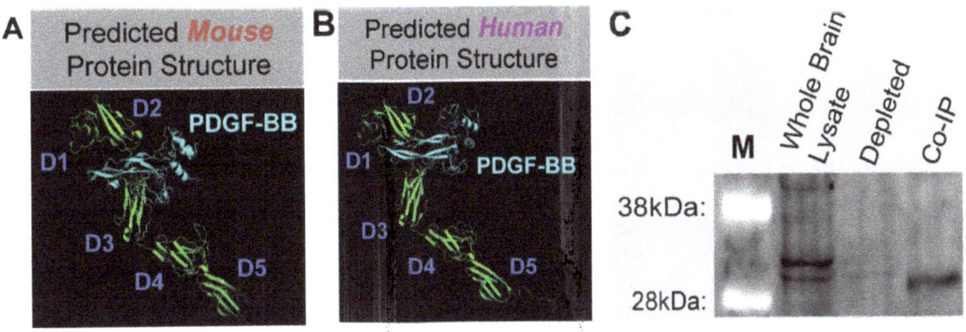

Figure 3. Soluble PDGFRβ Engagement with PDGF-BB Ligand was Supported by Predicted Protein Models of Mouse and Human Splice Variants and sPDGFRβ Co-Immunoprecipitation using PDGF-BB Capture. Protein models of mouse (**A**) and human (**B**) sPDGFRβ variants including predictions for PDGF-BB (cyan) binding. "D" denotes major folding domains. (**C**) Immunoblot results (n = 1, P21 male) from the sPDGFRβ co-IP, with "M" denoting size marker. Whole Lysate shown in Lane 2, the depleted fraction in Lane 3, and the co-IP elution labeled for PDGFRβ in Lane 4. Note the distinct bands at ~32 kDa in the Whole Lysate and co-IP lanes.

3.3. *Pdgfrβ Splice Variant Transcripts Localize near Pecam1-Positive Cells in the Murine Brain with PDGFRβ Protein Associated with Vessels and Interstitial ECM*

In addition to facilitating construction of predicted protein models, sequence identification for the murine i10 *Pdgfrβ* splice variant enabled the design and synthesis of custom RNA-Scope® probes to label mRNA transcripts in situ. We hypothesized that *sPdgfrβ* transcripts in the murine brain would spatially correlate with transcripts of full-length *Pdgfrβ* and with cerebrovascular endothelial cells positive for platelet-endothelial cell adhesion molecule-1 (PECAM-1). To test this hypothesis, we collected normal, healthy WT (c57BL/6) P21 murine brains for snap freezing and coronal cryosectioning. Brain cryo-slices were fixed, dehydrated, and prepared for the hybridization of RNA probes against (i) the i10 *sPdgfrβ* variant sequence (see Figure 2 for regions targeted by custom probes), (ii) full-length *Pdgfrβ* sequence corresponding to the intracellular kinase domain, and (iii) *Pecam1*. Positive and negative control probes were applied to parallel samples. Probes were labeled by corresponding dyes with signal amplification, and imaged by high resolution confocal microscopy. Tiled images were collected to spatially assess *sPdgfrβ* transcript distribution relative to full-length *Pdgfrβ* and *Pecam-1* transcripts (Figure 4).

The vast majority of fluorescent signals associated with *sPdgfrβ* transcripts overlapped with full-length *Pdgfrβ* transcript signals, although not entirely, as distinct *sPdgfrβ*-associated signals were noted. These overlapping signals demonstrated a strong localization with the *Pecam1*-positive signals throughout the brain as well, consistent with the spatial configuration of endothelial cells and pericytes within the brain microvasculature (Figure 4), though overlap could represent a subset of double-positive vascular cells derived from hematopoietic lineages [45]. The distribution of these signals also appeared to be fairly conserved across different brain regions, including the cortex, the hypothalamus, and along the floor of the third ventricle, adjacent to the thalamus (e.g., neighboring the paraventricular nucleus of the thalamus) (Figure 4). In addition, we did not observe any notable differences in the spatial distribution of these signals across sexes.

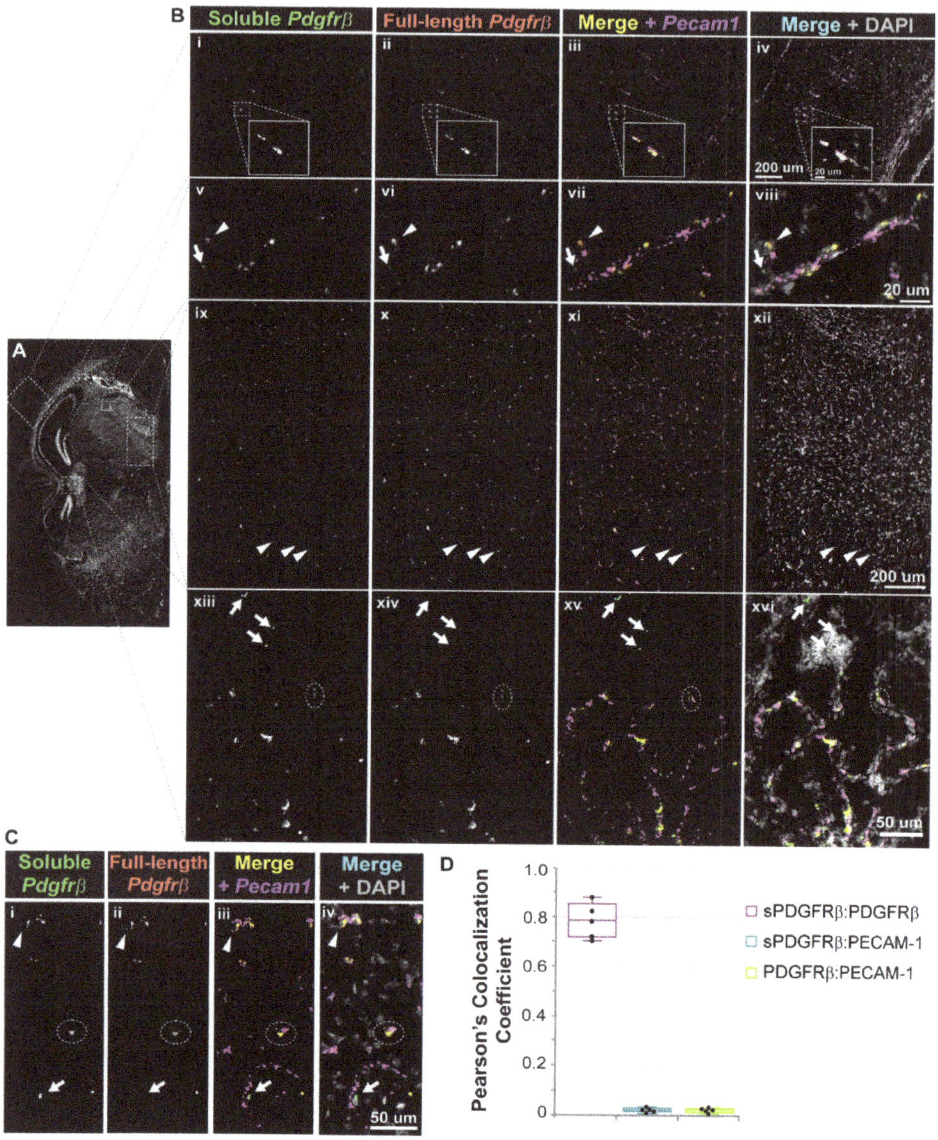

Figure 4. RNA-Scope® Probes Facilitated Detection of Soluble and Full length *Pdgfrβ* Transcripts Adjacent to *Pecam1* mRNA in the Adult Mouse Brain. (**A**) Representative coronal section of an adult mouse brain (from n = 4 biological replicates, both sexes) labeled for cell nuclei (DAPI) with white, dotted boxes indicating regions visualized at a higher-power magnification shown in (**B**). (**B**) High-power magnification images of mouse cerebral cortex and white matter (i–iv with associated inset), thalamus (v–viii), hypothalamus (ix–xii), and along the floor of the third ventricle, adjacent to the thalamus (xiii–xvi) labeled for soluble *Pdgfrβ* transcripts (i, v, ix, xiii; green in iv, viii, xii, xvi), and full-length *Pdgfrβ* transcripts (ii, vi, x, xiv; red in iv, viii, xii, xvi). A merge of these two signals appears yellow in iii, vii, xi, and xv. While most of these signals overlap, we found instances where the signal was stronger for full-length *Pdgfrβ* (arrowheads, ix–xii) or soluble *Pdgfrβ* (arrows, xiii–xvi).

Pecam1 transcripts were also detected (magenta in iii, iv, vii, and viii), along with cell nuclei (DAPI, white in iv and viii), with many locations where *Pecam1*-positive signals spatially correlated with *Pdgfrβ* transcript signals (see dashed white oval, xiii–xvi). Scale bars are 200 microns in iv and xii, 20 microns in the inset in iii and in viii, and 50 microns in xvi. (**C**) Representative high-power images of a mouse brain section labeled with RNA-Scope® probes against soluble *Pdgfrβ* (i, and green in iii and iv), full-length *Pdgfrβ* (ii, and red in iii and iv), and *Pecam1* (magenta in iii and iv), along with cell nuclei (DAPI, white in iv). Most signals converged, though we noted instances of more prominent full-length *Pdgfrβ* (arrowhead) or soluble *Pdgfrβ* (arrow), with signal also corresponding to *Pecam1*-positive signals (dashed white oval). Scale bar is 50 microns. (**D**) Graph of Pearson's Colocalization Coefficient for each of the signal combinations indicated, specifically sPDGFRβ and full-length PDGFRβ (magenta box, sPDGFRβ:PDGFRβ), sPDGFRb and PECAM-1 (cyan box, sPDGFRb: PECAM-1), and full-length PDGFRb and PECAM-1 (green box, PDGFRb:PECAM-1).

Having mapped the spatial distribution of *sPdgfrβ* transcripts relative to full length *Pdgfrβ* mRNA and *Pecam1*-expressing cells within murine brain sections, we then wanted to assess the distribution and localization of PDGFRβ protein. We have previously immunostained early postnatal murine brains for PDGFRβ (P1 [46], and P5 [47]) to observe pericytes and their interactions with the developing cerebrovasculature. In addition to observing distinct signals on the cell surface of pericytes within these brain slices, we noted regions of diffuse signals with varying intensity associated with the PDGFRβ staining. Here, we replicated those findings by immunolabeling PDGFRβ protein alongside visualization of endothelial cells (anti-PECAM-1 antibody labeling) and glial fibrillary acidic protein (GFAP)-positive astrocytes (Figure 5). In brain sections from normal, healthy P1 and P21 WT (c57BL6) mice, we found PDGFRβ-positive pericytes along brain microvessels located in several areas neighboring the caudothalamic groove in the floor of the lateral ventricle, posterior to the interventricular foramen. However, we also observed regions of PDGFRβ signal located beyond vessel walls that were not associated with PECAM-1-positive endothelial cells, presumably distributed in the surrounding extracellular matrix (ECM) (Figure 5). Additionally, we observed increased cellular density corresponding with ECM-associated PDGFRβ signal. Though we could not exclude the possibility of direct overlap between the PDGFRβ and GFAP signals, obvious convergence of these signals appeared relatively infrequent, suggesting minimal association with astrocytes (Figure 5), which have been proposed to express full-length *Pdgfrβ* under certain conditions [48]. Overall, these observations were consistent with non-cell associated PDGFRβ that likely represents a soluble variant that can tether to the ECM.

3.4. *Aging and Acute Hypoxia Induce an Increase in Transcript and Protein Levels of sPDGFRβ*

Cognitive and behavioral abnormalities such as memory loss and dementia arise in part from neurodegeneration occurring in the context of aging and during hypoxic stress [49]. These factors likely contribute to neuronal damage through, among others, cerebrovascular dysfunction, which is presumably driven by aberrant regulation of cells within the blood vessel wall. Because pericyte regulation by the PDGF-BB signaling axis might be involved in these vascular defects, we sought to address the question of how soluble PDGFRβ variants might be affected by these factors, namely aging and hypoxia. To do this, we collected whole murine brains at P5, P21, P90 (3 months old), and 20 months old. Following RNA isolation and cDNA synthesis, we conducted quantitative RT-PCR (Figure 6) using probes against full-length *Pdgfrβ* transcripts (targeting the exon 13–14 boundary) and the soluble i10 *Pdgfrβ* variant (via custom-designed probes—see Figure 2 for regions targeted by custom probes). We found that full-length *Pdgfrβ* transcripts were more abundant than *sPdgfrβ* transcripts at earlier ages (P5 and P21) (Figure 6); however, by P90, *sPdgfrβ* transcript abundance increased by nearly five-fold relative to the P5 timepoint, while full-length *Pdgfrβ* mRNA stayed relatively constant. At 20 months of age, full-length *Pdgfrβ* expression appeared to diminish relative to P5 levels, while *sPdgfrβ* transcript levels continued to increase to nearly 10-11-fold higher relative quantitation than at P5 (Figure 6). This trend was reflected in the relative protein levels of each PDGFRβ isoform at each time point as well. Protein lysates from whole P7, P21, and P90 murine brains were analyzed by Western blot with anti-PDGFRβ and anti-α-tubulin antibodies used for detection. For P7 brains, a distinct band at ~180 kDa was detected, corresponding to full-length PDGFRβ; in contrast, faint signals around the 32 kDa molecular weight were observed, suggesting a relatively low abundance of sPDGFRβ isoforms at this early age (Figure 6). Bands appeared more prominently in this size range for P21 murine brain lysates relative to P7, and signal associated with full-length PDGFRβ also appeared more intense at this time-point. Signal strength for bands in the 32 kDa range intensified further for P90 brain lysates, indicating a potential increase in sPDGFRβ abundance. Bands at the higher molecular weight (~180 kDa) appeared to decrease in density at the P90 age, mirroring the mRNA relative quantitation data (Figure 6).

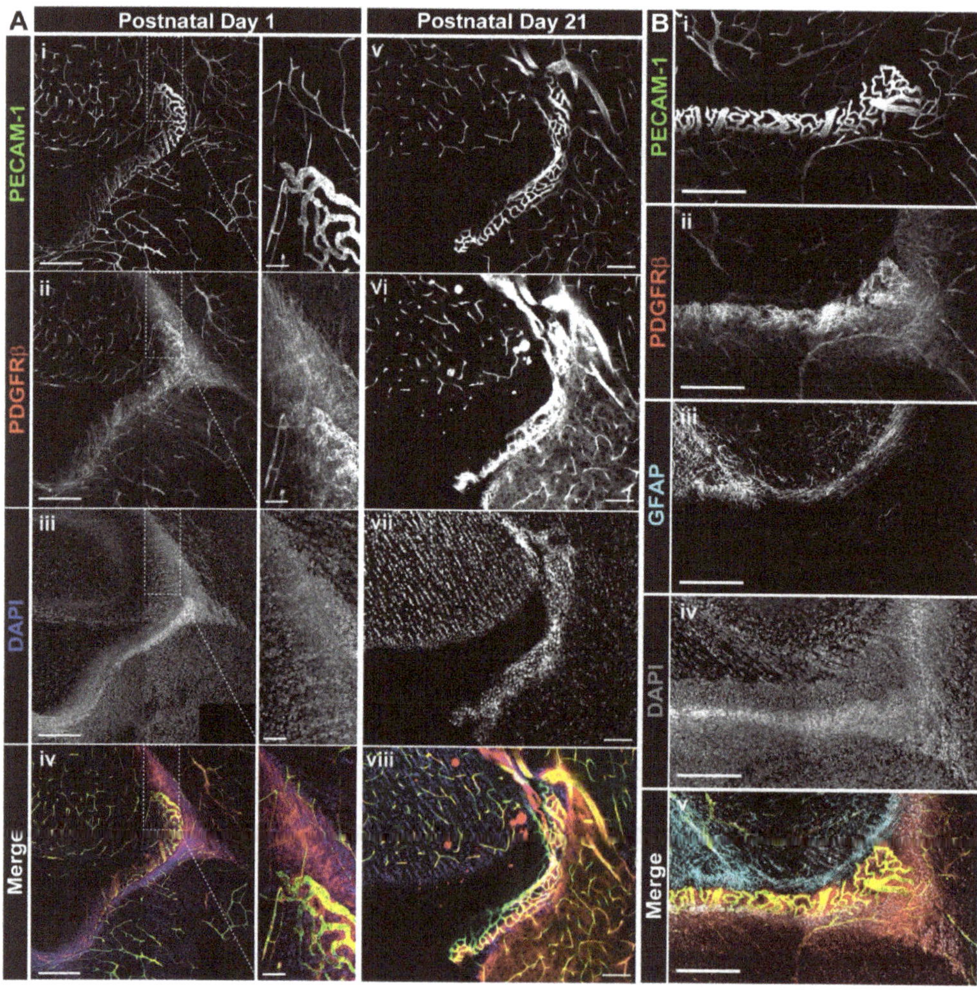

Figure 5. Immunofluorescent Labeling of PDGFRβ in the Mouse Brain Revealed Signals Alongside PECAM-1-Positive Vessels but also Diffusely Distributed throughout the Parenchyma. (**A**) Representative confocal images of mouse brain sections immunohistochemically labeled for PECAM-1 (i, v; green in iv, viii) and PDGFRβ (ii, vi; red in iv, viii) with cell nuclei stained with DAPI

(iii, vii; blue in iv, viii) in P1 (from n = 4 biological replicates, both sexes) and P21 murine brains (n = 3 biological replicates, both sexes—see *Supplemental Figure S1 for images from additional biological replicates*). Scale bars are 250 microns. Note the diffuse signal associated with PDGFRβ immunostaining in addition to signals associated with vessels in ii, iv, vi, and viii. A select region of the P1 vessels is shown at higher magnification to visualize PDGFRβ signals associated with vessel walls and those in the parenchyma. Scale bar is 50 microns. (**B**) Representative confocal images of immunohistochemical labeling of PECAM-1 (i, green in v), PDGFRβ (ii, red in v), and GFAP (iii, cyan in v) with cell nuclei stained with DAPI (iv, white in v) in P1 murine brain (from n = 5 biological replicates, both sexes). Scale bars are 200 microns. Note the diffuse signal associated with PDGFRβ immunostaining in addition to signals associated with vessels in ii and iv.

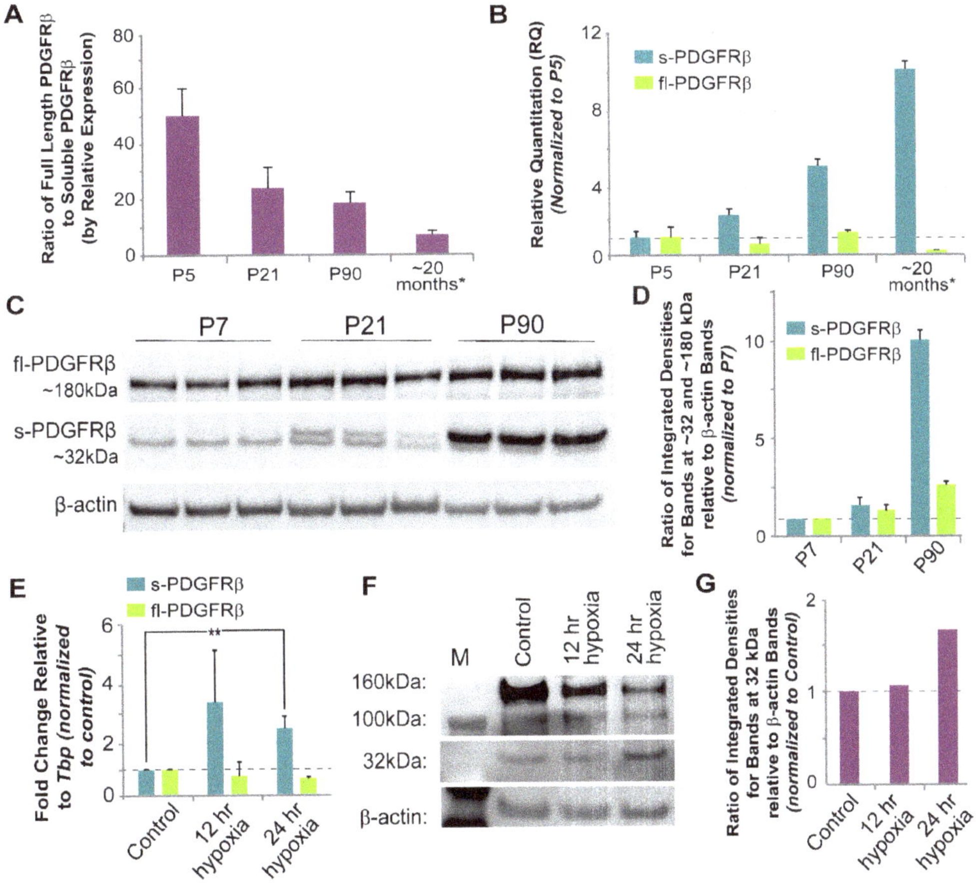

Figure 6. Soluble PDGFRβ Transcript and Protein Levels Increased in the Murine Brain with Age and in a Cell-Based Model of Acute Hypoxia. (**A**) Graph of the ratio of full-length *Pdgfrβ* transcripts relative to the soluble i10 isoform by relative expression using qRT-PCR to evaluate mRNA levels in murine brains at postnatal day 5 (P5), P21, P90, and at ~20 months. The housekeeping gene was *Tbp*. The asterisk (*) denotes samples taken from non-induced *Vhl* control littermates instead of WT, as described in the Methods section. Values are averages with error bars representing standard error of

the mean. Biological replicates were n = 4–5 for all ages and represent both sexes. (**B**) Graph of the relative quantitation (RQ) of soluble i10 (teal) and full-length (green) *Pdgfrβ* in mouse brain at P5, P21, P90, and ~20 months, with values referenced to P5 for relative comparison across time (i.e., indicative of fold change). The asterisk (*) denotes samples taken from non-induced *Vhl* control littermates instead of WT, as described in the Methods section. Values are averages with error bars representing standard error of the mean. Biological replicates were n = 4–5 for all ages and represent both sexes. (**C**) Representative Western blot images of full-length PDGFRβ (fl-PDGFRβ, ~180 kDa) and soluble PDGFRβ (s-PDGFRβ, ~32 kDa) with β-actin shown for reference. Protein lysates were collected from murine brains at P7 (lanes 1–3, n = 3, both sexes), P21 (lanes 4–6, n = 3, both sexes), and P90 (lanes 7–9, n = 3, both sexes). (**D**) Graph of the ratios of integrated densities for 32 kDa (s-PDGFRβ) and 180 kDa (fl-PDGFRβ) bands relative to β-actin bands, with values referenced to P7 for relative comparison across time (i.e., indicative of fold change). (**E**) Relative quantitation of soluble i10 (teal) and full-length (green) *Pdgfrβ* referenced to control samples using qRT-PCR to evaluate mRNA levels from mouse embryonic stem cell (ESC)-derived vessels exposed to hypoxia 12 or 24 h before collection at day 10 differentiation (*Tbp* housekeeping gene). Values are averages with error bars representing standard deviation. n = 3–5 biological replicates for each condition. ** $p \leq 0.01$ by ordinary one-way ANOVA followed by unpaired Tukey's *t*-test (outlier data points removed by Grubbs' test, or the extreme studentized deviate method, with $\alpha = 0.05$). (**F**) Representative Western blot images of full-length PDGFRβ (fl-PDGFRβ, ~160 kDa) and soluble PDGFRβ (s-PDGFRβ, ~32 kDa) with β-actin (~50 kDa) shown for reference. Protein lysates were collected from differentiated ESCs under normoxia (control) or exposed to 12 or 24 h of hypoxia. n = 3–5 biological replicates for each condition. (**G**) Graph of the ratios of integrated densities for 32 kDa (s-PDGFRβ) bands relative to β-actin bands, with values referenced to Control (normoxia) for relative comparison across experimental groups.

To test our hypothesis that sPDGFRβ variant levels are oxygen-sensitive, we took advantage of a recently developed cell-based model containing reporter-labeled endothelial cell and pericyte lineages that would permit a high degree of experimental control over the oxygen environment. Specifically, mouse embryonic stem cells (ESCs) were differentiated into endothelial cell and pericyte precursors that give rise to primitive vascular structures composed of these cell types [43]. Following visual confirmation of early vessel formation, we exposed these cultures to hypoxic conditions for 12 and 24 h, and then collected mRNA for quantification of the soluble i10 *Pdgfrβ* variant and full-length *Pdgfrβ*. After 12 and 24 h of exposure to 3% O_2, soluble *Pdgfrβ* isoforms displayed a nearly three-fold increase in transcript levels relative to the control (i.e., cultures exposed to ambient O_2 levels until collection) (Figure 6). In contrast, the expression of full-length *Pdgfrβ* remained relatively constant, with a slight decrease at the 24-h time point. Protein levels of a ~32 kDa isoform that stained positive for PDGFRβ remained fairly constant at 12 h of hypoxia (relative to controls), but increased after 24 h of hypoxic conditions (Figure 6). These data suggest that *sPdgfrβ* levels are likely sensitive to variations in oxygen levels, with acute hypoxia inducing a marked increase.

4. Discussion

The PDGF-BB pathway is a critical regulator of microvascular pericytes, facilitating their investment and retention within the cerebral microcirculation. Similar RTKs such as VEGF-A are modulated by soluble isoforms of cognate receptors that maintain signaling within a physiological range. Soluble variants of PDGFRβ have been reported, with their origins being attributed to enzymatic cleavage from the surface of vascular mural cells, and specifically cerebral pericytes [31,32]. Here, we present evidence that sPDGFRβ is also generated by pre-mRNA alternative splicing, terminating prior to the transmembrane and intracellular domains. We detected truncated PDGFRβ isoforms in a range of murine tissues, including the blood (serum), kidney, and brain. Given the importance of PDGFRβ signaling for cerebral pericytes, we utilized brain tissue for follow-on analysis, including sequence identification and protein model predictions. Corresponding sequences and protein models were also found in human cells, suggesting a conservation across species. Co-immunoprecipitation of truncated/soluble PDGFRβ with PDGF-BB corroborated protein-protein docking model predictions and crystal structure data (human PDGF-BB with PDGFRβ domains 1–3), supporting the likely essential functional relevance of sPDGFRβ. In situ hybridization and fluorescent labeling of *sPdgfrβ* transcripts in murine brain revealed a strong, though not complete, colocalization with full-length *Pdgfrβ* transcripts. Soluble *Pdgfrβ* transcripts were also in close proximity to transcripts labeled for the endothelial cell marker *Pecam1*, consistent with cerebrovascular mural cells as a source of sPDGFRβ under normal physiological conditions. Visualization of immunofluorescently labeled PDGFRβ protein distribution throughout certain brain regions suggested the presence of ECM-associated PDGFRβ protein in the brain parenchyma, in addition to vessel-associated pericytes. We also took initial steps to understand factors that might regulate the production of *sPdgfrβ* variants, and found that aging and hypoxia were associated with an increased abundance of i10 soluble PDGFRβ isoforms relative to full-length *Pdgfrβ* expression. Overall, our data are consistent with the idea that (i) soluble PDGFRβ isoforms are present in the normal murine brain, likely generated by pre-mRNA alternative splicing, and (ii) corresponding *sPdgfrβ* transcripts, which were similarly detected in human cells, are sensitive to aging and hypoxia–two factors that can fuel cerebrovascular dysfunction, neurodegeneration, and in turn, memory loss and dementia.

Truncated isoforms of PDGFRβ have been previously reported, with a range of molecular weights reported for different physiological and pathological contexts [5,30–33,44]. For instance, in the setting of amyloid-beta exposure or severe hypoxia (1% O_2), human pericytes are proposed to shed a soluble isoform of PDGFRβ in the size range of ~160 kDa [5], suggested to be an indicator of A Disintegrin And Metalloproteinase domain-containing protein 10 (ADAM10) activity, BBB breakdown, and cognitive decline [32]. Cultured primary mouse pericytes from normal, healthy postnatal (P5) and adult (6 months) brains, however, have also been reported to generate 110 and 160 kDa species of PDGFRβ, while cortical slices from mouse brain generated a 60 kDa PDGFRβ isoform after 2 weeks in organotypic culture [33]. Considering current sequence data from both humans and mice, it appears that the entire extracellular portion of PDGFRβ–post signal cleavage–is around 55 kDa without glycosylation. Therefore, truncated forms of PDGFRβ larger than ~60 kDa (allowing for glycosylation) likely extend through the transmembrane domain (<3 kDa) and into the intracellular region. As we learn more about truncated PDGFRβ isoforms, we will also need to expand our understanding of the potential mechanisms that may give rise to these unique growth factor receptor species. In the current study, we repeatedly detected an ~32 kDa PDGFRβ variant from normal, healthy murine brains, with double bands potentially arising from variable glycosylation. Our sequence analysis found a correspondence between this protein and a *Pdgfrβ* splice variant generated by truncation in intron 4 (i4 splice variant–identified in human cells, and a predicted sequence in mice cells still being resolved). The i10 *Pdgfrβ* splice variant likely corresponds to larger protein isoforms (~50–60 kDa) previously detected by our group [44] and others [33]. We have demonstrated here that sPDGFRβ i10 is expressed during normal physiological conditions, but is altered during hypoxia and aging in the brain. Thus, these findings provide a new perspective for future investigations into PDGFRβ isoforms that are essential for tissue homeostasis versus those associated with pathology, as well as for studies identifying their context-dependent regulatory mechanisms.

The spatial distribution of soluble growth factor receptors often depends on the presence of unique motifs that facilitate binding to the ECM. Soluble Flt-1/VEGFR1, for instance, includes a heparin-binding domain in the fourth Ig-like loop of its protein structure [50], which can be targeted for sFlt-1/sVEGFR1 release from the ECM by exposure to unfractionated heparin in tissues such as the placenta [51]. Sequence analysis in the current study suggests that the previously unreported i10 PDGFRβ splice variant (but not the i4 isoform) retains a heparin binding domain capable of facilitating physical association with the ECM via heparin sulfate interactions. This property is consistent with our observation of a diffuse signal for PDGFRβ immunostaining in murine brain slices in addition to the expected labeling of full-length PDGFRβ on pericyte cell walls. We noted an enrichment of this more dispersed signal in periventricular brain regions. These areas displayed a relatively higher density of cells, as seen from the corresponding nuclear labeling, suggesting an increased concentration of ECM proteins as well. PDGFRβ that may be present on the astrocyte cell wall did not appear to substantially contribute to this diffuse signal, though we could not fully exclude contributions from astrocytes or all possible cell types that may be expressing PDGFRβ in these brain regions. Another phenomenon potentially contributing to elevated sPDGFRβ in these sub-regions may be an increased exposure to PDGF-BB ligands. Soluble receptors associated with RTKs are often generated as a feedback mechanism to maintain growth factor signaling within an appropriate range [22,26,35–38]. Therefore, their levels frequently increase when cells are exposed to high levels of ligand, and this response may be partially responsible for the noted increase in sPDGFRβ immunostaining signal in periventricular areas. Previous studies have suggested that the production of specific PDGFRβ isoforms may indeed be ligand-sensitive [33], but testing this response for the isoforms identified herein was beyond the scope of the current study. Additional insight into how PDGFRβ isoform levels and distribution may be differentially regulated throughout the brain will be critical for understanding not only their upstream regulation, but also their functional relevance for cerebrovascular homeostasis.

Growth factor receptor isoforms can perform a number of unique functions, depending on the signaling pathway and the context. The soluble epidermal growth factor receptor (sEGFR), for example, can inhibit intracellular signaling in target cells by (i) acting as a "decoy" receptor to prevent ligand binding of surface receptors, or (ii) binding the extracellular domain of full-length EGFRs. Soluble variant interaction with transmembrane receptors can create inactive heterodimers or prevent internalization [52], thereby interfering with initial signaling events. Previous studies have utilized full-length sequence information to artificially create truncated, soluble versions of PDGFRβ [41,53]. These dominant-negative constructs have been introduced into a variety of settings, but their primary effect appears to be antagonizing signaling via full-length PDGFRβ and accordingly altering cell activation and behavior. Truncated PDGFRβ isoforms generated by proteolytic cleavage have been primarily considered biomarkers of pericyte injury or death [31,32], with potential functional roles remaining relatively unexplored. Soluble PDGFRβ variants produced by alternative splicing also remain to be characterized with regard to their functional relevance, but, supported by ligand-binding capacity demonstrated herein, it is likely that they play similar roles as those within other RTK pathways, such as sFlt-1/VEGFR1 [22,26,35–38]. In pathological scenarios, dysregulated sPDGFRβ interference with signaling via the full-length receptor, particularly for cerebrovascular pericytes, is likely to lead to deleterious effects akin to those reported for sFlt-1/sVEGFR1 [34]. More specifically, sPDGFRβ-mediated attenuation of pericyte signaling via full-length receptors may undermine pericyte investment within cerebral capillaries, leading to BBB and/or vasomotor defects, neuronal dysfunction and degeneration, and ultimately, memory loss and cognitive decline [13]. Thus, as we continue to learn more about how sPDGFRβ isoforms are created and regulated in the brain and beyond, it will be essential to also establish their contributions to both physiological and pathological processes, to better understand their utility as biomarkers of disease and also as potential drug targets.

Supplementary Materials: The following supporting information can be downloaded at: https://www.mdpi.com/article/10.3390/biom13040711/s1, Figure S1: Immunofluorescent Labeling of PDGFRβ in the Mouse Brain Revealed Signals alongside PECAM-1-Positive Vessels but also Distributed Diffusely throughout the Parenchyma; File S1: Images of Raw Immunoblots in Support of the Indicated Datasets within the Manuscript; Table S1: Nucleotide sequence information for mouse and human *Pdgfrβ* variants; Table S2: Detailed amino acid predictions for the mouse and human i4 and i10 PDGFRβ splice variants.

Author Contributions: Conceptualization, L.B.P. and J.C.C.; methodology, L.B.P., H.A., M.H., A.B., Z.M., C.E.W., J.D., C.H., M.W.S., S.R.J. and J.C.C.; formal analysis, L.B.P., H.A., M.H., A.B., Z.M., C.E.W., J.D., C.H., M.W.S., S.R.J. and J.C.C.; investigation, L.B.P., H.A., M.H., A.B., Z.M., C.E.W., J.D., C.H., M.W.S., S.R.J. and J.C.C.; resources, L.B.P., S.R.J. and J.C.C.; data curation, L.B.P. and J.C.C.; writing—original draft preparation, L.B.P. and J.C.C.; writing—review and editing, L.B.P., S.R.J. and J.C.C.; visualization, L.B.P. and J.C.C.; supervision, L.B.P. and J.C.C.; project administration, L.B.P. and J.C.C.; funding acquisition, L.B.P., S.R.J. and J.C.C. All authors have read and agreed to the published version of the manuscript.

Funding: This work was supported in part by funding from the National Institutes of Health (R01HL146596 and R21HL165296 to J.C.C.), the National Science Foundation (CAREER Award 1752339 to J.C.C.), and the American Heart Association (Career Development Award #19CDA34630036 to S.R.J.).

Institutional Review Board Statement: All experiments involving animal use were performed following review and approval from the Institutional Animal Care and Use Committee (IACUC) at Virginia Tech. All experimental protocols were reviewed and approved by Virginia Tech Veterinary Staff and the IACUC. The Virginia Tech NIH/PHS Animal Welfare Assurance Number is A-3208-01 (Expires: 31 July 2025).

Informed Consent Statement: Informed consent was not required for the current study, as the human cells used herein were acquired from a commercial source.

Data Availability Statement: The data presented in this study are available on request from the corresponding author.

Acknowledgments: We would like to thank all of the members of the Chappell lab for their support and valued assistance on this project, both materially and intellectually.

Conflicts of Interest: The authors declare no conflict of interest.

Abbreviations

BBB—Blood-Brain Barrier, PDGFRβ—Platelet-Derived Growth Factor Receptor-b, VEGF-A—Vascular Endothelial Growth Factor-A, Receptor Tyrosine Kinase—RTK, ECM—Extracellular Matrix, P#—Postnatal Day #, PECAM-1—Platelet-Endothelial Cell Adhesion Molecule-1.

References

1. Coelho-Santos, V.; Shih, A.Y. Postnatal development of cerebrovascular structure and the neurogliovascular unit. *Wiley Interdiscip. Rev. Dev. Biol.* **2020**, *9*, e363. [CrossRef]
2. Iadecola, C. The Neurovascular Unit Coming of Age: A Journey through Neurovascular Coupling in Health and Disease. *Neuron* **2017**, *96*, 17–42. [CrossRef]
3. Miners, J.S.; Schulz, I.; Love, S. Differing associations between Aβ accumulation, hypoperfusion, blood-brain barrier dysfunction and loss of PDGFRB pericyte marker in the precuneus and parietal white matter in Alzheimer's disease. *J. Cereb. Blood Flow Metab.* **2018**, *38*, 103–115. [CrossRef]
4. Charkviani, M.; Muradashvili, N.; Lominadze, D. Vascular and non-vascular contributors to memory reduction during traumatic brain injury. *Eur. J. Neurosci.* **2019**, *50*, 2860–2876. [CrossRef]
5. Montagne, A.; Barnes, S.R.; Sweeney, M.D.; Halliday, M.R.; Sagare, A.P.; Zhao, Z.; Toga, A.W.; Jacobs, R.E.; Liu, C.Y.; Amezcua, L.; et al. Blood-brain barrier breakdown in the aging human hippocampus. *Neuron* **2015**, *85*, 296–302. [CrossRef] [PubMed]
6. Armulik, A.; Genove, G.; Betsholtz, C. Pericytes: Developmental, physiological, and pathological perspectives, problems, and promises. *Dev. Cell* **2011**, *21*, 193–215. [CrossRef]
7. Abdelazim, H.; Payne, L.B.; Nolan, K.; Paralkar, K.; Bradley, V.; Kanodia, R.; Gude, R.; Ward, R.; Monavarfeshani, A.; Fox, M.A.; et al. Pericyte heterogeneity identified by 3D ultrastructural analysis of the microvessel wall. *Front. Physiol.* **2022**, *13*, 1016382. [CrossRef] [PubMed]
8. Armulik, A.; Genové, G.; Mäe, M.; Nisancioglu, M.H.; Wallgard, E.; Niaudet, C.; He, L.; Norlin, J.; Lindblom, P.; Strittmatter, K.; et al. Pericytes regulate the blood-brain barrier. *Nature* **2010**, *468*, 557–561. [CrossRef] [PubMed]
9. Daneman, R.; Zhou, L.; Kebede, A.A.; Barres, B.A. Pericytes are required for blood-brain barrier integrity during embryogenesis. *Nature* **2010**, *468*, 562–566. [CrossRef] [PubMed]

10. Lindblom, P.; Gerhardt, H.; Liebner, S.; Abramsson, A.; Enge, M.; Hellström, M.; Bäckström, G.; Fredriksson, S.; Landegren, U.; Nyström, H.C.; et al. Endothelial PDGF-B retention is required for proper investment of pericytes in the microvessel wall. *Genes Dev.* **2003**, *17*, 1835–1840. [CrossRef]
11. Ando, K.; Shih, Y.-H.; Ebarasi, L.; Grosse, A.; Portman, D.; Chiba, A.; Mattonet, K.; Gerri, C.; Stainier, D.Y.R.; Mochizuki, N.; et al. Conserved and context-dependent roles for pdgfrb signaling during zebrafish vascular mural cell development. *Dev. Biol.* **2021**, *479*, 11–22. [CrossRef]
12. Shen, J.; Xu, G.; Zhu, R.; Yuan, J.; Ishii, Y.; Hamashima, T.; Matsushima, T.; Yamamoto, S.; Takatsuru, Y.; Nabekura, J.; et al. PDGFR-β restores blood-brain barrier functions in a mouse model of focal cerebral ischemia. *J. Cereb. Blood Flow Metab.* **2019**, *39*, 1501–1515. [CrossRef]
13. Smyth, L.C.D.; Highet, B.; Jansson, D.; Wu, J.; Rustenhoven, J.; Aalderink, M.; Tan, A.; Li, S.; Johnson, R.; Coppieters, N.; et al. Characterisation of PDGF-BB:PDGFR-β signalling pathways in human brain pericytes: Evidence of disruption in Alzheimer's disease. *Commun. Biol.* **2022**, *5*, 235. [CrossRef]
14. Brown, L.S.; Foster, C.G.; Courtney, J.-M.; King, N.E.; Howells, D.W.; Sutherland, B.A. Pericytes and Neurovascular Function in the Healthy and Diseased Brain. *Front. Cell. Neurosci.* **2019**, *13*, 282. [CrossRef] [PubMed]
15. Langen, U.H.; Ayloo, S.; Gu, C. Development and Cell Biology of the Blood-Brain Barrier. *Annu. Rev. Cell Dev. Biol.* **2019**, *35*, 591–613. [CrossRef] [PubMed]
16. Ayloo, S.; Lazo, C.G.; Sun, S.; Zhang, W.; Cui, B.; Gu, C. Pericyte-to-endothelial cell signaling via vitronectin-integrin regulates blood-CNS barrier. *Neuron* **2022**, *110*, 1641–1655.e6. [CrossRef]
17. Hall, C.N.; Reynell, C.; Gesslein, B.; Hamilton, N.B.; Mishra, A.; Sutherland, B.A.; O'Farrell, F.M.; Buchan, A.M.; Lauritzen, M.; Attwell, D. Capillary pericytes regulate cerebral blood flow in health and disease. *Nature* **2014**, *508*, 55–60. [CrossRef] [PubMed]
18. Hill, R.A.; Tong, L.; Yuan, P.; Murikinati, S.; Gupta, S.; Grutzendler, J. Regional Blood Flow in the Normal and Ischemic Brain Is Controlled by Arteriolar Smooth Muscle Cell Contractility and Not by Capillary Pericytes. *Neuron* **2015**, *87*, 95–110. [CrossRef]
19. Bennett, H.C.; Kim, Y. Pericytes Across the Lifetime in the Central Nervous System. *Front. Cell. Neurosci.* **2021**, *15*, 627291. [CrossRef]
20. Wu, A.; Sharrett, A.R.; Gottesman, R.F.; Power, M.C.; Mosley, T.H., Jr.; Jack, C.R., Jr.; Knopman, D.S.; Windham, B.G.; Gross, A.L.; Coresh, J. Association of Brain Magnetic Resonance Imaging Signs With Cognitive Outcomes in Persons With Nonimpaired Cognition and Mild Cognitive Impairment. *JAMA Netw. Open* **2019**, *2*, e193359. [CrossRef]
21. Andrae, J.; Gallini, R.; Betsholtz, C. Role of platelet-derived growth factors in physiology and medicine. *Genes Dev.* **2008**, *22*, 1276–1312. [CrossRef]
22. Vorlová, S.; Rocco, G.; Lefave, C.V.; Jodelka, F.M.; Hess, K.; Hastings, M.L.; Henke, E.; Cartegni, L. Induction of antagonistic soluble decoy receptor tyrosine kinases by intronic polyA activation. *Mol. Cell* **2011**, *43*, 927–939. [CrossRef]
23. Fong, G.-H.; Rossant, J.; Gertsenstein, M.; Breitman, M.L. Role of the Flt-1 receptor tyrosine kinase in regulating the assembly of vascular endothelium. *Nature* **1995**, *376*, 66–70. [CrossRef] [PubMed]
24. Hiratsuka, S.; Minowa, O.; Kuno, J.; Noda, T.; Shibuya, M. Flt-1 lacking the tyrosine kinase domain is sufficient for normal development and angiogenesis in mice. *Proc. Natl. Acad. Sci. USA* **1998**, *95*, 9349–9354. [CrossRef] [PubMed]
25. Sela, S.; Itin, A.; Natanson-Yaron, S.; Greenfield, C.; Goldman-Wohl, D.; Yagel, S.; Keshet, E. A novel human-specific soluble vascular endothelial growth factor receptor 1: Cell-type-specific splicing and implications to vascular endothelial growth factor homeostasis and preeclampsia. *Circ. Res.* **2008**, *102*, 1566–1574. [CrossRef] [PubMed]
26. Chappell, J.C.; Taylor, S.M.; Ferrara, N.; Bautch, V.L. Local guidance of emerging vessel sprouts requires soluble Flt-1. *Dev. Cell* **2009**, *17*, 377–386. [CrossRef]
27. Wu, F.T.H.; Stefanini, M.O.; Gabhann, F.M.; Kontos, C.D.; Annex, B.H.; Popel, A.S. A systems biology perspective on sVEGFR1: Its biological function, pathogenic role & therapeutic use. *J. Cell. Mol. Med.* **2009**, *14*, 528–552. [CrossRef]
28. Ancot, F.; Foveau, B.; Lefebvre, J.; Leroy, C.; Tulasne, D. Proteolytic cleavages give receptor tyrosine kinases the gift of ubiquity. *Oncogene* **2009**, *28*, 2185–2195. [CrossRef] [PubMed]
29. Raikwar, N.S.; Liu, K.Z.; Thomas, C.P. Protein kinase C regulates FLT1 abundance and stimulates its cleavage in vascular endothelial cells with the release of a soluble PlGF/VEGF antagonist. *Exp. Cell Res.* **2013**, *319*, 2578–2587. [CrossRef]
30. Mendelson, K.; Swendeman, S.; Saftig, P.; Blobel, C.P. Stimulation of platelet-derived growth factor receptor β (PDGFRβ) activates ADAM17 and promotes metalloproteinase-dependent cross-talk between the PDGFRβ and epidermal growth factor receptor (EGFR) signaling pathways. *J. Biol. Chem.* **2010**, *285*, 25024–25032. [CrossRef]
31. Sagare, A.P.; Sweeney, M.D.; Makshanoff, J.; Zlokovic, B.V. Shedding of soluble platelet-derived growth factor receptor-β from human brain pericytes. *Neurosci. Lett.* **2015**, *607*, 97–101. [CrossRef] [PubMed]
32. Nation, D.A.; Sweeney, M.D.; Montagne, A.; Sagare, A.P.; D'Orazio, L.M.; Pachicano, M.; Sepehrband, F.; Nelson, A.R.; Buennagel, D.P.; Harrington, M.G.; et al. Blood-brain barrier breakdown is an early biomarker of human cognitive dysfunction. *Nat. Med.* **2019**, *25*, 270–276. [CrossRef]
33. Hutter-Schmid, B.; Humpel, C. Platelet-derived Growth Factor Receptor-beta is Differentially Regulated in Primary Mouse Pericytes and Brain Slices. *Curr. Neurovasc. Res.* **2016**, *13*, 127–134. [CrossRef]
34. Lizano, P.L.; Yao, J.K.; Tandon, N.; Mothi, S.S.; Montrose, D.M.; Keshavan, M.S. Association of sFlt-1 and worsening psychopathology in relatives at high risk for psychosis: A longitudinal study. *Schizophr. Res.* **2017**, *183*, 75–81. [CrossRef]

35. Kendall, R.L.; Thomas, K.A. Inhibition of vascular endothelial cell growth factor activity by an endogenously encoded soluble receptor. *Proc. Natl. Acad. Sci. USA* **1993**, *90*, 10705–10709. [CrossRef]
36. Kendall, R.L.; Wang, G.; Thomas, K.A. Identification of a natural soluble form of the vascular endothelial growth factor receptor, FLT-1, and its heterodimerization with KDR. *Biochem. Biophys. Res. Commun.* **1996**, *226*, 324–328. [CrossRef] [PubMed]
37. Ebos, J.M.; Bocci, G.; Man, S.; Thorpe, P.E.; Hicklin, D.J.; Zhou, D.; Jia, X.; Kerbel, R.S. A naturally occurring soluble form of vascular endothelial growth factor receptor 2 detected in mouse and human plasma. *Mol. Cancer Res.* **2004**, *2*, 315–326. [CrossRef] [PubMed]
38. Albuquerque, R.J.; Hayashi, T.; Cho, W.G.; Kleinman, M.E.; Dridi, S.; Takeda, A.; Baffi, J.Z.; Yamada, K.; Kaneko, H.; Green, M.G.; et al. Alternatively spliced vascular endothelial growth factor receptor-2 is an essential endogenous inhibitor of lymphatic vessel growth. *Nat. Med.* **2009**, *15*, 1023–1030. [CrossRef]
39. Lemmon, M.A.; Schlessinger, J. Cell signaling by receptor tyrosine kinases. *Cell* **2010**, *141*, 1117–1134. [CrossRef]
40. Krueger, J.; Liu, D.; Scholz, K.; Zimmer, A.; Shi, Y.; Klein, C.; Siekmann, A.; Schulte-Merker, S.; Cudmore, M.; Ahmed, A.; et al. Flt1 acts as a negative regulator of tip cell formation and branching morphogenesis in the zebrafish embryo. *Development* **2011**, *138*, 2111–2120. [CrossRef]
41. Duan, D.S.; Pazin, M.J.; Fretto, L.J.; Williams, L.T. A functional soluble extracellular region of the platelet-derived growth factor (PDGF) beta-receptor antagonizes PDGF-stimulated responses. *J. Biol. Chem.* **1991**, *266*, 413–418. [CrossRef]
42. Arreola, A.; Payne, L.B.; Julian, M.H.; de Cubas, A.A.; Daniels, A.B.; Taylor, S.; Zhao, H.; Darden, J.; Bautch, V.L.; Rathmell, W.K.; et al. Von Hippel-Lindau mutations disrupt vascular patterning and maturation via Notch. *JCI Insight* **2018**, *3*, e92193. [CrossRef] [PubMed]
43. Payne, L.B.; Tewari, B.P.; Dunkenberger, L.; Bond, S.; Savelli, A.; Darden, J.; Zhao, H.; Willi, C.; Kanodia, R.; Gude, R.; et al. Pericyte Progenitor Coupling to the Emerging Endothelium During Vasculogenesis via Connexin 43. *Arter. Thromb. Vasc. Biol.* **2022**, *42*, e96–e114. [CrossRef]
44. Darden, J.; Payne, L.B.; Zhao, H.; Chappell, J.C. Excess vascular endothelial growth factor-A disrupts pericyte recruitment during blood vessel formation. *Angiogenesis* **2018**, *22*, 167–183. [CrossRef]
45. Yamamoto, S.; Muramatsu, M.; Azuma, E.; Ikutani, M.; Nagai, Y.; Sagara, H.; Koo, B.-N.; Kita, S.; O'donnell, E.; Osawa, T.; et al. A subset of cerebrovascular pericytes originates from mature macrophages in the very early phase of vascular development in CNS. *Sci. Rep.* **2017**, *7*, 3855. [CrossRef]
46. Payne, L.B.; Darden, J.; Suarez-Martinez, A.D.; Zhao, H.; Hendricks, A.; Hartland, C.; Chong, D.; Kushner, E.J.; Murfee, W.L.; Chappell, J.C. Pericyte migration and proliferation are tightly synchronized to endothelial cell sprouting dynamics. *Integr. Biol.* **2021**, *13*, 31–43. [CrossRef]
47. Payne, L.B.; Zhao, H.; James, C.C.; Darden, J.; McGuire, D.; Taylor, S.; Smyth, J.W.; Chappell, J.C. The pericyte microenvironment during vascular development. *Microcirculation* **2019**, *26*, e12554. [CrossRef] [PubMed]
48. Kyyriäinen, J.; Ndode-Ekane, X.E.; Pitkänen, A. Dynamics of PDGFRβ expression in different cell types after brain injury. *Glia* **2017**, *65*, 322–341. [CrossRef]
49. Livingston, G.; Huntley, J.; Sommerlad, A.; Ames, D.; Ballard, C.; Banerjee, S.; Brayne, C.; Burns, A.; Cohen-Mansfield, J.; Cooper, C.; et al. Dementia prevention, intervention, and care: 2020 report of the Lancet Commission. *Lancet* **2020**, *396*, 413–446. [CrossRef] [PubMed]
50. Park, M.; Lee, S.-T. The fourth immunoglobulin-like loop in the extracellular domain of FLT-1, a VEGF receptor, includes a major heparin-binding site. *Biochem. Biophys. Res. Commun.* **1999**, *264*, 730–734. [CrossRef]
51. Moore, K.H.; Chapman, H.; George, E.M. Unfractionated heparin displaces sFlt-1 from the placental extracellular matrix. *Biol. Sex Differ.* **2020**, *11*, 34. [CrossRef] [PubMed]
52. Maramotti, S.; Paci, M.; Manzotti, G.; Rapicetta, C.; Gugnoni, M.; Galeone, C.; Cesario, A.; Lococo, F. Soluble Epidermal Growth Factor Receptors (sEGFRs) in Cancer: Biological Aspects and Clinical Relevance. *Int. J. Mol. Sci.* **2016**, *17*, 593. [CrossRef] [PubMed]
53. Borkham-Kamphorst, E.; Herrmann, J.; Stoll, D.; Treptau, J.; Gressner, A.M.; Weiskirchen, R. Dominant-negative soluble PDGF-β receptor inhibits hepatic stellate cell activation and attenuates liver fibrosis. *Lab. Investig.* **2004**, *84*, 766–777. [CrossRef] [PubMed]

biomolecules

MDPI

Article

Elevated Soluble TNF-Receptor 1 in the Serum of Predementia Subjects with Cerebral Small Vessel Disease

Kaung H. T. Salai [1,2], Liu-Yun Wu [1,3], Joyce R. Chong [1,3], Yuek Ling Chai [1,3], Bibek Gyanwali [1,3], Caroline Robert [1], Saima Hilal [1,3,4,5], Narayanaswamy Venketasubramanian [6], Gavin S. Dawe [1,2,7,8], Christopher P. Chen [1,2,3] and Mitchell K. P. Lai [1,2,3,*]

1 Department of Pharmacology, Yong Loo Lin School of Medicine, National University of Singapore, Singapore 117600, Singapore
2 Healthy Longevity Translational Research Programme, Yong Loo Lin School of Medicine, National University of Singapore, Singapore 117456, Singapore
3 Memory Aging and Cognition Centre, National University Health System, Singapore 117600, Singapore
4 Saw Swee Hock School of Public Health, National University of Singapore, Singapore 117597, Singapore
5 Departments of Epidemiology and Radiology & Nuclear Medicine, Erasmus University Medical Center, 3015 GD Rotterdam, The Netherlands
6 Raffles Neuroscience Centre, Raffles Hospital, Singapore 188770, Singapore
7 Precision Medicine Translational Research Programme, Yong Loo Lin School of Medicine, National University of Singapore, Singapore 117596, Singapore
8 Neurobiology Programme, Life Sciences Institute, Centre for Life Sciences, National University of Singapore, Singapore 117456, Singapore
* Correspondence: mitchell.lai@dementia-research.org

Abstract: Tumor necrosis factor-receptor 1 (TNF-R1)-mediated signaling is critical to the regulation of inflammatory responses. TNF-R1 can be proteolytically released into systemic blood circulation in a soluble form (sTNF-R1), where it binds to circulating TNF and functions to attenuate TNF-mediated inflammation. Increases of peripheral sTNF-R1 have been reported in both Alzheimer's disease (AD) dementia and vascular dementia (VaD). However, the status of sTNF-R1 in predementia subjects (cognitive impairment, no dementia, CIND) is unknown, and putative associations with cerebral small vessel disease (CSVD), as well as with longitudinal changes in cognitive functions are unclear. We measured baseline serum sTNF-R1 in a longitudinally assessed cohort of 93 controls and 103 CIND, along with neuropsychological evaluations and neuroimaging assessments. Serum sTNF-R1 levels were increased in CIND compared with controls ($p < 0.001$). Higher baseline sTNF-R1 levels were specifically associated with lacunar infarcts (rate ratio = 6.91, 95% CI 3.19–14.96, $p < 0.001$), as well as lower rates of cognitive decline in the CIND subgroup. Our data suggest that sTNF-R1 interacts with vascular cognitive impairment in a complex manner at predementia stages, with elevated levels associated with more severe CSVD at baseline, but which may subsequently be protective against cognitive decline.

Keywords: Alzheimer's disease; biomarker; cerebral small vessel diseases; predementia; serum; TNF-receptor 1; tumor necrosis factor; vascular cognitive impairment

1. Introduction

Alzheimer's disease (AD) and vascular dementia (VaD) are the two most common forms of dementia [1]. AD is a neurodegenerative disease characterized by the accumulation of extracellular amyloid plaques and intracellular neurofibrillary tangles [1]. In contrast, VaD falls within the vascular cognitive impairment (VCI) spectrum and is associated with cerebrovascular disease, particularly cerebral small vessel disease (CSVD) [2]. CSVDs manifest on magnetic resonance imaging (MRI) scans as lacunes, white matter hyperintensities (WMHs) and cerebral microbleeds (CMBs) [2]. The presence of CSVD has been associated with cognitive impairment and an increased risk of dementia [2], and also acts additively or synergistically with AD pathology [3]. Therefore, in certain populations where the baseline CSVD burden is relatively high, such as those in Asia, the cerebrovascular contribution to cognitive decline and dementia may be more prominent than in Western populations [4,5], and more research efforts should be expended on uncovering the common pathophysiological mechanisms underlying CSVD and AD, as well as identifying potential biomarkers and therapeutic targets.

One pathogenic factor observed in both CSVD and AD is dysregulated neuroinflammation, characterized by chronic gliosis within compromised brain regions [6,7]. In AD, a proinflammatory state with widespread upregulation of cytokines and chemokines has been observed, and neuroinflammation is also implicated in the pathogenesis of amyloid plaques and neurofibrillary tangles [8]. In patients with CSVD, microglia and peripheral macrophages may play both protective and detrimental roles in maintaining cerebral vasculature integrity [6]. Furthermore, intracerebral ventricular administration of tumor necrosis factor (TNF), an acute-phase proinflammatory cytokine known to be secreted by both microglia and macrophages [9], dose-dependently increased infarct volume in rodent models of brain ischemia [10]. Studies also showed increased cerebrospinal fluid (CSF) levels of TNF in patients with VaD [11]. Interestingly, the high-affinity subtype of the receptor for TNF, TNF-receptor 1, can be cleaved from its membrane-bound state and secreted into the systemic blood circulation in a soluble form (sTNF-R1) as a response to various upstream stimuli, including TNF [12,13]. The release of sTNF-R1 may in turn attenuate TNF-induced inflammatory processes by acting as decoy receptors for circulating TNF [12]. As blood sTNF-R1 is detectable for prolonged periods, it has been proposed to be a more reliable indicator of circulating TNF concentrations than TNF itself [12], suggesting potential utility as a promising inflammatory biomarker for individuals with CSVD and dementia.

In AD and VaD subjects, plasma sTNF-R1 has previously been reported to be increased compared with healthy controls [14,15]. In patients with pre-clinical stages of dementia, CSF sTNF-R1 was positively correlated with CSF β-amyloid and tau [16]. However, while increased plasma levels of sTNF-R1 were associated with memory functions in healthy individuals [17], the putative links between sTNF-R1 and cognitive performance, as well as structural brain changes in predementia subjects, remain unclear. Within the spectrum of cognitive impairment and dementia, predementia (which may include MCI or cognitive impairment, no dementia (CIND)) represents a clinically important stage when disease-modifying interventions may still be possible [18,19]. Furthermore, associations between sTNF-R1 and neuroimaging features of CSVD await investigation. Therefore, we aimed to examine serum sTNF-R1 associations in a prospectively assessed cohort of predementia subjects with neuroimaging markers of CSVD and cerebral atrophy at baseline, as well as the relationship between sTNF-R1 and longitudinal changes in cognitive function in the cohort of patients with CIND over 3 years of follow-up.

2. Materials and Methods

2.1. Study Cohort

The selection and assessment of subjects for this study have been previously described [20]. Briefly, patients with subjective complaints of memory loss were recruited from memory clinics at Singapore's National University Hospital and Saint Luke's Hospital. Subjects underwent clinical, physical and neuropsychological assessments, together with neuroimaging at the National University of Singapore. Important demographic and medical covariates, including cardiovascular risk factors (see Section 2.4 on covariates below) and exclusion factors (previous head trauma, psychiatric illnesses, thyroid disease and non-dementia neurodegenerative conditions), were collected by detailed questionnaires and reviews of medical records. The inclusion criteria of this study were as follows: (i) patients diagnosed as no cognitive impairment (NCI, used as control group) and CIND at baseline, (ii) complete neuropsychological data at baseline and $\geq$2 follow-up visits and (iii) sufficient baseline serum available for sTNF-R1 measurement. The diagnoses of all study subjects were made at regular consensus meetings attended by study clinicians and neuropsychologists. NCI was defined as those found to be cognitively normal and without functional loss based on the standard neuropsychological test battery (see below). CIND cases were defined as not meeting the Diagnostic and Statistical Manual of Mental Disorders, Fourth Edition (DSM-IV) diagnostic criteria for dementia but showing impairment in at least one domain of the neuropsychological battery (that is, education-adjusted cognitive scores ≥1.5 standard deviations (SDs) below cut-off scores on any test) [21]. As this study is focused on predementia, patients diagnosed with AD dementia or VaD based on current criteria [22,23] were excluded.

2.2. Standard Protocol Approvals, Registrations and Patient Consent

This study was performed in accordance with the Declaration of Helsinki and approved by the Singapore National Healthcare Group Domain-Specific Review Board (NHG-DSRB) (reference: 2010/00017; study protocol number: DEM4233). Written informed consent was obtained from all participants or their caregivers.

2.3. Neuropsychological Battery

All patients were administered an annual comprehensive neuropsychological battery which consists of six cognitive domains (see Supplementary Table S1 for the component tests of each domain) [24]. Raw scores of individual tests were transformed into standardized z-scores using the mean and standard deviation (SD) of the control group. The score for each cognitive domain was created by averaging the z-scores of individual component tests and standardized using the composite mean and SD of the NCI. In order to obtain the global cognitive score for each patient, the domain z-scores were averaged and standardized using the mean and SD of the NCI group, as previously described [25]. The follow-up annual visits' global and domain-specific cognitive z-scores were obtained using the means and SDs of the NCI group at baseline.

2.4. Covariates

Together with the demographic variables, vascular risk factors (hypertension, hyperlipidemia, diabetes mellitus and cardiovascular disease) were collected and classified as absent or present. Hypertension was determined by a systolic blood pressure ≥140 mmHg and/or diastolic blood pressure ≥90 mmHg, or the patient being on medications for hypertension. Hyperlipidemia was determined by total cholesterol levels of ≥4.14 mM, or the patient being on medications for hyperlipidemia. Diabetes mellitus was defined by glycated hemoglobin (HbA1c) of ≥6.5%, or the patient being on medications for diabetes mellitus. Medications for hypertension, hyperlipidemia and diabetes mellitus were also recorded. Cardiovascular disease was determined by a previous history of atrial fibrillation, congestive heart failure or myocardial infarction. Apolipoprotein E (APOE) genotyping was performed as previously described [26] to determine positive APOE ε4 carrier status (presence of at least one APOE ε4 allele).

2.5. Neuroimaging

Using a 3T Siemens Magnetom Trio Tim scanner with a 32-channel head coil, MRI scans were obtained at the Clinical Imaging Research Center of the National University of Singapore. For each subject, the following MRI markers were determined as previously described [27,28]. Lacunar infarcts (lacunes) were identified as lesions involving the subcortical regions, 3–15 mm in diameter, with a low signal on the T1-weighted image and a fluid attenuated inversion recovery (FLAIR) sequence giving a high signal on the T2-weighted image, as well as a hyperintense rim with a center following CSF intensity on FLAIR. WMHs were identified as hyperintense regions on T2 and FLAIR sequences without cavitation and hypointense on T1-weighted images. WMHs were graded using the Age-Related White Matter Changes (ARWMC) scale [29]. Cerebral microbleeds (CMBs) were identified as small focal rounded hypointense lesions, graded based on susceptibility-weighted imaging sequences using the Brain Observer MicroBleed Scale [30]. Medial temporal lobe atrophy (MTA) was graded by coronal sections on a 5-point Scheltens' scale (0—normal, 1—mild, 2—mild to moderate, 3—moderate, 4—severe), which considers the widening of the choroid fissure and temporal horn and loss of hippocampal height [28]. The presence vs. absence of significant neurodegeneration predictive of AD [31] was defined by MTA scores of 2–4 vs. scores of 0–1. Furthermore, the presence vs. absence of significant CSVD burden was defined as: confluent WMHs (ARWMC score $\geq$ 8 vs. <8), $\geq$2 lacunes vs. <2 lacune or $\geq$2 CMBs vs. <2 CMBs.

2.6. *Serum sTNF-R1 Measurements*

Non-fasting blood was drawn from patients into serum-separating tubes; centrifuged at 2000$\times$ g for 10 min at 4 °C. Serum samples were mixed well, aliquoted, and stored at −80 °C for future use. All samples underwent only one freeze–thaw cycle. Samples were diluted by 10$\times$ volume with the kit's diluent, then assayed for sTNF-R1 concentrations by quantitative sandwich enzyme immunoassay technique (Quantikine® ELISA Kit, R&D Systems Inc., Minneapolis, MN, USA), as per the manufacturer's instructions. sTNF-R1 was detectable and linear from 7.8 pg/mL to 500 pg/mL.

2.7. *Statistical Analyses*

The baseline serum biomarker (sTNF-R1) level and clinical data were used for cross-sectional analyses, while the global and cognitive domain-specific *z*-scores data were obtained at baseline, as well as up to three years follow-up (mean 2.9 $\pm$ SD 0.3 years) for longitudinal analyses of cognitive trajectories. Analyses of data were performed using the SPSS (version 26, IBM Inc., Armonk, NY, USA) and R (version 4.0.5, R Foundation) software. Independent samples t-tests, Mann–Whitney U tests or chi-square tests were used to compare demographic characteristics and sTNF-R1 levels between NCI and CIND. For regression analyses, sTNF-R1 levels were stratified into tertiles (lowest tertile: $\leq$1054.04 pg/mL, middle tertile: 1054.05–1333.69 pg/mL, highest tertile: $\geq$1333.70 pg/mL) as sTNFR1 concentrations were not normally distributed in this study (Shapiro–Wilk test $p < 0.001$, skewness = 1.34, kurtosis = 2.36), then examined for associations with CSVD markers at baseline. Poisson regression modelling was performed for counts of lacunes, while negative binomial regression was used for over-dispersed CMB counts and expressed as rate ratios (RR) with 95% confidence intervals (95% CI). For WMH grading, linear regression models were performed with measures of associations by mean differences (MD) with 95% CI. Binary logistic regression was used for significant MTA and expressed as odds ratios (OR) with 95% CI. Models were then adjusted with various covariates (see Section 2.4. Covariates above).

Lastly, we used linear mixed-effect models to examine the associations between baseline serum sTNF-R1 levels (independent variable) and changes in both global and domain-specific cognitive *z*-scores (dependent variables) of CIND subjects over time. All models included random effects for subjects and fixed effects for baseline levels of sTNF-R1 (lowest, middle and highest tertiles) and time, as well as interactions between baseline sTNF-R1 and time. Covariates added to the models included age, gender, education, APOE4 carrier status, hypertension, diabetes mellitus, cardiovascular disease, and hyperlipidemia. The model estimates were produced using the maximum likelihood method with random intercept and slope.

3. Results

3.1. *Study Participants*

Of the 273 participants diagnosed as NCI or CIND in the cohort, 186 subjects had available collected blood samples, as well as complete neuroimaging data at baseline (see Supplementary Figure S1). Similar to previous observations for our cohort [32–35], CIND subjects were significantly older ($p < 0.001$) with fewer years of education ($p = 0.006$) than NCI subjects. Therefore, both demographic variables (age and education) are included as covariates for subsequent regression analyses. Table 1 shows that sTNF-R1 levels in CIND were significantly higher than NCI ($p < 0.001$).

Table 1. Demographic characteristics of cases and controls at baseline.

Characteristics	NCI (n = 93)	CIND (n = 103)	p-Value
Age, mean (SD)	68.9 (6.4)	74.1 (6.9)	<0.001 *
Female, n (%)	54 (58.1)	49 (47.6)	0.142
Years of education, mean (SD)	9.7 (4.7)	7.9 (4.4)	0.006 *
APOE4 carrier, n (%)	22 (23.7)	27 (26.2)	0.680
Hypertension, n (%)	53 (57.0)	69 (67.0)	0.149
Diabetes mellitus, n (%)	20 (21.5)	33 (32.0)	0.097
Cardiovascular disease, n (%)	5 (5.4)	14 (13.6)	0.052
Hyperlipidemia, n (%)	66 (71.0)	80 (77.7)	0.282
sTNF-R1, median (IQR), pg/mL	1126.5 (381.3)	1236.5 (587.4)	<0.001 *

CIND = cognitive impairment, no dementia; IQR = interquartile range; NCI = no cognitive impairment; n = number of cases; SD = standard deviation; sTNF-R1 = soluble tumor necrosis factor-receptor 1. * $p < 0.05$; Student's t-test or Mann–Whitney U-test.

3.2. Associations between sTNF-R1 and Neuroimaging Markers

In order to investigate potential associations between sTNF-R1 levels and CSVD, we compared sTNF-R1 levels between subjects with and without significant WMH, lacunes and cerebral microbleeds (total n = 186). sTNF-R1 concentrations were significantly higher in subjects with confluent WMH (ARWMC score $\geq$ 8, $p = 0.034$) or the presence of lacunes ($p = 0.004$) (Figure 1). No significant differences were observed in subjects with or without significant cerebral microbleeds ($p = 0.621$). In unadjusted univariate regression models on the entire cohort, higher sTNF-R1 (analyzed as tertiles) was significantly associated with the severity of WMH (as measured by ARWMC scores) and the number of lacunes and cerebral microbleeds (Table 2). After adjusting for covariates, sTNF-R1 remained significantly associated with lacunes but not with WMH and cerebral microbleeds (Table 2). Interestingly, sTNF-R1 was not associated with significant MTA after adjusting for covariates (Supplementary Table S2), suggesting that sTNF-R1 may be a biomarker of CSVD rather than the pattern of neurodegeneration which is characteristic of AD.

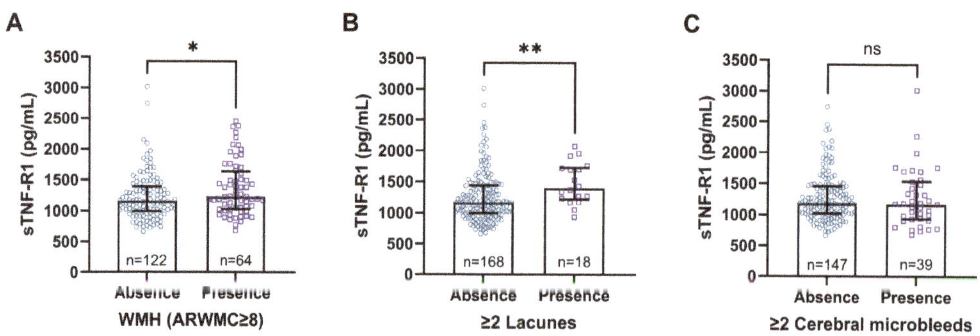

Figure 1. Serum sTNF-R1 concentrations in predementia subjects with CSVD. Bar graphs are median sTNF-R1 values ± interquartile range (IQR) stratified by the absence vs. presence of significant (**A**) WMH, (**B**) lacunes and (**C**) cerebral microbleeds (see Section 2.5 on neuroimaging for definitions). Datapoints represent available individual measures. * $p < 0.05$, ** $p < 0.01$, ns: not significant, pairwise comparisons using independent-samples Mann–Whitney U-tests. ARWMC = age-related white matter changes; n = number of cases; WMH = white matter hyperintensities.

Table 2. Association between serum sTNF-R1 tertiles and CSVD markers at baseline [1].

sTNF-R1 (Tertiles)	WMH (ARWMC) MD (95% CI)	Number of Lacunes RR (95% CI)	Number of CMBs RR (95% CI)
Model I			
Lowest	0	1	1
Middle	0.34 (−0.90, 1.59) $p = 0.590$	1.68 (0.79, 3.58) $p = 0.182$	1.59 (1.00, 2.53) $p = 0.051$
Highest	1.43 (0.20, 2.65) $p = 0.022$ *	4.16 (2.15, 8.04) $p < 0.001$ *	2.02 (1.29, 3.16) $p = 0.002$ *
Model II			
Lowest	0	1	1
Middle	−0.21 (−1.43, 1.02) $p = 0.738$	2.08 (0.95, 4.56) $p = 0.068$	1.56 (0.91, 2.68) $p = 0.110$
Highest	0.18 (−1.22, 1.58) $p = 0.800$	6.91 (3.19, 14.96) $p < 0.001$ *	1.73 (0.97, 3.08) $p = 0.061$

ARWMC = age-related white matter changes; CSVD = cerebral small vessel disease; CI = confidence interval; CMBs = cerebral microbleeds; WMH = white matter hyperintensities; MD = mean difference; RR = rate ratio. CSVD markers were available for 186 subjects. * $p < 0.05$ using linear, Poisson or negative binomial regression. *Model I*: unadjusted; *Model II*: adjusted for age, gender, APOE4 carrier status, hypertension, diabetes mellitus, cardiovascular disease and hyperlipidemia. [1] Interpretation: the lowest sTNF-R1 tertile was set as a reference group. Linear regression models were used for WMHs by ARWMC scores: when MD > 0, subjects with the middle/highest sTNF-R1 tertile will have MD units higher (lower when MD < 0) in the ARWMC scores compared with those with the lowest sTNF-R1 tertile. Poisson and negative binomial regression models were used for lacunes and CMBs: subjects with the middle/highest sTNF-R1 tertile were RR times more likely to have as many lacunes or CMBs compared with those with the lowest sTNF-R1 tertile.

3.3. Associations between sTNF-R1 and Cognitive Trajectories

Interestingly, while no significant difference in baseline cognitive z-scores was observed among the tertiles of sTNF-R1 (Supplementary Figure S2), we found that sTNF-R1 was associated with the cognitive trajectories of CIND subjects, in that significant declines in scores over time were observed within specific sTNF-R1 tertiles for global cognition (lowest: $\beta = -0.24$, 95% CI = −0.35 to −0.13; highest: $\beta = -0.09$, 95% CI = −0.19 to −0.01, Figure 2A), executive function (lowest: $\beta = -0.26$, 95% CI = −0.42 to −0.10; highest: $\beta = -0.15$, 95% CI = −0.28 to −0.03, Figure 2B), language (lowest: $\beta = -0.26$, 95% CI = −0.47 to −0.05, Figure 2D), visuomotor speed (lowest: $\beta = -0.16$, 95% CI = −0.24 to −0.08; highest: $\beta = -0.11$, 95% CI = −0.17 to −0.04, Figure 2F) and memory (lowest: $\beta = -0.12$, 95% CI = −0.22 to −0.02, Figure 2G). However, subjects with higher sTNF-R1 showed slower cognitive declines compared to those with the lowest sTNF-R1 tertile in global cognition (highest vs. lowest: $\beta' = 0.15$, 95% CI = 0.01 to 0.29; middle vs. lowest: $\beta' = 0.19$, 95% CI = 0.03 to 0.35, Figure 2A), language (highest vs. lowest: $\beta' = 0.29$, 95% CI = 0.02 to 0.57, Figure 2D) and visuomotor speed (middle vs. lowest: $\beta' = 0.18$, 95% CI = 0.07 to 0.29, Figure 2F).

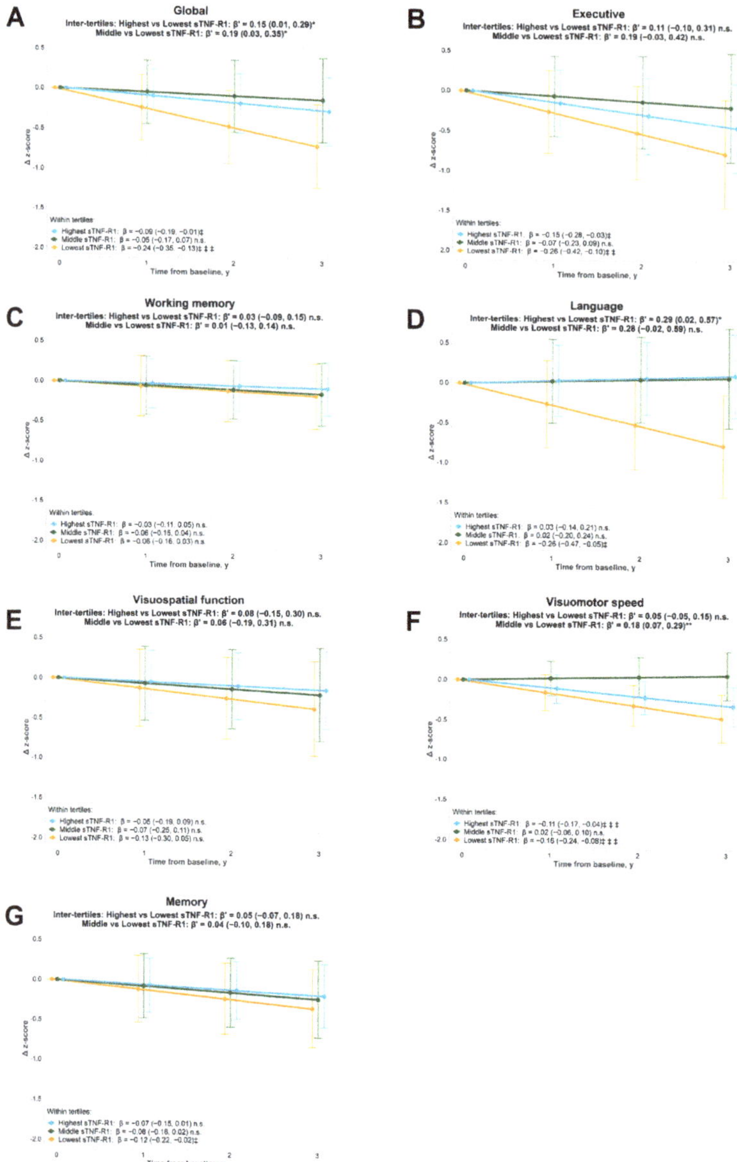

Figure 2. Associations between baseline sTNF-R1 tertiles and the cognitive trajectories of CIND subjects. Estimated mean change in cognitive domain scores (Δ z-scores) with 95% CI are stratified by the tertiles of sTNF-R1 (yellow: lowest tertile (n = 30), green: middle tertile (n = 27), blue: highest tertile (n = 46)) for (**A**) global cognition as well as (**B–G**) each of the six cognitive domains (see Section 2.3 on neuropsychological battery). β-coefficients with 95% CIs are derived from linear mixed-effect models adjusted for age, education, gender, APOE4 carrier status, hypertension, diabetes mellitus, cardiovascular disease and hyperlipidemia. Interpretation: Within tertiles, β represents the decline in cognitive z-scores over time for each tertile of sTNF-R1; ‡ $p < 0.05$, ‡‡ $p < 0.01$, ‡‡‡ $p < 0.001$, n.s.: not significant. Between tertiles, β′ represents the difference in the rate of change in cognitive z-scores between the highest/middle tertiles of sTNF-R1 vs. the lowest sTNF-R1 tertile; * $p < 0.05$, ** $p < 0.01$, n.s.: not significant. CIND = cognitive impairment, no dementia; n = number of cases.

4. Discussion

This study reports significantly higher levels of serum sTNF-R1 in CIND compared to NCI patients, and extends previous findings of elevated sTNF-R1 in dementia patients [14,15] to the predementia (also known as minimal cognitive impairment, MCI) stages. Furthermore, we investigated potential associations between sTNF-R1 and CSVD, showing that higher levels of sTNF-R1 were specifically linked to baseline measurements of lacunes independent of vascular risk factors. While the precise mechanisms linking sTNF-R1 with lacunes are unknown, postmortem and in vivo studies have reported the migration of microglia toward regions of infarcts after an ischemic episode, whereupon the secretion of chemoattractants, including TNF [9], initiates the recruitment of macrophages, monocytes and granulocytes, as well as the perpetuation of neuroinflammatory responses [36,37]. Interestingly, the production of TNF under chronic inflammatory conditions has been known to induce sTNF-R1 secretion into the systemic blood circulation [38], and data from mouse models suggest that sTNF-R1 upregulation binds to peripheral free TNF, thus attenuating TNF-mediated inflammatory responses and ameliorating increases in infarct size after middle cerebral artery occlusion [10,39]. Considering the current findings of sTNF-R1 in association with lacunes, we postulate that sTNF-R1 might be an adaptive response to the increased TNF and may play a protective role in TNF-associated CSVD processes, thus underscoring the biological relevance of sTNF-R1 as a biomarker for CSVD. Regarding lacunes, some studies have reported associations with worse cognitive outcomes [40], while others have not [41]. Nevertheless, successful validation of peripheral sTNF-R1 as a biomarker for the early identification of individuals with cerebral vascular pathologies may allow clinicians to prioritize timely management of vascular risk factors in these patients, thereby lowering the incidence of VCI, and potentially preventing concomitant AD progression [42]. Our data may have added clinical significance for Asian populations which are known to have a relatively high CSVD burden [3,4]. Moreover, compared with expensive MRI scans, peripheral biomarkers may be a cost-effective alternative for at-risk populations, especially those from developing countries [43]. Hence, sTNF-R1 may be a useful, economical blood-based diagnostic tool for the early identification of subjects with CSVD before progression to dementia.

Besides neuroimaging markers, the second main aim of the study was to investigate sTNF-R1's associations with cognitive decline. Although we did not find significant differences in the baseline z-scores among sTNF-R1 tertiles (Supplementary Figure S2), longitudinal analyses showed that CIND subjects with a higher sTNF-R1 declined more slowly in terms of global cognition than those at the lowest sTNF-R1 tertile. These results complement a recent study that reported elevated levels of CSF markers, including sTNF-R1, associated with decreased cognitive decline and a risk of conversion to AD in MCI subjects [44]. Since sTNF-R1 acts as a decoy receptor to inhibit TNF-induced signaling [13], it is possible that predementia subjects with higher sTNF-R1 levels may be protected from cognitive decline resulting from TNF-mediated neuroinflammatory or neurovascular injuries, as higher CSF levels of TNF were found to be associated with conversion to dementia [45]. Taken together, the present findings of higher sTNF-R1 being associated with more severe baseline lacunes but also a slower cognitive decline over the follow-up period, suggest a complex picture where robust responses to more severe CSVD-induced neuroinflammation resulted in better protection against cognitive decline in the longer term. Interestingly, the protective effects of sTNF-R1 may be specific to VCI, as sTNF-R1 changes were unrelated to the characteristic brain perturbations of AD (using MTA as the neuroimaging marker).

This study has several limitations. Firstly, our results on memory clinic patients might not be generalizable to the population at-large as there may be non-memory clinic patients with risk factor profiles that are not considered in our analyses. Secondly, as associations between baseline serum sTNF-R1 and CSVD neuroimaging markers were cross-sectionally analyzed, investigations into the temporal relationship between sTNF-R1 and the progression of the markers are needed. Thirdly, we were not able to detect any sTNF-R1 associations with clinical conversion (NCI to CIND or CIND to dementia) or with cognitive trajectories within the NCI group, due to the small sample size and relatively short follow-up period. Furthermore, whilst our finding of higher sTNF-R1 levels correlating with lacunes supports sTNF-R1 elevation as a putatively protective response to TNF-associated CSVD progression, we were not able to detect TNF directly due to platform and sample factors (data not shown), and correlations between blood TNF and sTNF-R1 levels are needed in future investigations to validate the functions of TNF-mediated inflammation and responses in VCI and dementia. Lastly, the relatively small sample size of the current study necessitates follow-up research with larger cohorts and a longer follow-up duration to more adequately study sTNF-R1 effects on cognitive trajectories stratified by CVSD severity.

5. Conclusions

To the best of our knowledge, this study is the first to examine serum sTNF-R1 associations with both neuroimaging CSVD markers, as well as cognitive trajectories in predementia subjects. Since vascular risk factors may influence the onset and progression of dementia [42], our multiple regression models controlled for potential confounding effects of vascular risk factors, and still found that the higher levels of serum sTNF-R1 in CIND subjects were associated with more severe lacunar infarcts but also with slower cognitive decline, suggesting a complex interaction between sTNF-R1 and CSVD-associated processes. Our study proposes sTNF-R1 as a prognostic peripheral biomarker for identifying patients with concomitant cerebrovascular pathologies, as well as for early identification of individuals at risk for VCI. However, further clinical studies using larger cohorts with longitudinal design are needed to validate the clinical utility of sTNF-R1.

Supplementary Materials: The following supporting information can be downloaded at: https://www.mdpi.com/article/10.3390/biom13030525/s1, Table S1: Summary of neuropsychological battery and component tests [22,46–51]; Table S2: Association between serum sTNF-R1 tertiles and MTA at baseline; Figure S1: Selection of study participants; Figure S2: Baseline global and specific cognitive domains z-scores of CIND subjects stratified by sTNF-R1 tertiles.

Author Contributions: Conceptualization, J.R.C., G.S.D. and M.K.P.L.; formal analysis, K.H.T.S., L.-Y.W., J.R.C., Y.L.C. and M.K.P.L.; funding acquisition, S.H., G.S.D., C.P.C. and M.K.P.L.; investigation, K.H.T.S., L.-Y.W., J.R.C., Y.L.C., B.G., C.R., S.H., N.V. and C.P.C.; supervision, G.S.D., C.P.C. and M.K.P.L.; validation, K.H.T.S., L.-Y.W., J.R.C., Y.L.C., G.S.D. and M.K.P.L.; writing—original draft, K.H.T.S. and L.-Y.W.; writing—review and editing, J.R.C., Y.L.C., B.G., C.R., S.H., N.V., G.S.D., C.P.C. and M.K.P.L. All authors have read and agreed to the published version of the manuscript.

Funding: This work was funded by the Singapore National Medical Research Council (grants MOH-000500-01, MOH-000707-01 and MOH-001086-00), the Healthy Longevity Translational Research Programme, Yong Loo Lin School of Medicine (grant HLTRP/2022/PS-01) and the National University Health System (grant A-0006090-00-00). Y.L.C. is a recipient of a post-doctoral fellowship award from the Yong Loo Lin School of Medicine (NUSMED/2021/PDF/05). Research in the laboratory of G.S.D. is supported by the Singapore Ministry of Education (grant MOE2017-T3-1-002).

Institutional Review Board Statement: This study was conducted in accordance with the Declaration of Helsinki, and approval was given by the Singapore National Healthcare Group Domain-Specific Review Board (NHG-DSRB) (reference: 2010/00017; study protocol number: DEM4233).

Informed Consent Statement: Participants or their caregivers gave informed consent to participate in this study before taking part.

Data Availability Statement: Anonymized data derived from this study may be provided by the corresponding author upon reasonable request.

Acknowledgments: The authors are grateful to the study participants and to the clinical staff who were involved in this study. We acknowledge Boon Yeow Tan, St Luke's Hospital, for his contribution to patient recruitment.

Conflicts of Interest: The authors declare no conflict of interest. The funders had no role in the design of the study; in the collection, analyses, or interpretation of data; in the writing of the manuscript; or in the decision to publish the results.

References

1. Prince, M.; Albanese, E.; Guerchet, M.; Prina, M. *World Alzheimer Report 2014: Dementia and Risk Reduction: An Analysis of Protective and Modifiable Risk Factors*; Alzheimer's Disease International: London, UK, 2014; p. 104.
2. Zanon Zotin, M.C.; Sveikata, L.; Viswanathan, A.; Yilmaz, P. Cerebral small vessel disease and vascular cognitive impairment: From diagnosis to management. *Curr. Opin. Neurol.* **2021**, *34*, 246–257. [CrossRef]
3. Chen, C.; Homma, A.; Mok, V.C.; Krishnamoorthy, E.; Alladi, S.; Meguro, K.; Abe, K.; Dominguez, J.; Marasigan, S.; Kandiah, N.; et al. Alzheimer's disease with cerebrovascular disease: Current status in the Asia-Pacific region. *J. Intern. Med.* **2016**, *280*, 359–374. [CrossRef] [PubMed]
4. Lam, B.Y.K.; Yiu, B.; Ampil, E.; Chen, C.L.-H.; Dikot, Y.; Dominguez, J.C.; Ganeshbhai, P.V.; Hilal, S.; Kandiah, N.; Kim, S.; et al. High burden of cerebral white matter lesion in 9 Asian cities. *Sci. Rep.* **2021**, *11*, 11587. [CrossRef] [PubMed]
5. Hilal, S.; Mok, V.; Youn, Y.C.; Wong, A.; Ikram, M.K.; Chen, C.L. Prevalence, risk factors and consequences of cerebral small vessel diseases: Data from three Asian countries. *J. Neurol. Neurosurg. Psychiatry* **2017**, *88*, 669–674. [CrossRef] [PubMed]
6. Koizumi, T.; Kerkhofs, D.; Mizuno, T.; Steinbusch, H.W.M.; Foulquier, S. Vessel-Associated Immune Cells in Cerebrovascular Diseases: From Perivascular Macrophages to Vessel-Associated Microglia. *Front. Neurosci.* **2019**, *13*, 1291. [CrossRef]
7. Pasqualetti, G.; Brooks, D.J.; Edison, P. The role of neuroinflammation in dementias. *Curr. Neurol. Neurosci. Rep.* **2015**, *15*, 17. [CrossRef]
8. Kinney, J.W.; Bemiller, S.M.; Murtishaw, A.S.; Leisgang, A.M.; Salazar, A.M.; Lamb, B.T. Inflammation as a central mechanism in Alzheimer's disease. *Alzheimers Dement (N. Y.)* **2018**, *4*, 575–590. [CrossRef]
9. Liu, L.-R.; Liu, J.-C.; Bao, J.-S.; Bai, Q.-Q.; Wang, G.-Q. Interaction of Microglia and Astrocytes in the Neurovascular Unit. *Front. Immunol.* **2020**, *11*, 1024. [CrossRef]
10. Barone, F.C.; Arvin, B.; White, R.F.; Miller, A.; Webb, C.L.; Willette, R.N.; Lysko, P.G.; Feuerstein, G.Z. Tumor necrosis factor-alpha. A mediator of focal ischemic brain injury. *Stroke* **1997**, *28*, 1233–1244. [CrossRef]
11. Tarkowski, E.; Blennow, K.; Wallin, A.; Tarkowski, A. Intracerebral Production of Tumor Necrosis Factor-α, a Local Neuroprotective Agent, in Alzheimer Disease and Vascular Dementia. *J. Clin. Immunol.* **1999**, *19*, 223–230. [CrossRef]
12. Cheng, X.; Yang, L.; He, P.; Li, R.; Shen, Y. Differential activation of tumor necrosis factor receptors distinguishes between brains from Alzheimer's disease and non-demented patients. *J. Alzheimer's Dis.* **2010**, *19*, 621–630. [CrossRef]
13. Diez-Ruiz, A.; Tilz, G.P.; Zangerle, R.; Baier-Bitterlich, G.; Wachter, H.; Fuchs, D. Soluble receptors for tumour necrosis factor in clinical laboratory diagnosis. *Eur. J. Haematol.* **1995**, *54*, 1–8. [CrossRef]
14. Zhang, J.; Jia, J.; Qin, W.; Wang, S. Combination of plasma tumor necrosis factor receptors signaling proteins, beta-amyloid and apolipoprotein E for the detection of Alzheimer's disease. *Neurosci. Lett.* **2013**, *541*, 99–104. [PubMed]
15. Zhang, J.; Peng, M.; Jia, J. Plasma amyloid-β oligomers and soluble tumor necrosis factor receptors as potential biomarkers of AD. *Curr. Alzheimer Res.* **2014**, *11*, 325–331.
16. Buchhave, P.; Zetterberg, H.; Blennow, K.; Minthon, L.; Janciauskiene, S.; Hansson, O. Soluble TNF receptors are associated with Aβ metabolism and conversion to dementia in subjects with mild cognitive impairment. *Neurobiol. Aging* **2010**, *31*, 1877–1884. [CrossRef]
17. West, N.A.; Kullo, I.J.; Morris, M.C.; Mosley, T.H. Sex-specific associations of inflammation markers with cognitive decline. *Exp. Gerontol.* **2020**, *138*, 110986. [CrossRef]
18. Grandy, J.K. Melatonin: Therapeutic intervention in mild cognitive impairment and Alzheimer disease. *J. Neurol. Neurophysiol.* **2013**, *4*, 148. [CrossRef]
19. Soininen, H.; Solomon, A.; Visser, P.J.; Hendrix, S.B.; Blennow, K.; Kivipelto, M.; Hartmann, T.; Hallikainen, I.; Hallikainen, M.; Helisalmi, S. 24-month intervention with a specific multinutrient in people with prodromal Alzheimer's disease (LipiDiDiet): A randomized, double-blind, controlled trial. *Lancet Neurol.* **2017**, *16*, 965–975. [PubMed]
20. Chai, Y.L.; Hilal, S.; Chong, J.P.; Ng, Y.X.; Liew, O.W.; Xu, X.; Ikram, M.K.; Venketasubramanian, N.; Richards, A.M.; Lai, M.K. Growth differentiation factor-15 and white matter hyperintensities in cognitive impairment and dementia. *Medicine* **2016**, *95*, e4566.
21. *Diagnostic and Statistical Manual of Mental Disorders*, 4th ed.; American Journal of Psychiatry: Washington, DC, USA, 1994. [CrossRef]
22. McKhann, G.; Drachman, D.; Folstein, M.; Katzman, R.; Price, D.; Stadlan, E.M. Clinical diagnosis of Alzheimer's disease: Report of the NINCDS-ADRDA Work Group under the auspices of Department of Health and Human Services Task Force on Alzheimer's Disease. *Neurology* **1984**, *34*, 939–944. [CrossRef]
23. Román, G.C.; Tatemichi, T.K.; Erkinjuntti, T.; Cummings, J.; Masdeu, J.; Garcia, J.; Amaducci, L.; Orgogozo, J.-M.; Brun, A.; Hofman, A. Vascular dementia: Diagnostic criteria for research studies: Report of the NINDS-AIREN International Workshop. *Neurology* **1993**, *43*, 250.

24. Xu, X.; Chan, Q.L.; Hilal, S.; Ikram, M.K.; Venketasubramanian, N.; Tan, B.Y.; Dong, Y.; Chen, C.L.; Collinson, S.L. The Diagnostic Utility of the NINDS-CSN Neuropsychological Battery in Memory Clinics. *Dement. Geriatr. Cogn. Dis. Extra* **2016**, *6*, 276–282. [CrossRef] [PubMed]
25. Wu, L.-Y.; Kan, C.N.; Cheah, I.K.; Chong, J.R.; Xu, X.; Vrooman, H.; Hilal, S.; Venketasubramanian, N.; Chen, C.P.; Halliwell, B.; et al. Low Plasma Ergothioneine Predicts Cognitive and Functional Decline in an Elderly Cohort Attending Memory Clinics. *Antioxidants* **2022**, *11*, 1717. [PubMed]
26. Chai, Y.L.; Yeo, H.K.; Wang, J.; Hilal, S.; Ikram, M.K.; Venketasubramanian, N.; Wong, B.S.; Chen, C.L. Apolipoprotein e4 is Associated with Dementia and Cognitive Impairment Predominantly Due to Alzheimer's Disease and Not with Vascular Cognitive Impairment: A Singapore-Based Cohort. *J. Alzheimer's Dis.* **2016**, *51*, 1111–1118. [CrossRef]
27. Hilal, S.; Chai, Y.L.; Ikram, M.K.; Elangovan, S.; Yeow, T.B.; Xin, X.; Chong, J.Y.; Venketasubramanian, N.; Richards, A.M.; Chong, J.P.; et al. Markers of cardiac dysfunction in cognitive impairment and dementia. *Medicine* **2015**, *94*, e297. [CrossRef]
28. Saridin, F.N.; Chew, K.A.; Reilhac, A.; Gyanwali, B.; Villaraza, S.G.; Tanaka, T.; Scheltens, P.; van der Flier, W.M.; Chen, C.L.H.; Hilal, S. Cerebrovascular disease in suspected non-Alzheimer's pathophysiology and cognitive decline over time. *Eur. J. Neurol.* **2022**, *29*, 1922–1929. [CrossRef] [PubMed]
29. Wahlund, L.O.; Barkhof, F.; Fazekas, F.; Bronge, L.; Augustin, M.; Sjogren, M.; Wallin, A.; Ader, H.; Leys, D.; Pantoni, L.; et al. A new rating scale for age-related white matter changes applicable to MRI and CT. *Stroke* **2001**, *32*, 1318–1322.
30. Gyanwali, B.; Shaik, M.A.; Venketasubramanian, N.; Chen, C.; Hilal, S. Mixed-Location Cerebral Microbleeds: An Imaging Biomarker for Cerebrovascular Pathology in Cognitive Impairment and Dementia in a Memory Clinic Population. *J. Alzheimer's Dis.* **2019**, *71*, 1309–1320. [CrossRef]
31. Convit, A.; de Asis, J.; de Leon, M.J.; Tarshish, C.Y.; De Santi, S.; Rusinek, H. Atrophy of the medial occipitotemporal, inferior, and middle temporal gyri in non-demented elderly predict decline to Alzheimer's disease. *Neurobiol. Aging* **2000**, *21*, 19–26. [CrossRef]
32. Wu, L.Y.; Cheah, I.K.; Chong, J.R.; Chai, Y.L.; Tan, J.Y.; Hilal, S.; Vrooman, H.; Chen, C.P.; Halliwell, B.; Lai, M.K.P. Low plasma ergothioneine levels are associated with neurodegeneration and cerebrovascular disease in dementia. *Free Radic. Biol. Med.* **2021**, *177*, 201–211. [CrossRef]
33. Chong, J.R.; Ashton, N.J.; Karikari, T.K.; Tanaka, T.; Saridin, F.N.; Reilhac, A.; Robins, E.G.; Nai, Y.H.; Vrooman, H.; Hilal, S.; et al. Plasma P-tau181 to Aβ42 ratio is associated with brain amyloid burden and hippocampal atrophy in an Asian cohort of Alzheimer's disease patients with concomitant cerebrovascular disease. *Alzheimer's Dement. J. Alzheimer's Assoc.* **2021**, *17*, 1649–1662. [CrossRef]
34. Chai, Y.L.; Chong, J.R.; Raquib, A.R.; Xu, X.; Hilal, S.; Venketasubramanian, N.; Tan, B.Y.; Kumar, A.P.; Sethi, G.; Chen, C.P.; et al. Plasma osteopontin as a biomarker of Alzheimer's disease and vascular cognitive impairment. *Sci. Rep.* **2021**, *11*, 4010. [CrossRef]
35. Chua, X.Y.; Chai, Y.L.; Chew, W.S.; Chong, J.R.; Ang, H.L.; Xiang, P.; Camara, K.; Howell, A.R.; Torta, F.; Wenk, M.R.; et al. Immunomodulatory sphingosine-1-phosphates as plasma biomarkers of Alzheimer's disease and vascular cognitive impairment. *Alzheimer's Res. Ther.* **2020**, *12*, 122. [CrossRef] [PubMed]
36. Schilling, M.; Besselmann, M.; Leonhard, C.; Mueller, M.; Ringelstein, E.B.; Kiefer, R. Microglial activation precedes and predominates over macrophage infiltration in transient focal cerebral ischemia: A study in green fluorescent protein transgenic bone marrow chimeric mice. *Exp. Neurol.* **2003**, *183*, 25–33. [CrossRef]
37. Tanaka, R.; Komine-Kobayashi, M.; Mochizuki, H.; Yamada, M.; Furuya, T.; Migita, M.; Shimada, T.; Mizuno, Y.; Urabe, T. Migration of enhanced green fluorescent protein expressing bone marrow-derived microglia/macrophage into the mouse brain following permanent focal ischemia. *Neuroscience* **2003**, *117*, 531–539. [CrossRef]
38. Dri, P.; Gasparini, C.; Menegazzi, R.; Cramer, R.; Albéri, L.; Presani, G.; Garbisa, S.; Patriarca, P. TNF-induced shedding of TNF receptors in human polymorphonuclear leukocytes: Role of the 55-kDa TNF receptor and involvement of a membrane-bound and non-matrix metalloproteinase. *J. Immunol.* **2000**, *165*, 2165–2172. [CrossRef] [PubMed]
39. Works, M.G.; Koenig, J.B.; Sapolsky, R.M. Soluble TNF receptor 1-secreting ex vivo-derived dendritic cells reduce injury after stroke. *J. Cereb. Blood Flow Metab. Off. J. Int. Soc. Cereb. Blood Flow Metab.* **2013**, *33*, 1376–1385. [CrossRef] [PubMed]
40. Benjamin, P.; Lawrence, A.J.; Lambert, C.; Patel, B.; Chung, A.W.; MacKinnon, A.D.; Morris, R.G.; Barrick, T.R.; Markus, H.S. Strategic lacunes and their relationship to cognitive impairment in cerebral small vessel disease. *NeuroImag. Clin.* **2014**, *4*, 828–837. [CrossRef] [PubMed]
41. Gyanwali, B.; Lui, B.; Tan, C.S.; Chong, E.J.Y.; Vrooman, H.; Chen, C.; Hilal, S. Cerebral Microbleeds and White Matter Hyperintensities are Associated with Cognitive Decline in an Asian Memory Clinic Study. *Curr. Alzheimer Res.* **2021**, *18*, 399–413. [CrossRef] [PubMed]
42. Alagiakrishnan, K.; McCracken, P.; Feldman, H. Treating vascular risk factors and maintaining vascular health: Is this the way towards successful cognitive ageing and preventing cognitive decline? *Postgrad Med. J.* **2006**, *82*, 101–105. [CrossRef]
43. Hosoki, S.; Tanaka, T.; Ihara, M. Diagnostic and prognostic blood biomarkers in vascular dementia: From the viewpoint of ischemic stroke. *Neurochem. Int.* **2021**, *146*, 105015. [CrossRef] [PubMed]
44. Hu, W.T.; Ozturk, T.; Kollhoff, A.; Wharton, W.; Christina Howell, J.; Weiner, M.; Aisen, P.; Petersen, R.; Jack, C.R.; Jagust, W.; et al. Higher CSF sTNFR1-related proteins associate with better prognosis in very early 'Alzheimer's disease. *Nat. Commun.* **2021**, *12*, 4001. [CrossRef] [PubMed]

45. Fu, P.; Peng, F.; Initiative, A.s.D.N. CSF TNF α levels were associated with conversion from mild cognitive impairment to dementia. *PLoS ONE* **2022**, *17*, e0274503. [CrossRef]
46. D'Elia, L.F.; Satz, P.; Uchiyama, C.L.; White, T. *Color Trails Test*; PAR: Odessa, FL, USA, 1996. Available online: https://www.parinc.com/Products/Pkey/77 (accessed on 9 March 2023).
47. Wechsler, D. *Subtest Administration and Scoring. WAIS–IV: Administration and Scoring Manual*; The Psychological Corporation: San Antonio, TX, USA, 2009; pp. 87–93.
48. Mack, W.J.; Freed, D.M.; Williams, B.W.; Henderson, V.W. Boston Naming Test: Shortened versions for use in Alzheimer's disease. *J. Gerontol.* **1992**, *47*, P154–P158. [CrossRef] [PubMed]
49. Meyers, J. *Meyers Scoring System for the Rey Complex Figure Test and the Recognition Trial*; Psychological Assessment Resources: Odessa, FL, USA, 1994. Available online: https://www.parinc.com/Products/Pkey/355 (accessed on 9 March 2023).
50. Smith, A. *Symbol Digit Modalities Test*; Western Psychological Services: Los Angeles, CA, USA, 1973. Available online: https://www.wpspublish.com/sdmt-symbol-digit-modalities-test (accessed on 9 March 2023).
51. Brandt, J. The Hopkins Verbal Learning Test: Development of a new memory test with six equivalent forms. *Clin. Neuropsychol.* **1991**, *5*, 125–142. [CrossRef]

biomolecules

MDPI

Communication

PHLPP Inhibitor NSC74429 Is Neuroprotective in Rodent Models of Cardiac Arrest and Traumatic Brain Injury

Travis C. Jackson [1,2,*], Cameron Dezfulian [3,4,5], Vincent A. Vagni [3,5], Jason Stezoski [3,5], Keri Janesko-Feldman [3,5] and Patrick M. Kochanek [3,5]

1 Department of Molecular Pharmacology & Physiology, Morsani College of Medicine, University of South Florida, 12901 Bruce B Downs BLDV, Tampa, FL 33612, USA
2 USF Health Heart Institute, Morsani College of Medicine, University of South Florida, 560 Channelside Dr, Tampa, FL 33602, USA
3 Safar Center for Resuscitation Research, UPMC Children's Hospital of Pittsburgh, Rangos Research Center—6th Floor, Pittsburgh, PA 15224, USA
4 Department of Pediatrics, Baylor College of Medicine, 6651 Main Street, Houston, TX 77030, USA
5 Department of Critical Care Medicine, University of Pittsburgh School of Medicine, 4401 Penn Avenue, Pittsburgh, PA 15224, USA
* Correspondence: tcjackson@usf.edu

Abstract: Pleckstrin homology domain and leucine rich repeat protein phosphatase (PHLPP) knock-out mice have improved outcomes after a stroke, traumatic brain injury (TBI), and decreased mal-adaptive vascular remodeling following vascular injury. Thus, small-molecule PHLPP inhibitors have the potential to improve neurological outcomes in a variety of conditions. There is a paucity of data on the efficacy of the known experimental PHLPP inhibitors, and not all may be suited for targeting acute brain injury. Here, we assessed several PHLPP inhibitors not previously explored for neuroprotection (NSC13378, NSC25247, and NSC74429) that had favorable predicted chemistries for targeting the central nervous system (CNS). Neuronal culture studies in staurosporine (apoptosis), glutamate (excitotoxicity), and hydrogen peroxide (necrosis/oxidative stress) revealed that NSC74429 at micromolar concentrations was the most neuroprotective. Subsequent testing in a rat model of asphyxial cardiac arrest, and in a mouse model of severe TBI, showed that serial dosing of 1 mg/kg of NSC74429 over 3 days improved hippocampal survival in both models. Taken together, NSC74429 is neuroprotective across multiple insult mechanisms. Future pharmacokinetic and pharmacodynamic (PK/PD) studies are warranted to optimize dosing, and mechanistic studies are needed to determine the percentage of neuroprotection mediated by PHLPP1/2 inhibition, or potentially from the modulation of PHLPP-independent targets.

Keywords: PHLPP; PHLPP1; PHLPP2; neuroprotection; NSC74429; brain

1. Introduction

Studies in mice show that knockout (KO) of the pleckstrin homology domain and leucine rich repeat protein phosphatase (PHLPP) gene improves CNS outcomes in mouse models of stroke, and traumatic brain injury (TBI), and decreases detrimental vascular remodeling after damage to the carotid artery [1–3]. PHLPPs promote cell death by inhibiting or activating a variety of signaling molecules. For instance, protein kinase B (AKT) is a potent pro-survival kinase and is inactivated by PHLPP-dependent dephosphorylation at Serine 473 [4]. Conversely, mammalian sterile 20-like kinase 1 (Mst1) is a potent pro-apoptotic effector and is activated by PHLPP-dependent dephosphorylation at Threonine 387 [5]. Thus, PHLPP antagonists, by blocking phosphatase activity, have great clinical potential to promote brain health by modulating neuronal survival and vascular homeostasis. Germane to that possibility, Sierecki et al. identified ~116 small molecule experimental PHLPP inhibitors [6]. All had IC50s ranging from 1–100 μM to inhibit the

PHLPP2-PP2C phosphatase domain in a cell-free assay, and also blocked the PP2C domain of the PHLPP1 isoform with varying potencies [6]. Most of those inhibitors have not been characterized further and their research utility or translational potential is unknown. The exceptions are PHLPP inhibitors NSC117079 and NSC45586. Sierecki et al. showed that both compounds inhibited PHLPP signaling at ~50 μM in mammalian cells in vitro and were cytoprotective in a model of etoposide-induced cell death [6]. Consistent with that finding, we reported that 50 μM NSC117079 or NSC45586 altered PHLPP signaling in primary rat cortical neurons in vitro and were neuroprotective following staurosporine injury [7]. However, we also observed that NSC45586 strongly bound to albumin in the culture media, raising questions on its potential translatability for in vivo use [7]. In addition, both NSC117079 and NSC45586 are poor candidates to cross the blood-brain barrier (BBB) based on their physiochemical properties. Here, we used a predictive chemistry-guided approach to select three additional PHLPP inhibitors for testing that have not been further explored among the 116 known compounds, and to use three in vitro assays and two in vivo models to assess neuroprotection.

2. Materials and Methods

Reagents. Experimental PHLPP Inhibitors: NSC13378, NSC25247, and NSC74429 were obtained from the Developmental Therapeutics Program (DTP) of the National Cancer Institute (NCI) upon completion of a Material Transfer Agreement. Compounds were provided as powder in opaque glass vials and dissolved in 100% dimethylsulfoxide (DMSO) before diluting to the desired final concentration. DMSO, 30% Hydrogen Peroxide Solution, and Glutamate were purchased from Sigma (St. Louis, MO, USA). Staurosporine was purchased from Tocris (Bristol, UK). Cell Titer Blue was purchased from Promega (Madison, WI, USA). Neurobasal/B27 culture media was purchased from ThermoFisher Scientific/Life Technologies (Waltham, MA, USA).

Animals. All studies were approved by the IACUC of the University of Pittsburgh. Methods of euthanasia adhered with the AVMA Guidelines for the Euthanasia of Animals. Timed pregnant female Sprague Dawley rats were purchased from Charles River (Wilmington, MA, USA). Male adult Sprague Dawley rats (~12 weeks old) weighing 350–375 g were purchased from Charles River. Adult male C57BL/6 mice (~12 weeks old) weighing ~28 g were also purchased from Charles River. All animals were given ad lib access to food and water and housed on a 12 h light/dark cycle.

Preparation of Experimental PHLPP Inhibitors for In Vitro and In Vivo Testing. For in vitro studies, NSC13378, NSC25247, and NSC74429 were prepared in 10 mM stocks in 100% DMSO and diluted to a final concentration of 25 μM or 50 μM in cell culture media. For in vivo studies, pilot testing indicated low water solubility of NSC74429. Therefore, a combination of DMSO, heating via a water bath, and vigorous vortexing was used to enhance solubilization. Two drug doses were tested in rats and in mice (1 or 3 mg/kg). For studies in rats, NSC74429 was prepared in 24% DMSO/D5W (dextrose 5% in water) and administered at a volume of 1.25 mL/kg (intravenous [IV] or intraperitoneal [IP]). For studies in mice, NSC74429 was prepared in 5% DMSO/D5W (vehicle) and a 250 μL bolus was administered IV or IP.

Primary Neuronal Culture. Neuron culture was done using our established methods [7,8]. In brief, timed-pregnant Sprague Dawley rats were purchased from Charles River and cortices isolated from E18-19 embryos under a dissecting microscope (Leica M651, Buffalo Grove, IL, USA). Mixed-sex neurons were dissociated and seeded onto 96-well plates at 1.5×10^5/well in Neurobasal/B27 media. The culture media was supplemented with 8 μM cytosine β-D-arabinofuranoside hydrochloride on day in vitro (DIV) 3 to reduce glial proliferation. A second feeding ($\frac{1}{2}$ media exchange) was given on DIV6 and neurons were subjected to injury and/or drug therapy on DIV9.

Staurosporine Injury Model. On DIV9, half of the media from each well was collected and pooled (conditioned media) and mixed with an equal volume fresh neurobasal/B27 to make "24 h treatment media". Treatment media was prepared with or without 150 nM

STS and with or without experimental PHLPP inhibitors at 25 μM or 50 μM. The final concentration of DMSO in all groups was ~0.5%. Each well received 100 μL of treatment media per well and incubated for 24 h at 37 °C. Cell viability was measured on DIV10.

Hydrogen Peroxide (H_2O_2) Injury Model. On DIV9, 2/3 conditioned media was prepared and set aside for the post-injury phase. H_2O_2 was prepared fresh. Specifically, 30% concentrated H_2O_2 with stabilizers (Sigma) was first diluted in sterile ddH_2O. The diluted H_2O_2 stock (Stock A) was immediately added to a balanced salt solution containing 5 mM glucose (BSS-G) and diluted to a final concentration of 40 μM H_2O_2 (Stock B). Cells were quickly washed once with plain BSS-G and then incubated with 40 μM H_2O_2 for 35 min. Controls received identical manipulations without injury. After 35 min, the Stock B injury solution was removed and 100 μL of conditioned media with or without experimental PHLPP inhibitors was added to the cells. The final DMSO concentration was ~0.5%.

Glutamate-Excitotoxicity Model. On DIV9, 2/3 conditioned media was prepared and set aside for the post-injury phase. Cells were quickly washed once with plain BSS-G and then incubated in BSS-G containing 10 μM Glutamate for 5 min. Controls received identical manipulations without injury. After 5 min, the glutamate solution was removed and 100 μL of conditioned media with or without experimental PHLPP inhibitors was added to the cells. The final DMSO concentration was ~0.5%.

Cell Viability Assay. In vitro neuronal injury was assessed using the CellTiter-Blue method as reported by our group [7,8]. Briefly, 20 μL of CellTiter-Blue reagent (Promega; Madison, WI, USA) was added to each well of a 96-well plate at 24 h post-injury. Plates were incubated in the dark for ~1 h and absorbance was measured on a plate reader.

Asphyxial Cardiac Arrest Rat Model. Rats were injured by ACA as described by our group [9]. Rats were fasted overnight prior to surgery. Anesthetized rats (4% isoflurane in a 30:70 mix of oxygen/nitrogen) were intubated and mechanically ventilated (~45–50 breaths/min). Positive end-expiratory pressure (PEEP) was set to 3 cm H_2O and tidal volume to 7 mL/kg. Temperature was monitored with a rectal thermometer and a tympanic probe (Physitemp Instruments LLC, Clifton, New Jersey, USA) and adjusted via a heating pad/lamp to target ~36.5–37 °C. Electrocardiographic leads were attached to the limbs for continuous ECG monitoring (Powerlab, ADinstruments, CO, USA). Femoral artery and venous catheters (PE50) were surgically inserted in the left hind limb for continuous blood pressure/heart rate monitoring (arterial) and for drug injections (venous). Ventilator rate, tidal volume and FiO2 were adjusted based on arterial blood gas data to achieve goal PaCO2 of 35–40 mm Hg and PaO2 of 80–150 mm Hg. To induce ACA, isoflurane was discontinued just prior to cisatracurium (2 mg/kg) administration to induce neuromuscular blockade, and rats were then disconnected from the mechanical ventilator, "No-flow" was considered achieved when mean arterial blood pressure (MAP) reached ≤10 mmHg and was maintained for 5 min in all subjects prior to starting resuscitation efforts. The total asphyxia insult time was ~8–9 min and = (a) the time of the hypoxic period starting from the discontinuation of ventilation until no-flow is reached + (b) 5 min no-flow time + (c) the time to achieve return of spontaneous circulation (ROSC) after starting resuscitation maneuvers. Subjects that were not able to be resuscitated within 2 min of the onset of chest compressions were out of protocol and excluded from the study. Resuscitation was achieved by reinitiating mechanical ventilation, starting mechanical chest compressions using a rat thumper at a rate of 275 beats per minute, and IV administration of epinephrine (20 μg/kg), and sodium bicarbonate (1 mg/kg). At 5 min post-restoration of spontaneous circulation (ROSC) rats received vehicle or NSC74429 (1 or 3 mg/kg) IV, and booster doses were administered via IP injections at 1 d, 2 d, and 3 d post-ROSC. At 15 min post-ROSC, rats received 5 mg/kg of ketoprofen to treat surgical pain, and booster IP injections followed deficit scoring (5 mg/kg ketoprofen) on days 1, 2, and 3 post-injury. A MiniMitter (Philips Respironics, Murrysville, PA, USA) temperature probe was inserted into the abdominal cavity for telemetric monitoring of temperature to ensure normothermia for the first ~18 h of recovery. Femoral catheters were removed, wounds sutured, and rats received 10 mL subcutaneous Dextrose 5% (D5NS) to ensure hydration. Outcome performance category (OPC; 1 = normal, 2 = mild disability, 3 = moderate disability, 4 = severe disability, and 5 = death) and neurological deficit score (NDS; 0–500, 0 = normal and 500 = brain dead) were assessed daily for the first 72 h and on 7 d as described by our group [9,10]. After 72 h, rats were returned to the primary animal care facility until euthanasia for histological endpoints on day 7. Shams received surgical manipulations and vehicle injections. Rats were randomized to treatments and technicians were blind to groups (e.g., de-identified tubes labeled "A", "B", or "C" were provided to the technician that administered the experimental therapy after ACA injury and for subsequent booster doses). Rats that died before reaching the 7 d endpoint were removed from statistical analysis. Three rats failed to achieve ROSC (death before drug therapy), and 4 were excluded for death after ROSC but before day 7 (1 vehicle, 1 low dose NSC74429, 2 high dose NSC74429).

Traumatic Brain Injury Mouse Model. Mice were injured by controlled cortical impact (CCI) followed by a secondary hemorrhagic shock (HS) insult (to mimic the commonly observed clinical polytrauma scenario) as described by our group [11,12]. In brief, anesthetized mice (4% isoflurane in 70% N_2O/30% O_2) were secured inside a stereotaxic frame. Body temperature was recorded via a rectal probe (targeted to ~37 °C), and catheters inserted into the left femoral artery and vein. A dental drill was used to perform a craniotomy over the left parietal cortex. A pressure-driven pneumatic impactor, with a 3 mm steel flat tip, was advanced into the brain parenchyma (velocity 5/ms; depth 1.0 mm). Five minutes after the CCI, isoflurane was reduced to 0.5%, and blood was removed by the femoral venous catheter connected to a syringe containing citrate anticoagulant. Blood was withdrawn over 15 minutes until MAP reached 25 to 27 mmHg. Hemorrhagic shock was pressure-controlled for 35 mins by removing or administering shed blood in 50 μL aliquots. After HS, volume resuscitation was initiated with 20 mL/kg lactated ringers (LR) solution, plus additional bolus injections of 10 mL/kg LR as needed to sustain a target MABP of 70 mmHg for 90 min (Pre-Hospital Care Phase). The final 'Hospital Phase' was initiated by returning shed blood to mice over 15 mins. After re-infusion of the shed blood, MAP generally returns to near baseline levels (~90 mm Hg). Immediately following the Hospital Phase, mice were administered the experimental therapy IV (vehicle, 1 mg/kg, or 3 mg/kg). Mice were weaned from isoflurane, catheters were removed, and they were allowed to recover on supplemental oxygen for 30 mins before returning to the animal housing facility. In the acute study cohort (48 h endpoint) mice received an additional IP bolus injection of the experimental therapy at 24 h post-injury. In the subacute study cohort (7 d endpoint) mice received additional IP bolus injections at 1 d, 2 d, and 3 d post-injury. Shams received craniotomy, catheterization, and vehicle injections. Mice were randomized to treatments and technicians were blind to groups (e.g., de-identified tubes labeled "A", "B", or "C" were provided to the technician that administered the experimental therapy after CCI + HS injury and for subsequent booster injections).

Hematoxylin and Eosin (H&E) Staining and Cell Counts. Histological staining and analysis were done as described by our group [2]. In brief, at the determined study endpoint, rodents were anesthetized and flushed with heparinized ice-cold saline by transcardial perfusion into the left ventricle with drainage from a right atrial incision until all blood was cleared. Subsequently, the animal was immediately flushed with 10% buffered formalin. Brains were removed and postfixed in 10% buffered formalin for 72 h. Brains were cut, dehydrated, cleared, embedded into paraffin wax, and sectioned on a microtome. Sections were deparaffinized, treated with hematoxylin stain (ThermoScientific, Waltham, MA, USA), washed, treated with Bluing reagent (ThermoScientific), washed, and treated with Eosin Y (ThermoScientific), mounted with a coverslip on glass slides, and imaged on a Nikon Eclipse 90i microscope (Nikon, Melville, NY, USA). Live cells (CA1 and CA3) in the stratum pyramidale hippocampus were counted, divided by the length of the CA1/CA3, and reported as cells/0.1 mm. The technician quantifying CA1/CA3 cell counts was blind to treatment.

Statistics. Fluorescent units (CellTiter-Blue viability) were analyzed by one-way ANOVA followed by a post hoc Newman-Keuls multiple comparison adjustment. CA1 cell counts from ACA injured rats were analyzed by one-way ANOVA followed by Dunnett's post hoc comparison with vehicle-injury established as the control group for group comparisons. The CA1 cell count for a single subject was eliminated from statistical analysis because it was a statistical outlier as determined by the ROUT method. No other subjects across all groups were identified as outliers in the ROUT analysis. The one outlier animal received a ~9 min cardiac arrest in the vehicle treated group but had the highest CA1 cell count across all groups including when compared to uninjured sham rats. OPC and NDS data were analyzed by two-way ANOVA followed by Dunnett's post hoc comparison with vehicle-injury established as the control group for group comparisons. In the 48 h CCI + HS study, cell count data between groups had unequal variance. Therefore, 48 h CA1/CA3 cell count data were analyzed by Welch's ANOVA followed by Dunnett's T3 post hoc comparison test with vehicle-injury established as the control group for group comparisons. For the 7 d CCI + HS cohort, descriptive statistics confirmed all groups had equal variance. Therefore, a standard two-way ANOVA followed by Dunnett's post hoc comparison was performed, with vehicle-injury established as the control group for group comparisons. Scatter plots were graphed in GraphPad Prism (GraphPad Software Inc., La Jolla, CA, USA, Version 8.1.2). Data were significant at $p < 0.05$.

3. Results

A five-step screening strategy (Figure 1A) was used to identify a potential CNS permeable PHLPP inhibitor for neuroprotection. Polar surface area (PSA) and calculated partition coefficient (CLogP) are critical chemical properties that influence BBB penetration of compounds [13,14]. PSA and CLogP were calculated for each of the 116 known PHLPP inhibitors using Chemicalize software (Chemaxon; Boston, MA, USA) (Figure 1B). The upper limit of the PSA for BBB penetrating compounds is <90 Å2 [13]. This led us to eliminate 65 PHLPP inhibitors from further consideration. The remaining 51/116 compounds were screened for CLogP. The optimal mean CLogP of BBB penetrating drugs was reported to be 2.1 [13,15], 2.5 [13,16], or 3.4 [13,17]. However, CLogP of BBB penetrating drugs is normally distributed on a bell curve ranging from ~1 to ~5 [14]. To support our goal of refining the number of hits, we set 2.1 as the optimal CLogP then instituted cutoffs by adding or subtracting 1.3 from that value (i.e., a minimum of 0.8 and a maximum of 3.4). Thus, we selected for compounds that cluster near the optimal CLogP mean of 2.1–3.4 and which led us to eliminate an additional 29 inhibitors from further consideration.

To further rank the 22 remaining PHLPP inhibitors, we calculated the BBB Score for each compound. The BBB Score methodology developed by Gupta et al. [18] considers five physicochemical descriptors including: (a) number of aromatic rings, (b) heavy atoms, (c) a descriptor comprising molecular weight, hydrogen bond donor, and hydrogen bond acceptors (i.e., the MWHBN), (d) the PSA, and (e) pKa. The utility of the BBB Score was verified by Gupta et al. against 270 CNS active and 720 non-CNS FDA-approved drugs [18]. In their study no compounds with a calculated BBB Score of ≤ 2 were CNS active. A total of 12.8% of drugs with a BBB Score between 2 and 3 were CNS active; 21.9% of drugs with a BBB Score between 3 and 4 were CNS active; 51.5% of drugs with a BBB Score between 4 and 5 were CNS active; and 90.3% of drugs with a BBB Score between 5 and 6 were CNS active. Germane to the 22 PHLPP inhibitors deemed of interest, 21/22 had BBB Scores ≥ 3, 14/22 had scores > 4, and 4/22 had scores ≥ 5. (Figure 1C). Notably, the two most prolifically investigated PHLPP inhibitors have a BBB score < 2, and the frontrunner for potential translation (NSC177079 [6,7,19,20]) had the lowest value of 0.86. The 22 unexplored inhibitors were then grouped and ordered by IC50 to block the PP2C domain of PHLPPs [6]. Only compounds with the lowest IC50 to inhibit PHLPP (≤ 5 μM) were considered for empirical testing of neuroprotection. This led us to consider four compounds for in vitro studies (NSC13378, NSC25247, NSC48961, and NSC74429), Ultimately, NSC48961 was excluded due to having a CLogP furthest outside the optimal range (2.1–3.4) and because better options were identified. This led us to obtain from the DTP/NCI three highly promising compounds to test in cell culture (Figure 1D): NSC13378 (BBB Score = 5.06), NSC25247 (BBB Score = 4.82), and NSC74429 (BBB Score = 3.60).

A

Screening Scheme

Level 1: PSA <90 Å² + CLogP 0.8 ≥ and ≤ 3.4

↓

Level 2: BBB Score ≥ 3 + IC50 to Inhibit PHLPP ≤ 5 μM

↓

Level 3: Positive Hits in Staurosporine Injury Assay

↓

Level 4: Most Neuroprotective in Glutamate & H2O2

↓

Level 5: Best Compound Advanced to *In Vivo* Testing

B

116 Total Experimental PHLPP Inhibitors

NSC#	CLogP	PSA	NSC#	CLogP	PSA	NSC#	CLogP	PSA
117079	1.2	186.7	49692	3.2	66.4	60043	2.5	78.0
125908	2.3	170.8	134149	2.1	118.0	48961	1.0	72.3
117274	4.1	108.7	77393	1.5	171.8	24479	−2.1	36.8
115448	6.7	143.0	47716	0.5	156.5	270156	4.1	95.1
69343	−1.6	204.8	65537	3.0	120.1	601364	−4.4	189.4
170008	4.4	186.5	37031	−0.8	149.9	327704	3.3	76.0
23173	4.2	78.8	45208	−0.1	176.8	327705	5.2	72.8
125910	2.7	207.0	134196	0.7	114.0	41663	5.9	51.8
23128	14.9	126.5	361672	2.4	92.1	11241	2.8	68.8
13987	7.8	63.2	338517	1.9	100.7	106505	−2.1	138.7
255980	0.7	153.8	75140	0.5	133.7	130813	4.7	60.9
16211	2.8	71.4	134148	1.8	138.9	321518	3.8	45.8
117949	3.8	167.7	4895	2.0	124.6	131734	5.4	42.4
634663	−0.9	113.5	45586	3.1	155.5	13378	2.1	46.2
366218	2.1	155.7	79714	4.6	234.6	130796	3.7	157.9
119910	2.5	124.3	45618	6.5	210.2	67091	6.6	53.4
47924	3.7	41.5	51532	5.2	234.6	639174	−0.5	134.9
109248	3.7	60.8	65541	2.7	167.1	116708	3.5	44.1
114609	2.4	73.1	75913	3.1	228.7	210627	8.5	103.2
47949	5.1	43.7	134145	3.1	137.8	47106	1.0	88.5
67608	1.8	115.1	47754	5.0	243.1	117198	3.3	54.4
88852	4.1	43.7	47760	6.7	190.1	354961	3.6	137.8
103779	−2.2	102.4	114428	2.1	137.8	1013	4.7	91.0
375985	3.1	42.2	79729	6.7	190.1	48949	−0.2	50.9
120917	3.3	32.6	51536	3.3	275.5	136469	6.5	50.2
88850	3.9	43.7	45619	3.2	187.0	134244	4.4	68.7
11404	3.2	71.5	27236	0.7	138.9	12509	3.5	91.7
88915	6.2	77.5	12129	3.4	232.8	25247	1.2	55.6
93352	9.1	89.5	326196	3.2	157.9	37173	2.1	109.5
75541	2.4	91.7	134146	3.05	137.8	3323	5.6	52.3
72254	4.7	80.7	348401	1.6	165.2	13161	0.7	132.9
48010	2.9	60.4	42384	4.3	69.6	99550	8.0	23.9
109131	5.0	77.7	154829	−2.8	190.4	80126	2.3	52.3
121137	4.6	52.6	74429	3.4	81.4	117241	2.7	55.4
88135	2.9	118.2	207895	1.1	103.8	105444	4.8	71.6
119913	2.1	124.3	26699	4.0	122.4	72938	1.2	16.1
119886	10.5	76.0	133896	3.0	108.7	103019	−0.9	87.4
31076	4.5	84.2	175650	1.1	93.0	117438	1.1	87.1
114885	2.8	123.5	360215	−0.8	94.9	Most Characterized To Date		

C

22 Potential CNS Active PHLPP Inhibitors

IC50 to Block PP2C	NSC#	MW	BBB Score
≤ 5μM	117079	473.47	0.86
	45586	413.39	1.69
	13378	285.18	5.06
	25247	266.11	4.82
	48961	246.26	4.65
	74429	308.36	3.60
10-25 μM	80126	391.48	5.00
	114609	461.03	4.41
	48010	374.49	4.39
	16211	336.36	4.04
>25 μM	117241	303.38	5.20
	116708	294.74	5.07
	120917	217.15	5.02
	117198	273.30	4.89
	72938	775.23	4.58
	375985	249.27	4.39
	49692	279.30	4.33
	117438	249.27	3.60
	47106	312.33	3.54
	11404	415.27	3.53
	327704	321.35	3.25
	361672	288.29	3.04
	60043	317.34	3.00
	11241	930.94	2.98

- IC50 Values To Block PHLPP-PP2C Domain According to Sierecki et al. 2010
- BBB Score (CNS Penetration Probability): 1-2 (0%), 2-3 (12.2%), 3-4 (21.9%), 4-5 (54.%), 5-6 (90.3%).
- References: (1) Sierecki et al. 2010, (2) Gupta et al. 2019

D

3 Top Candidates Selected for Emperical Testing

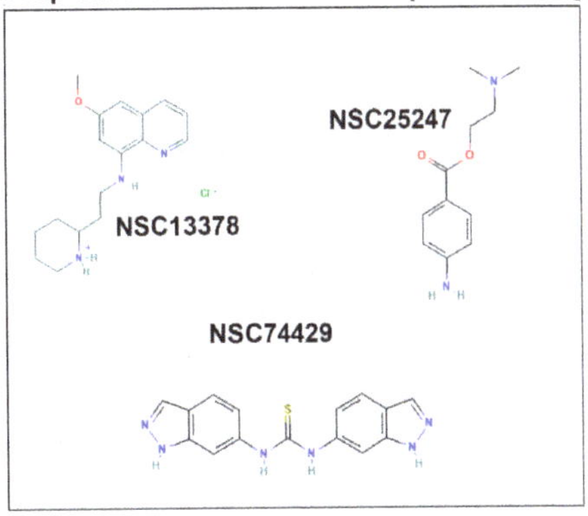

Figure 1. Physicochemical Properties of PHLPP Inhibitors and the Prediction of Their CNS Activity. (**A**) An overview of the top-down screening process employed in this study for drug selection. (**B**) The predicted polar surface aera (PSA) and calculated partition coefficient (CLogP) for each PHLPP inhibitor calculated in Chemicalize (Chemaxon). Compounds highlighted in purple denote the two PHLPP inhibitors that have received the most empirical study/characterization to date. Compounds highlighted in yellow denote unexplored PHLPP inhibitors that have chemistries supportive of BBB penetration. (**C**) The 2 most studied compounds and the 22 unexplored PHLPP inhibitors organized by known IC50s to block the PP2C domain of PHLPP2. Additionally, the BBB Score (BBB-S) was calculated for each of the compounds and arranged in each subgroup highest to lowest. The red highlighted BBB Scores indicate values < 2 and consistent with non-CNS drugs [6,18]. (**D**) The chemical structures of 3 unexplored PHLPP inhibitors chosen for testing in neurons is depicted.

NSC13378 did not increase neuronal survival in primary rat cortical neurons after a 24 h staurosporine insult (Figure 2A). In contrast, NSC25247 (at 50 μM) and NSC74429 (at 25 and 50 μM) were both neuroprotective in the STS assay (Figure 2A,B). NSC13378 was eliminated from further consideration while the other two compounds were advanced to testing in models of excitotoxicity and oxidative stress. NSC25247 increased 24 h neuronal survival after glutamate-induced injury (at 50 μM) but had no effect on survival in the H_2O_2 injury assay (Figure 2C,D). In contrast, NSC74429 increased 24 h neuronal survival in both the glutamate (at 50 μM) and H_2O_2 injury assay (at 25 and 50 μM) (Figure 2C,D). At the highest dose tested NSC25247 was neuroprotective in 2/3 assays and was a promising candidate for a CNS-active PHLPP inhibitor. However, it was eliminated from further consideration because NSC74429 demonstrated superior characteristics (i.e., it protected in 3/3 assays at one or both concentrations).

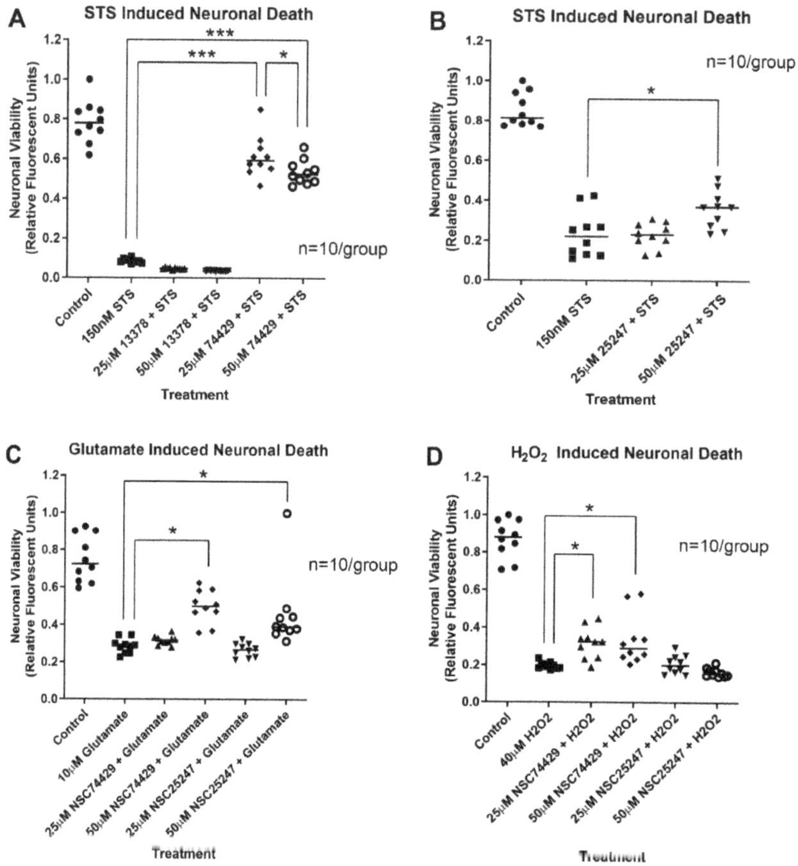

Figure 2. Neuroprotective Efficacy of NSC13378, NSC25247, and NSC74429 in Cultured Primary Rat Cortical Neurons. Neurons were injured on day in vitro (DIV) 9 and treated with vehicle or NSC compounds. (**A**) A scatter plot showing 24 h viability in staurosporine injured neurons treated with NSC13378 or NSC74429. (**B**) A scatter plot showing 24 h viability in staurosporine injured neurons treated with NSC25247. (**C**) A scatter plot showing 24 h viability in glutamate injured neurons treated with NSC25247 or NSC74429. (**D**) A scatter plot showing 24 h viability in hydrogen peroxide injured neurons treated with NSC25247 or NSC74429. All groups include n = 10/group. Data were analyzed by ANOVA and post hoc significance was detected using the Newman-Keuls multiple comparison test. Data were significant at $p < 0.05$. (*) = $p < 0.05$, (***) = $p < 0.001$.

Rats underwent an ~8–9 min ACA and were administered post-insult vehicle, 1 mg/kg NSC74429, or 3 mg/kg NSC74429. Both doses of the experimental PHLPP inhibitor increased 7 d neuronal survival in the CA1 subregion of the hippocampus (Figures 3A–F and S1). The duration of the total ischemia time did not differ compared to vehicle (IQR [min:sec]: vehicle 8:39–9:07, 1 mg/kg 8:34–8:56, 3 mg/kg 8:39–9:02; post hoc Dunnett's $p = 0.856$ for vehicle vs. 1 mg/kg and $p = 0.982$ for vehicle vs. 3 mg/kg), indicating that differences in 7 d neuron survival were not due to differences in insult severity. OPC and NDS testing did not reveal significant differences in neurological recovery at 1 d, 2 d, 3 d, or 7 d post-injury in vehicle versus NSC74429 treated rats (Figure 3G,H).

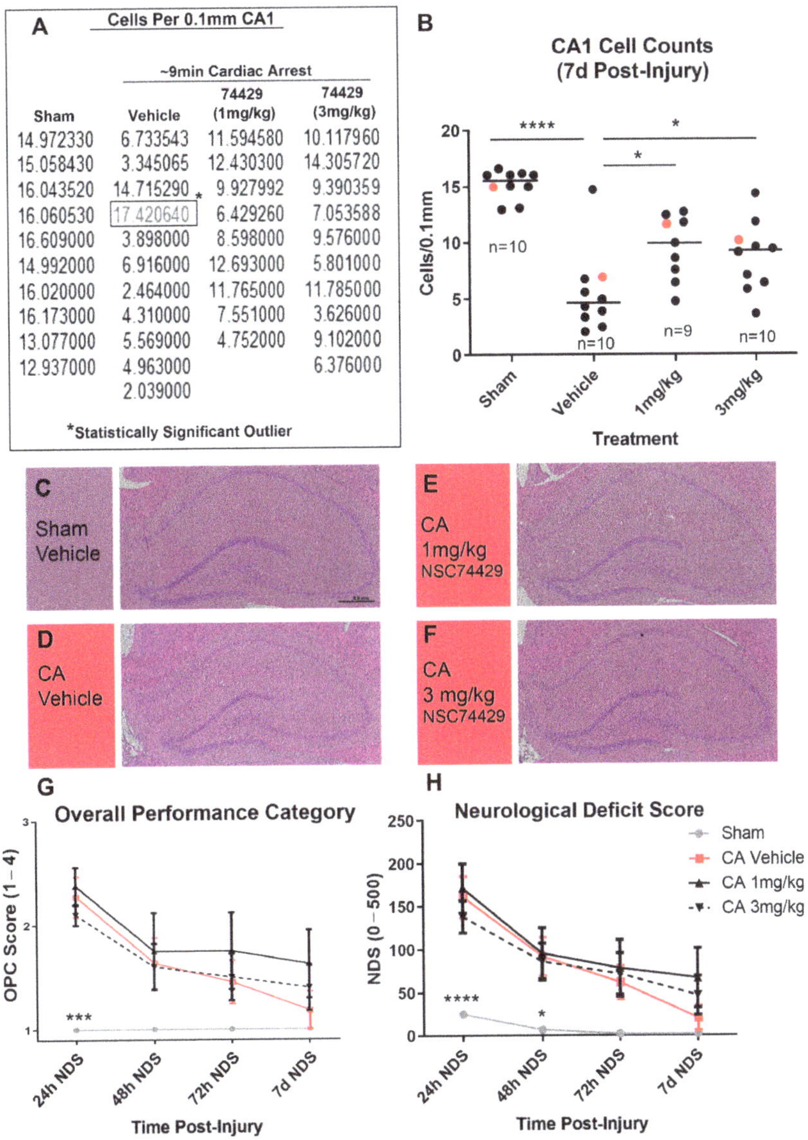

A Cells Per 0.1mm CA1

	~9min Cardiac Arrest		
Sham	Vehicle	74429 (1mg/kg)	74429 (3mg/kg)
14.972330	6.733543	11.594580	10.117960
15.058430	3.345065	12.430300	14.305720
16.043520	14.715290	9.927992	9.390359
16.060530	17.420640 *	6.429260	7.053588
16.609000	3.898000	8.598000	9.576000
14.992000	6.916000	12.693000	5.801000
16.020000	2.464000	11.765000	11.785000
16.173000	4.310000	7.551000	3.626000
13.077000	5.569000	4.752000	9.102000
12.937000	4.963000		6.376000
	2.039000		

*Statistically Significant Outlier

Figure 3. Evaluation of NSC74429 Mediated Neuroprotection in a Rat Model of Cardiac Arrest. (**A**,**B**) Normalized CA1 cell count values are shown along with the associated scatter plot. A single

CA1 cell count outlier was confirmed by the ROUT method (indicated by the *) and removed prior to statistical analysis of histology. Individual data points highlighted in red in the scatter plot correspond to the rat brains selected as representative images. **(C–F)** Representative images of H&E-stained brains used to assess CA1 cell loss. **(G,H)** Line graphs (mean + SEM) show changes in overall performance category (OPC) and neurological deficit score (NDS) testing measured in the first 72 h post-injury and at 7 d post-injury. Asterisks indicate significance compared to vehicle-injured rats. Data were significant at $p < 0.05$. (*) = $p < 0.05$, (***) = $p < 0.001$, (****) = $p < 0.0001$.

Finally, to begin to assess the scope of hippocampal neuroprotection in vivo, we also tested NSC74429 in an established mouse model of severe TBI. Vehicle treated mice subjected to a CCI + HS had significant CA1 and CA3 neuronal loss in the hippocampus at 48 h post-injury (Figure 4A,B). By 7 d post-injury, neuronal loss persisted in the CA1 subregion but was not statistically different versus shams in the CA3 subregion (Figure 4C,D). One mg/kg NSC74429 significantly increased CA1 neuronal survival at 48 h post-injury but not at 7 d post-injury (Figure 4A,C). Additionally, 1 mg/kg NSC74429 was associated with a trend toward improved CA3 cell counts at both the 48 h and 7 d post-injury time points but this was not statistically significant (Figure 4B,D). CA1/CA3 neuronal survival in the 3 mg/kg treatment group was also not statistically different compared to vehicle treated mice (Figure 4A,B). However, CA1 cell counts at 48 h post-injury displayed a biomodal pattern in the high dose group—one subgroup appeared to show signs of neuroprotection, whereas the other subgroup appeared worse than vehicle treated mice (Figure 4A). Mortality was highest overall in the 3 mg/kg group (in the 48 h study) and had the only case of a surviving mouse with brain hemorrhage that prevented accurate hippocampal neuronal counting (Figure 4I–L). For that reason, 3 mg/kg was not further explored in the 7 d outcome study.

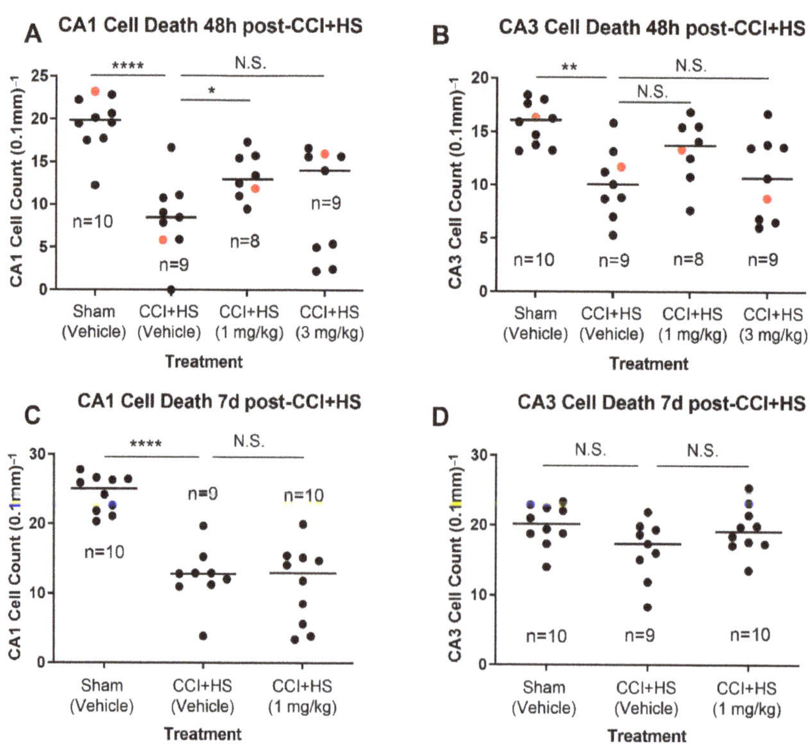

Figure 4. *Cont.*

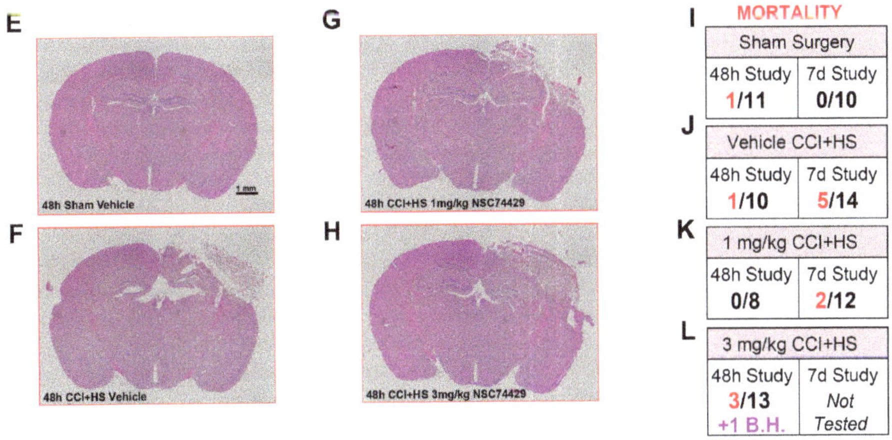

Figure 4. Evaluation of NSC74429 Mediated Neuroprotection in a Mouse Model of TBI. (**A**,**B**) Scatter plots of normalized CA1 and CA3 cell counts at 48 h post-injury. (**C**,**D**) Scatter plots of normalized CA1 and CA3 cell counts at 7 d post-injury. Individual data points highlighted in red in the scatter plot correspond to the mouse brains selected as representative images in panels E-H. (**E**–**H**) Representative images of H&E-stained brains used to assess CA1/CA3 cell loss. (**I**–**L**) Mortality for each group by treatment and study endpoint. Asterisks indicate significance compared to vehicle-injured rats. The red text indicates animals that died. The purple text indicates an animal that survived to the study endpoint but had bleeding in the brain that precluded cell counting. Data were significant at $p < 0.05$. Not Significant (N.S.), Brain Hemorrhage (B.H.), Controlled Cortical Impact (CCI), Hemorrhagic Shock (HS). (*) = $p < 0.05$, (**) = $p < 0.01$, (****) = $p < 0.0001$.

4. Discussion

Sierecki et al. used a combination of cell-free screening assays and in silico tools to identify ~116 small-molecules that selectively block the PP2C-phosphatase domain in PHLPPs [6]. These compounds are non-selective for PHLPP1 and PHLPP2. Two compounds (NSC117079 and NSC45586) were subsequently investigated by others but are unlikely candidates for CNS therapeutics [7,19,20]. Specifically, their chemical properties are not ideal for targeting the brain (e.g., their BBB score < 2) and some bind albumin.

Our objective was to employ an accelerated win/kill approach to quickly filter/identify, among the 114 unexplored PHLPP inhibitors, a single compound that merited in vivo testing for neuroprotection in multiple models of brain injury. Our rationale was based on the need for a PHLPP inhibitor that is more likely to target the brain, and to establish neuroprotection before advancing with elaborate follow up studies to comprehensively assess cell signaling in a variety of cell types, behavioral outcomes, pharmacokinetic and pharmacodynamic (PK/PD) studies, and brain disposition studies to determine dosing and clearance kinetics.

A five-staged algorithm was used to identify a promising PHLPP inhibitor candidate for CNS drug development (Figure 1A). First, we reviewed both key traditional chemical criteria (e.g., CLogP thresholds for CNS drugs [13]) and contemporary tools (e.g., calculation of the BBB Score [18]) to identify 22 compounds with chemical properties suggestive of BBB penetration (1st and 2nd level analysis). Next, we selected the top three drugs (NSC13378, NSC25247, and NSC74429) with optimal properties and the lowest IC50s to inhibit PHLPPs, and empirically tested them in neurons in a gold-standard model of STS-induced apoptosis (3rd level analysis). STS was the initial screening injury-assay because it models a pure apoptotic insult, and because apoptosis is the prototypical cell death pathway augmented by PHLPPs [4,21]. Furthermore, we previously reported that PHLPP gene knockdown or treating cells with NSC117079 is neuroprotective in STS, and therefore using the same assay here facilitated historical comparisons germane to the efficacy of the uncharacterized novel PHLPP inhibitors that were tested [7]. The most promising compounds advanced to testing in in vitro models of glutamate-induced excitotoxicity and H_2O_2-induced oxidative stress (4th level analysis). The insults produced by these agents activate a complex constellation of cell death mechanisms including apoptosis, necrosis, and autophagy [22–25]. These mechanisms are felt to be critical in the pathophysiology of clinical brain injury including in TBI, stroke and ACA [26–28]. NSC74429 performed best overall (on all three assays) and was advanced to animal testing (5th level analysis). We used a 1 mg/kg (low) and 3 mg/kg (high) dose for in vivo studies. Dosing was based on a similar dose range reported with NSC117079 and NSC45586 to induce a biological effect in animals [20]. To further strengthen our in vivo studies, we used two different brain injury models across two different species, to assess the generalizability and robustness of neuroprotection. Our rat ACA model results in a primary insult to the CA1 subregion of the hippocampus but does not damage the BBB [29,30]. In contrast, our mouse CCI + HS model is comparatively more severe, results in widespread necrosis, a cortical contusion, robust hippocampal CA1/CA3 neuron death, and the BBB is compromised as determined by Evans blue extravasation [11,31].

Low and/or high dose NSC74429 increased rat neuron survival after STS, glutamate, or H_2O_2 -induced cell death. High dose (50 μM) NSC25247 also showed promise and was neuroprotective in both the STS and glutamate injury models, but not in H_2O_2. Importantly, all assays employed a post-treatment strategy, except for STS in which NSC compounds were added at the same time as the injury agent. Additionally, the finding that low dose (25 μM) NSC74429 robustly protected in STS (apoptosis), modestly in H_2O_2 (necrosis), and had no effect in glutamate (excitotoxicity), precluded the possibility that neuroprotection was simply due to an artifact of the compound interacting with the viability assay. Similarly, cells treated with NSC13378 were visibly dead.

NSC74429 also decreased neuronal death in vivo in a rat model of ACA and a mouse model of TBI. Neuroprotection in the injured CA1 in the ACA model persisted for at least 7 d post-injury, which suggests that NSC74429 did not simply delay inevitable cell death—although even longer endpoints were not studied [32,33]. Additionally, the fact that BBB function is maintained in the ACA model, and that neuroprotection with NSC74429 was observed, supports the predictive chemistry suggesting that it is a promising candidate to target the brain. Nevertheless, future studies are needed to directly assess drug levels in brain tissue and/or to ascertain if peripheral effects could have contributed to CA1 neuroprotection via indirect mechanisms. In the CCI + HS model, CA1 neuroprotection was seen at 48 h but not 7 d post-injury. Thus, NSC74429 delayed cell death in CA1 in the TBI model but did not produce long-lasting neuroprotection in that region. The neuroprotective benefit on 48 h CA1 survival is nevertheless striking given that (a) the CCI + HS model produces a very severe insult which represents a high bar for neuroprotectants to show a benefit, and (b) the administration of NSC74429 was delayed by almost ~2 h after the initial CCI-TBI component of the insult (i.e., therapy was initiated after the hospital phase of the HS component). Interestingly, in the CA3, there was a trend toward a sustained increase in neuronal survival in the 1 mg/kg NSC74429 group at 48 h and 7 d post-injury; however, this effect was not significant, and our study was not powered to appropriately test that hypothesis. Higher mortality in the 3 mg/kg group merits additional study. While bolus dosing was a logical first step to assess neuroprotection, future PK/PD studies are needed to determine if sustained delivery at targeted serum doses may further improve histological outcomes and minimize adverse side effects. The mechanism(s) mediating increased mortality at the high bolus dose are unclear but will need to be elucidated prior to clinical translation, if warranted, or to potentially determine if modifications to the parent compound may mitigate some unwanted side effects.

We were surprised that NSC13378 failed to protect; it was our frontrunner based on its LogP, BBB Score, and IC50 to inhibit PHLPP. This unexpected result highlights the benefits of our multi-staged approach to screening. Conversely, we were surprised by the robustness of NSC74429. Indeed, the magnitude of neuroprotection appeared to be much greater than what we previously observed in the STS assay in primary rat neurons after PHLPP1 gene knockdown or by using NSC117079 [7]. Future cell culture studies are needed to test if protection with NSC74429 is lost or ameliorated after PHLPP1 or PHLPP2 knockdown, or after combined PHLPP1/PHLPP2 knockdown. Those experiments may shed light on the potential contribution of PHLPP-dependent versus -independent mechanisms mediating neuroprotection.

It is possible that NSC13378 failed to protect due to toxic off-target effects, whereas NSC74429 was beneficial due to neuroprotective off-target effects. Thus, it is not clear if hippocampal protection was fully or partially dependent on PHLPP inhibition and the specificity of NSC74429 remains to be determined. Importantly, contemporary thinking in neurotrauma has moved away from focusing solely on drugs that target a single mechanism. That is because the nature of brain damage seen in heterogenous TBI patients is multifactorial and has contributed, in part, to the lack of new interventions after >191 clinical trials [34]. Thus, from a translational perspective, we think it would be advantageous if future studies discover that NSC74429 targets additional neuroprotective mechanisms on top of PHLPP inhibition. We recognize however a disadvantage is that the more pathways targeted by NSC74429 (if confirmed) the less useful this compound may turn out to be as a research tool to selectively probe PHLPP biology in the brain, but whether this concern is justified remains to be seen. Additionally, it is equally possible that NSC74429 is a selective PHLPP inhibitor and that it performed the best because of its high specificity and lack of side effects at the doses tested.

Clinically, we think that our findings may be important. Pro-death PHLPP1 and PHLPP2, also referred to as suprachiasmatic nucleus circadian oscillatory proteins (SCOP), are encoded by two genes [4,35,36]. The PHLPP-PP2C domain is highly homologous between isoforms (58% identity), inhibits a variety of pro-survival signaling pathways, and is of interest germane to blocking PHLPP-PP2C activity for neuroprotection [1,36]. Indeed, PHLPP1 gene KO mice have decreased infarct volume after experimental stroke [1]. Additionally, PHLPP1 impairs hippocampal-dependent memory function, and gene KO mice have enhanced memory performance following a TBI [2,37]. PHLPP2 similarly promotes neuronal death and siRNA mediated gene knockdown decreased brain damage in a rat model of global cerebral ischemia [38]. More recently it was discovered that endothelium-specific PHLPP2 KO mice have decreased neointima formation in injured carotid arteries (i.e., decreased harmful vascular remodeling) [3]. Conversely, PHLPP2 levels are increased in the endothelium of human patients with atherosclerotic plaques [3]. Thus, PHLPPs are linked to a spectrum of disease-promoting cell signaling perturbations that could negatively impact chronic brain health, and pharmacological inhibitors targeting PHLPPs in the human brain may be useful.

Study limitations include sample size, the lack of an assessment of drug efficacy in females (i.e., only males were tested), lack of additional outcome metrics (e.g., behavior), and that our findings come from a single center, and lack of assessment of cognitive outcome and/or longer-term outcomes [11]. However, the high level of rigor in our study, both technical (randomization and blinding) and statistical, strengthens our overall findings. Additionally, assessing neuroprotective agents across models in a single study is uncommon and has many potential scientific advantages to better predict real-world efficacy and limitations of CNS drugs, as is being demonstrated by the multi-center drug testing consortium for TBI, Operation Brain Trauma Therapy, and/or in consortia in early stages of development for other forms of acute brain injury, including cardiac arrest [39–41]. Another limitation was our compound screening design. While logical, it may have incorporated biases that resulted in the elimination of drugs that ultimately warrant testing. For instance, changes in the thresholding rules could have led to other compounds being assessed. Additionally, we did not leverage other sophisticated methods to estimate BBB permeability (e.g., AI learning [42]). Thus, other molecules among the 111 PHLPP inhibitors yet to be characterized could have advantages over NSC74429 (e.g., more neuroprotective, greater ability to penetrate the BBB, and/or less side effects at the optimal dose). Nevertheless, we believe that our approach resulted in exciting new findings that expand the current armamentarium of research tools (NSC117079, NSC45586, NSC74429, and NSC25247) available to researchers to study PHLPPs. Additionally, our findings support the need for further tests on NSC74429 and NSC25247 to modulate PHLPP activity in the brain.

5. Patents

Travis C. Jackson and Patrick M. Kochanek are co-inventors on a pending patent application titled: "Compounds for the treatment of acute organ injury" (USPTO Application No. 16/584,314).

Supplementary Materials: The following supporting information can be downloaded at: https://www.mdpi.com/article/10.3390/biom12101352/s1, Figure S1: H&E Staining of CA1 in All Animals in the ACA Study.

Author Contributions: T.C.J. conceived the study. T.C.J., P.M.K. and C.D. contributed to the study design. T.C.J. drafted the manuscript. J.S., V.A.V., T.C.J. and K.J.-F. contributed to experiments and data acquisition. T.C.J., P.M.K. and C.D. contributed to data analysis. J.S., C.D., V.A.V. and K.J.-F. edited the draft and contributed to the final submitted version. All authors have read and agreed to the published version of the manuscript.

Funding: This research was funded by an American Heart Association (AHA) grant (14SDG20210000) to T.C.J., NIH/NINDS R01NS105721 to T.C.J., and by the Ake N. Grenvik Chair in Critical Care Medicine to P.M.K.

Institutional Review Board Statement: The animal study protocol was approved by the IACUC of the University of Pittsburgh (Protocol #: 14073890).

Informed Consent Statement: Not applicable.

Data Availability Statement: Data is contained within the article.

Acknowledgments: We thank the Developmental Therapeutics Program (DTP) of the National Cancer Institute (NCI) for generously providing the experimental compounds.

Conflicts of Interest: Travis C. Jackson and Patrick M. Kochanek are co-inventors on a pending patent application titled: "Compounds for the treatment of acute organ injury" (USPTO Application No. 16/584,314).

References

1. Chen, B.; Van Winkle, J.A.; Lyden, P.D.; Brown, J.H.; Purcell, N.H. PHLPP1 gene deletion protects the brain from ischemic injury. *J. Cereb. Blood Flow Metab.* **2013**, *33*, 196–204. [CrossRef]
2. Jackson, T.C.; Dixon, C.E.; Janesko-Feldman, K.; Vagni, V.; Kotermanski, S.E.; Jackson, E.K.; Kochanek, P.M. Acute Physiology and Neurologic Outcomes after Brain Injury in SCOP/PHLPP1 KO Mice. *Sci. Rep.* **2018**, *8*, 7158. [CrossRef] [PubMed]
3. Huang, J.; Cai, C.; Zheng, T.; Wu, X.; Wang, D.; Zhang, K.; Xu, B.; Yan, R.; Gong, H.; Zhang, J.; et al. Endothelial Scaffolding Protein ENH (Enigma Homolog Protein) Promotes PHLPP2 (Pleckstrin Homology Domain and Leucine-Rich Repeat Protein Phosphatase 2)-Mediated Dephosphorylation of AKT1 and eNOS (Endothelial NO Synthase) Promoting Vascular Remodeling. *Arterioscler. Thromb. Vasc. Biol.* **2020**, *40*, 1705–1721. [CrossRef]
4. Gao, T.; Furnari, F.; Newton, A.C. PHLPP: A phosphatase that directly dephosphorylates Akt, promotes apoptosis, and suppresses tumor growth. *Mol. Cell* **2005**, *18*, 13–24. [CrossRef]
5. Qiao, M.; Wang, Y.; Xu, X.; Lu, J.; Dong, Y.; Tao, W.; Stein, J.; Stein, G.S.; Iglehart, J.D.; Shi, Q.; et al. Mst1 is an interacting protein that mediates PHLPPs' induced apoptosis. *Mol. Cell* **2010**, *38*, 512–523. [CrossRef]
6. Sierecki, E.; Sinko, W.; McCammon, J.A.; Newton, A.C. Discovery of small molecule inhibitors of the PH domain leucine-rich repeat protein phosphatase (PHLPP) by chemical and virtual screening. *J. Med. Chem.* **2010**, *53*, 6899–6911. [CrossRef]
7. Jackson, T.C.; Verrier, J.D.; Drabek, T.; Janesko-Feldman, K.; Gillespie, D.G.; Uray, T.; Dezfulian, C.; Clark, R.S.; Bayir, H.; Jackson, E.K.; et al. Pharmacological inhibition of pleckstrin homology domain leucine-rich repeat protein phosphatase is neuroprotective: Differential effects on astrocytes. *J. Pharmacol. Exp. Ther.* **2013**, *347*, 516–528. [CrossRef]
8. Jackson, T.C.; Verrier, J.D.; Kochanek, P.M. Anthraquinone-2-sulfonic acid (AQ2S) is a novel neurotherapeutic agent. *Cell Death Dis.* **2013**, *4*, e451. [CrossRef]
9. Dezfulian, C.; Kenny, E.; Lamade, A.; Misse, A.; Krehel, N.; St Croix, C.; Kelley, E.E.; Jackson, T.C.; Uray, T.; Rackley, J.; et al. Mechanistic characterization of nitrite-mediated neuroprotection after experimental cardiac arrest. *J. Neurochem.* **2016**, *139*, 419–431. [CrossRef]

10. Neumar, R.W.; Bircher, N.G.; Sim, K.M.; Xiao, F.; Zadach, K.S.; Radovsky, A.; Katz, L.; Ebmeyer, E.; Safar, P. Epinephrine and sodium bicarbonate during CPR following asphyxial cardiac arrest in rats. *Resuscitation* **1995**, *29*, 249–263. [CrossRef]
11. Hemerka, J.N.; Wu, X.; Dixon, C.E.; Garman, R.H.; Exo, J.L.; Shellington, D.K.; Blasiole, B.; Vagni, V.A.; Janesko-Feldman, K.; Xu, M.; et al. Severe brief pressure-controlled hemorrhagic shock after traumatic brain injury exacerbates functional deficits and long-term neuropathological damage in mice. *J. Neurotrauma* **2012**, *29*, 2192–2208. [CrossRef] [PubMed]
12. Jackson, T.C.; Du, L.; Janesko-Feldman, K.; Vagni, V.A.; Dezfulian, C.; Poloyac, S.M.; Jackson, E.K.; Clark, R.S.; Kochanek, P.M. The nuclear splicing factor RNA binding motif 5 promotes caspase activation in human neuronal cells, and increases after traumatic brain injury in mice. *J. Cereb. Blood Flow Metab.* **2015**, *35*, 655–666. [CrossRef] [PubMed]
13. Pajouhesh, H.; Lenz, G.R. Medicinal chemical properties of successful central nervous system drugs. *NeuroRx* **2005**, *2*, 541–553. [CrossRef]
14. Ghose, A.K.; Herbertz, T.; Hudkins, R.L.; Dorsey, B.D.; Mallamo, J.P. Knowledge-Based, Central Nervous System (CNS) Lead Selection and Lead Optimization for CNS Drug Discovery. *ACS Chem. Neurosci.* **2012**, *3*, 50–68. [CrossRef]
15. Flynn, G.L. Substituent constants for correlation analysis in chemistry and biology. By Corwin Hansch and Albert Leo. Wiley, 605 Third Ave., New York, NY 10016. 1979. 339 pp. 21 × 28 cm. Price $24.95. *J. Pharm. Sci.* **1980**, *69*, 1109. [CrossRef]
16. Leeson, P.D.; Davis, A.M. Time-related differences in the physical property profiles of oral drugs. *J. Med. Chem.* **2004**, *47*, 6338–6348. [CrossRef]
17. Varma, M.V.; Sateesh, K.; Panchagnula, R. Functional role of P-glycoprotein in limiting intestinal absorption of drugs: Contribution of passive permeability to P-glycoprotein mediated efflux transport. *Mol. Pharm.* **2005**, *2*, 12–21. [CrossRef]
18. Gupta, M.; Lee, H.J.; Barden, C.J.; Weaver, D.F. The Blood-Brain Barrier (BBB) Score. *J. Med. Chem.* **2019**, *62*, 9824–9836. [CrossRef]
19. Hwang, S.M.; Feigenson, M.; Begun, D.L.; Shull, L.C.; Culley, K.L.; Otero, M.; Goldring, M.B.; Ta, L.E.; Kakar, S.; Bradley, E.W.; et al. Phlpp inhibitors block pain and cartilage degradation associated with osteoarthritis. *J. Orthop. Res.* **2018**, *36*, 1487–1497. [CrossRef]
20. Taylor, E.L.; Weaver, S.R.; Zars, E.L.; Turner, C.A.; Buhrow, S.A.; Reid, J.M.; Bradley, E.W.; Westendorf, J.J. Chondrocytic and pharmacokinetic properties of Phlpp inhibitors. *Osteoarthr. Cartil. Open* **2021**, *3*, 100190. [CrossRef]
21. Koh, J.Y.; Wie, M.B.; Gwag, B.J.; Sensi, S.L.; Canzoniero, L.M.; Demaro, J.; Csernansky, C.; Choi, D.W. Staurosporine-induced neuronal apoptosis. *Exp. Neurol.* **1995**, *135*, 153–159. [CrossRef]
22. Higgins, G.C.; Beart, P.M.; Nagley, P. Oxidative stress triggers neuronal caspase-independent death: Endonuclease G involvement in programmed cell death-type III. *Cell. Mol. Life Sci.* **2009**, *66*, 2773–2787. [CrossRef] [PubMed]
23. McNeill-Blue, C.; Wetmore, B.A.; Sanchez, J.F.; Freed, W.J.; Merrick, B.A. Apoptosis mediated by p53 in rat neural AF5 cells following treatment with hydrogen peroxide and staurosporine. *Brain Res.* **2006**, *1112*, 1–15. [CrossRef] [PubMed]
24. Ankarcrona, M.; Dypbukt, J.M.; Bonfoco, E.; Zhivotovsky, B.; Orrenius, S.; Lipton, S.A.; Nicotera, P. Glutamate-induced neuronal death: A succession of necrosis or apoptosis depending on mitochondrial function. *Neuron* **1995**, *15*, 961–973. [CrossRef]
25. Kumari, S.; Mehta, S.L.; Li, P.A. Glutamate induces mitochondrial dynamic imbalance and autophagy activation: Preventive effects of selenium. *PLoS ONE* **2012**, *7*, e39382. [CrossRef]
26. McGinn, M.J.; Povlishock, J.T. Pathophysiology of Traumatic Brain Injury. *Neurosurg. Clin. N. Am.* **2016**, *27*, 397–407. [CrossRef]
27. Abramov, A.Y.; Scorziello, A.; Duchen, M.R. Three distinct mechanisms generate oxygen free radicals in neurons and contribute to cell death during anoxia and reoxygenation. *J. Neurosci.* **2007**, *27*, 1129–1138. [CrossRef]
28. Mitsios, N.; Gaffney, J.; Kumar, P.; Krupinski, J.; Kumar, S.; Slevin, M. Pathophysiology of acute ischaemic stroke: An analysis of common signalling mechanisms and identification of new molecular targets. *Pathobiology* **2006**, *73*, 159–175. [CrossRef]
29. Tress, E.E.; Clark, R.S.; Foley, L.M.; Alexander, H.; Hickey, R.W.; Drabek, T.; Kochanek, P.M.; Manole, M.D. Blood brain barrier is impermeable to solutes and permeable to water after experimental pediatric cardiac arrest. *Neurosci. Lett.* **2014**, *578*, 17–21. [CrossRef]
30. Janata, A.; Drabek, T.; Magnet, I.A.; Stezoski, J.P.; Janesko-Feldman, K.; Popp, E.; Garman, R.H.; Tisherman, S.A.; Kochanek, P.M. Extracorporeal versus conventional cardiopulmonary resuscitation after ventricular fibrillation cardiac arrest in rats: A feasibility trial. *Crit. Care Med.* **2013**, *41*, e211–e222. [CrossRef]
31. Whalen, M.J.; Carlos, T.M.; Kochanek, P.M.; Clark, R.S.; Heineman, S.; Schiding, J.K.; Franicola, D.; Memarzadeh, F.; Lo, W.; Marion, D.W.; et al. Neutrophils do not mediate blood-brain barrier permeability early after controlled cortical impact in rats. *J. Neurotrauma* **1999**, *16*, 583–594. [CrossRef] [PubMed]
32. Dietrich, W.D.; Busto, R.; Alonso, O.; Globus, M.Y.; Ginsberg, M.D. Intraischemic but not postischemic brain hypothermia protects chronically following global forebrain ischemia in rats. *J. Cereb. Blood Flow Metab.* **1993**, *13*, 541–549. [CrossRef] [PubMed]
33. Elsersy, H.; Sheng, H.; Lynch, J.R.; Moldovan, M.; Pearlstein, R.D.; Warner, D.S. Effects of isoflurane versus fentanyl-nitrous oxide anesthesia on long-term outcome from severe forebrain ischemia in the rat. *Anesthesiology* **2004**, *100*, 1160–1166. [CrossRef]
34. Bragge, P.; Synnot, A.; Maas, A.I.; Menon, D.K.; Cooper, D.J.; Rosenfeld, J.V.; Gruen, R.L. A State-of-the-Science Overview of Randomized Controlled Trials Evaluating Acute Management of Moderate-to-Severe Traumatic Brain Injury. *J. Neurotrauma* **2016**, *33*, 1461–1478. [CrossRef] [PubMed]
35. Shimizu, K.; Okada, M.; Takano, A.; Nagai, K. SCOP, a novel gene product expressed in a circadian manner in rat suprachiasmatic nucleus. *FEBS Lett.* **1999**, *458*, 363–369. [CrossRef]
36. Brognard, J.; Sierecki, E.; Gao, T.; Newton, A.C. PHLPP and a second isoform, PHLPP2, differentially attenuate the amplitude of Akt signaling by regulating distinct Akt isoforms. *Mol. Cell* **2007**, *25*, 917–931. [CrossRef]

37. Shimizu, K.; Phan, T.; Mansuy, I.M.; Storm, D.R. Proteolytic degradation of SCOP in the hippocampus contributes to activation of MAP kinase and memory. *Cell* **2007**, *128*, 1219–1229. [CrossRef]
38. Wei, X.E.; Zhang, F.Y.; Wang, K.; Zhang, Q.X.; Rong, L.Q. Assembly of the FKBP51-PHLPP2-AKT signaling complex in cerebral ischemia/reperfusion injury in rats. *Brain Res.* **2014**, *1566*, 60–68. [CrossRef]
39. Kochanek, P.M.; Bramlett, H.M.; Dixon, C.E.; Dietrich, W.D.; Mondello, S.; Wang, K.K.W.; Hayes, R.L.; Lafrenaye, A.; Povlishock, J.T.; Tortella, F.C.; et al. Operation Brain Trauma Therapy: 2016 Update. *Mil. Med.* **2018**, *183*, 303–312. [CrossRef]
40. Bosetti, F.; Koenig, J.I.; Ayata, C.; Back, S.A.; Becker, K.; Broderick, J.P.; Carmichael, S.T.; Cho, S.; Cipolla, M.J.; Corbett, D.; et al. Translational Stroke Research: Vision and Opportunities. *Stroke* **2017**, *48*, 2632–2637. [CrossRef]
41. Lin, S.; Ramadeen, A.; Sundermann, M.L.; Dorian, P.; Fink, S.; Halperin, H.R.; Kiss, A.; Koller, A.C.; Kudenchuk, P.J.; McCracken, B.M.; et al. Establishing a multicenter, preclinical consortium in resuscitation: A pilot experimental trial evaluating epinephrine in cardiac arrest. *Resuscitation* **2022**, *175*, 57–63. [CrossRef] [PubMed]
42. Liu, L.; Zhang, L.; Feng, H.; Li, S.; Liu, M.; Zhao, J.; Liu, H. Prediction of the Blood-Brain Barrier (BBB) Permeability of Chemicals Based on Machine-Learning and Ensemble Methods. *Chem. Res. Toxicol.* **2021**, *34*, 1456–1467. [CrossRef] [PubMed]

biomolecules

MDPI

Review

Atrial Fibrillation and Dementia: Pathophysiological Mechanisms and Clinical Implications

Dimitrios Varrias [1,*], Tinatin Saralidze [1], Pawel Borkowski [1], Sumant Pargaonkar [1], Michail Spanos [2], George Bazoukis [3] and Damianos Kokkinidis [4,*]

1 Department of Medicine, Jacobi Medical Center, Albert Einstein College of Medicine, Bronx, NY 10461, USA; pawel.borkowski.md@gmail.com (P.B.); sumant.pargaonkar@gmail.com (S.P.)
2 Cardiovascular Research Center, Massachusetts General Hospital, Harvard Medical School, Boston, MA 02114, USA; mspanos1@mgh.harvard.edu
3 School of Medicine, European University Cyprus, 2417 Nicosia, Cyprus
4 Section of Cardiovascular Medicine, Yale University, New Haven, CT 06520, USA
* Correspondence: varriasdimitrios@gmail.com (D.V.); damianos.kokkinidis@yale.edu (D.K.)

Abstract: Numerous longitudinal studies suggest a strong association between cardiovascular risk factors and cognitive impairment. Individuals with atrial fibrillation are at higher risk of dementia and cognitive dysfunction, as atrial fibrillation increases the risk of cerebral hypoperfusion, inflammation, and stroke. The lack of comprehensive understanding of the observed association and the complex relationship between these two diseases makes it very hard to provide robust guidelines on therapeutic indications. With this review, we attempt to shed some light on how atrial fibrillation is related to dementia, what we know regarding preventive interventions, and how we could move forward in managing those very frequently overlapping conditions.

Keywords: dementia; atrial fibrillation; Alzheimer's; stroke; bleeding; anticoagulation

1. Introduction

Numerous longitudinal studies suggest a strong association between cardiovascular risk factors and cognitive impairment. Patients with atrial fibrillation (AF) are at higher risk of dementia and cognitive dysfunction as AF increases the risk of cerebral hypoperfusion, inflammation, and stroke [1]. However, the existing literature still lacks a thorough analysis of the observed association and complex relationship between these two diseases. In addition, no official guidelines exist to provide clear indications for treatment based on prospective intervention studies. This review will provide a comprehensive and detailed overview of the topic.

2. Epidemiology of AF and Dementia

AF is the most prevalent arrhythmia in adults, with around 37.5 million cases; at the same time, dementia accounts for a massive burden for the healthcare system, with an estimated prevalence of approximately 55 million cases. With the population ageing worldwide, epidemiological predictions foresee increased prevalence of both disorders as well as associated morbidity and mortality [2]. Saglietto et al. reported atrial fibrillation and dementia epidemiological data as analyzed from 1990 to 2019 [3]. The overall incidence rate of AF and dementia was 7–9 folds higher in higher sociodemographic index (SDI) countries compared to lower SDI countries. Remarkably, there is a significant overlap in the population affected by the two diseases, with Liu et al. reporting that AF was independently associated with dementia (HR = 1.34, 95% CI: 1.24–1.44) [4]. Unfortunately, the causal association is still being investigated [5].

3. Challenges in Associations

There are many potential links between AF and dementia. Firstly, a well-known risk factor for dementia is stroke, which is one of the most common complications of AF. Second, a decline in cardiac output secondary to AF leads to chronic cerebral hypoperfusion, which in turn causes central nervous system dysfunction. Although associating the two conditions appears obvious, significant gaps in the mechanistic etiology of AF-related cognitive consequences exist beyond aging and stroke. The shared etiology due to overlapping risk factors (including advanced age, hypertension, heart failure, diabetes, hyperlipidemia, sleep apnea, coronary artery disease, chronic kidney disease, obesity, and physical inactivity) as well as therapeutic options (DOACS) [2] acts as a significant confounder in existing observational studies. Furthermore, causal associations with dementia may be complicated to prove due to the slowly progressive nature of the disease. Still, it also has many different sub-categories (vascular, Alzheimer's, etc.) [6]. Recently, several longitudinal studies have suggested that AF correlates with all causes of dementia but also with Alzheimer's disease [6,7]. However, observational studies might be confounded by potential biases and reverse causation [8]. On the other hand, solid evidence from numerous prospective studies has shown AF to be associated with cognitive impairment, cognitive decline, and all causes of dementia [7,9–14]. However, most studies examined the relationship with all-cause dementia without separate investigation for vascular dementia and Alzheimer's disease [10–14]. Although there is evidence that AF correlates with Alzheimer's, there is no suggested mechanical relationship between the two, with studies concluding that it is likely related to confounding bias due to shared comorbidities [6]. Objective quantification of dementia can be challenging but vital when it comes to studies regarding associations. MRI is the gold standard as it can depict structural changes (lacunar infarcts, atrophy, and amyloid depositions) [15]. Computed tomography, despite being more cost effective, does not provide optimal results when it comes to differentiating types of dementia [16]. Recently, there has been a lot of progress in the field of neuro- imaging, with advanced metabolic and functional modalities. Identifying pathology before the radiographic evidence in T2 weighted MRI is a challenge, but single-photon emission computed tomography, fluorodeoxyglucose–positron emission tomography and blood oxygenation level–dependent MRI are state-of-the art and promising modalities [17].

4. Evidence of Association

Several well-established population-based longitudinal studies have demonstrated a higher risk of cognitive decline or dementia associated with AF. Nevertheless, most of the evidence comes from the data on the elderly population. Hence, the association might be related to underlying systemic vascular disease and the increased prevalence of both disorders with decreased age. Therefore, it is crucial to investigate whether the link arises from a common pathophysiological mechanism or whether AF plays a role in a causal pathway. Interestingly, in the Whitehall study [14], more than 10,000 patients (aged 45–69 years) who completed cognitive tests four times over 15 years were recruited. Compared to AF-free participants, patients with longstanding AF (5, 10, or 15 years) experienced a steeper curve of cognitive decline and a higher risk of dementia after adjusting confounders (sociodemographic, behavioral, and chronic diseases) [HR: 1.87; 95% CI: 1.37, 2.55]. Hence, in those with early AF onset, a more extended exposure period led to more significant neuronal injury and loss, possibly due to an enhancing effect of the degenerative disease on vascular changes. In addition, a study showed that even adults with incident AF at age 50–55 had accelerated cognitive decline [14]. Atrial fibrillation was associated with an increased risk of dementia (hazard ratio, 1.23; 95% confidence interval, 1.04–1.45), even after adjusting for cardiovascular risk factors, including ischemic stroke [18]. This association was strongest for younger patients who consistently had a longer duration of AF. A systematic review and meta-analysis by Kwork [19] identified 15 relevant studies, including 46,637 participants with a mean age of 71.7 years. Fourteen studies showed a significant increase in the risk of dementia associated with AF (OR 2.0, 95% CI 1.4 to 2.7,

$p < 0.0001$), with substantial heterogeneity (I(2) = 75%). After stratification by participants, the association was significant in studies focusing only on stroke patients (7 studies, OR 2.4, 95% CI 1.7 to 3.5, $p < 0.001$, I(2) = 10%), and of marginal significance in broader populations (7 studies, OR 1.6, 95% CI 1.0 to 2.7, $p = 0.05$, I(2) = 87%). Santangeli et al. comprised eight prospective studies including more than 77,000 patients, of whom 11,700 (17%) had AF [20–28]. The authors investigated whether there is an association between AF and dementia in patients with normal cognitive function at baseline who did not suffer an acute stroke. After adjusting for confounders, an independent risk of developing dementia was demonstrated in AF patients with an HR of 1.4 (95% CI 1.2–1.7) [7]. An even more extensive systematic review published by Kalantarian assessed the association of AF with cognitive decline, including prospective and non-prospective studies, mainly using MMSE and DSM-III or IV criteria. What was found was that the risk for cognitive impairment was more than doubled in AF patients with a history of stroke [RR 2.70 (95% CI 1.82–4.00)]. Additionally, there was a markedly increased risk for cognitive decline regardless of stroke history [RR 1.40 (95% CI 1.19–1.64)] [29]. There is limited data available regarding the prevalence of dementia within various treatment option populations for AF. The observed inconsistencies between studies that establish associations and those that don't could be due to the variations in the methodology used across studies, a lack of rigorous adjustment for potential confounder variables, and having highly selected populations. For instance, certain studies have small sample sizes or short follow-up periods. In addition, different approaches and age ranges were used to assess AF and dementia across the studies. In summary, current evidence from systematic reviews and population-based longitudinal studies suggests an independent association of dementia and cognitive decline with AF, especially among the age group of 64–74.

5. Proposed Mechanistic Associations between AF and Cognitive Dysfunction

Cerebral infarction, AF-related cerebral hypoperfusion, microbleeds, inflammation, brain atrophy, atherosclerosis, and others (Figure 1).

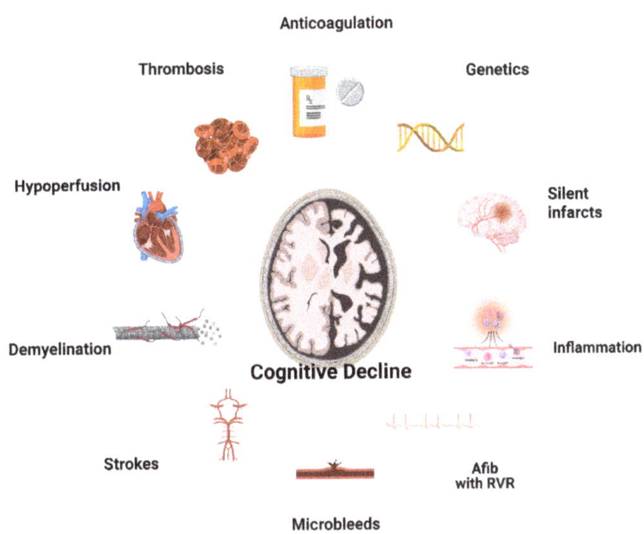

Figure 1. Relationship of Atrial Fibrillation with Dementia.

5.1. Silent Cerebral Infarcts

There have been many studies showing that silent cerebral infarcts are associated with dementia and stroke. In a meta-analysis, AF was associated with a 2.6-fold increased risk of silent cerebral infarcts [30]. Compared with controls in sinus rhythm, patients with AF had much lower cognitive function in an observational study that assessed patients with silent infarcts (validated by MRI) [31]. In studies using systematic brain imaging, 15% to 50% of patients with AF have a brain infarct [32]. What is more remarkable is that silent infarcts at the time of diagnosis of AF are up to five times more common compared to symptomatic infarcts [33]. As highlighted by Healy et al., frequently silent infarcts are miss-characterized as "innocent, underestimating their clinical significance and the risk of developing into fatal strokes [34]. Although this may often be regarded as a benign incidental finding without significant neurologic deficits at the time, it is associated with both concurrent cognitive performance and risk for cognitive decline [35,36]. The prospective SWISS-AF showed that the infarct size is crucial to delayed outcomes. It was described that only the large embolic silent noncortical or cortical infarcts were associated with lower cognitive scores in patients with AF, but not the smaller size infarcts [37].

5.2. Cerebral Microbleeds

Microbleeds in key brain locations (lobar locations) have been reported to be associated with poor cognition. What is remarkable is that even after adjusting for vascular risk factors and imaging markers of CSVD, this relationship persists [38]. On the other hand, in a large cohort of patients with AF, there was no association between cerebral microbleeds and cognitive dysfunction [38]. Whether microbleeds are directly causally related to cognitive impairment or they are bound to our therapeutic management of AF is still a mystery. There is a chance that microbleeds attributable to OAC pose an explanation for cognitive decline in patients with AF. A prospective MRI study found that only warfarin (not direct oral anticoagulants or antiplatelet agents) was associated with developing new microbleeds at one year in patients with AF. Still, we need larger cohorts to exclude that pathophysiologic mechanism from the list [22].

5.3. Cerebral Hypoperfusion

Transient or chronic cerebral hypoperfusion has been implicated in the development of dementia [39]. AF decreases cardiac output due to atrioventricular desynchrony, resulting in less stroke volume and lower blood pressure [7]. Another proposed mechanism is an interbeat volume variation, which may contribute to transient cerebral hypoperfusion [23]. Although cerebral auto-regulation is expected to keep blood flowing to the brain during a wide blood pressure range, several studies have noted decreased cerebral perfusion in patients suffering from AF [40]. In 358 patients with cognitive impairment and an age greater than 65 years, excluding those with a history of transient ischemic attack, stroke, or dementia, AF was an independent predictor for dementia, increasing the risk fourfold after a mean follow-up of 10 years [41]. Relevant to the cerebral hypoperfusion hypothesis, patients with either a low or high heart rate (<50 or >90) on 24-h Holter monitoring had a 7-fold risk for dementia compared to patients with a normal ventricular rate. These results are supported by computational data highlighting that faster rates relate to a progressive decrease in cerebral perfusion and hypotensive events in the cerebral circulation [42]. Relevant to the size of silent infarcts predicting cognitive impairment, the duration of AF was independently associated with worse outcomes. In a cross-sectional study, patients with sustained AF had reduced brain flow compared to those with paroxysmal AF or sinus rhythm [24].

5.4. Inflammation

Inflammation enhances hypercoagulability and potentiates the formation of thrombi, increasing the risk for stroke and malfunction of cerebrovascular regulation, which has been linked to Alzheimer's and vascular dementia [28,43]. AF is a pro-inflammatory condition [26,27]. Validating the above, studies showed that increased markers of inflammation correlate with cognitive impairment in patients with AF. Patients who developed dementia were shown to have impaired hemostatic ability, resulting in microbleeds, as discussed above. Continued microbleeds and inflammation form a vicious cycle that perpetuates, leading to a cascade of unfavorable physiologic changes. This association between hemostasis and vascular dementia is also expressed as an abundance of thrombin generation markers (D-dimer and prothrombin fragment 1 + 2) and endothelial dysfunction (von Willebrand factor and plasminogen activator inhibitor) [44]. In rodent models of Alzheimer's disease, dabigatran (a direct thrombin inhibitor) was shown to improve pathologic changes, reduce inflammation, and slow down cognitive dysfunction even in the absence of AF [45]. Thus, inflammatory markers could correlate with the degree of atherosclerosis associated with cognitive impairment and AF [46].

5.5. Systemic Atherosclerosis/CHADVASC

As discussed above, atherosclerosis is strongly associated with dementia. Preclinical markers of cardiovascular disease have been linked to both an increased risk of AF and cognitive impairment [46–48]. Some are subclinical atherosclerosis, aortic stiffness, and intima-media thickness of the carotid artery. Similarly, there is a direct correlation between higher CHADS2 and CHA2DS2VASc scores and the risk of developing cognitive dysfunction in patients with AF [49]. However, the number of studies on the relationship between those scores and cognitive impairment is limited [16,17]. Chou et al. investigated the correlation between vascular dementia/Alzheimer's disease and CHADS2 score in the Taiwan cohort, [16] concluding that the CHADS2 score is a valuable predictor. In another study using the Taiwan AF cohort of 332.665 patients with AF, Liao et al. showed that CHADS2 and CHA2DS2-VASc were independently associated with developing dementia during a 14-year follow-up. However, the CHA2DS2-VASc score was better at predicting dementia [17], with a score of four or greater having 1.5 times more chances of developing dementia than patients with less than that [11].

5.6. Genetics and Biomarkers

During the last ten years, significant progress has been made toward detecting biomarkers of diagnostic and prognostic value for dementia [50]. Identifying biomarkers predicting cognitive decline in AF patients could inform screening and management strategies. Biomarkers will undoubtedly help refine risk stratification [51], but as with most novel biomarkers (lncRNA BACE1-AS, MALAT1) [52], we must balance predictive ability against practicality [50]. In a Mendelian randomization study using 93 SNPs as the instruments, Pan et al. showed an insignificant association of genetically predicted AF with the risk of Alzheimer's disease. In detail, both a fixed-effect and a random-effect inverse-variance weighted Mendelian randomization (IVW-MR) method showed that genetically predicted AF was not associated with the risk of Alzheimer's disease (OR = 1.002, 95% CI: 0.996–1.009, $p = 0.47$; OR = 1.002, 95% CI: 0.995–1.010, $p = 0.52$). This study concludes that genetically predicted AF had no causal effect on the risk of Alzheimer's disease, with the findings being significant even when a sensitivity analysis was performed [6].

5.7. Amyloid $A\beta_{42}$

Atrial fibrillation compromises cerebral blood flow, which results in the development of senile plaques due to the long-term deposition of $A\beta_{42}$ [53]. What was known was that this type of cerebral amyloid deposition is associated with poor cognitive function and the development of dementia [54,55]. During the last decade, there have been tremendous advances in linking cerebral amyloid disease with low flow states, as in AF. Niwa et al. showed that atrial fibrillation leads to vascular dysfunction which aggravates the deposition of amyloid [56]. Hawkes et al. demonstrated how amyloid $A\beta_2$ hinders the clearance of cerebral amyloid, creating a vicious cycle [57]. Finally, Yao et al. elaborated on the role of hypoperfusion and cerebral amyloid disease, attributing it to activation of the tau phosphorylation enzymes glycogen synthase kinase-3 beta and cyclin-dependent kinase-5, which eventually induce the accumulation of $A\beta_{42}$.

6. Treatment

6.1. Rhythm Control

The potential impact of successful rhythm control on cognitive decline among patients with atrial fibrillation is an area of ongoing debate. Data originating from a recent study hint at the beneficial role of electric cardioversion in cerebral flow assessed by MRI [24]. Improved hemodynamics and increased brain perfusion resulting from rhythm control are hypothesized to delay cognitive decline. On the other hand, most of the existing literature reports that electric cardioversion increases the risk of cerebral microemboli. In a population of AF patients, there was no consistent evidence of an association between PVI (pulmonary vein isolation) and neurocognitive function [58]. In other studies, AF ablation has been associated with declining cognitive function and acute brain lesions [59–61]. However, a direct association between silent cerebral microembolism and cognitive decline is yet to be proven [62]. Findings from the EAST-AFNET 4 trial did not demonstrate a significant impact of early rhythm control on cognitive function [63]. Despite this, ongoing studies, such as the Comparison of Brain Perfusion in Rhythm Control and Rate Control of Persistent Atrial Fibrillation (NCT02633774) and AFCOG (NCT04033510), aim to investigate the role of rhythm control in preventing cognitive decline in AF. The results of these ongoing trials will inform treatment strategies to improve the cognitive outcomes of individuals with AF and potentially impact this patient population's management [1]. Since this is an issue of vital importance, the NOR-FIB2 (Fibrosis, Inflammation, and Brain Health in Atrial Fibrillation: The Norwegian Atrial Fibrillation and Stroke Study; URL: https://www.clinicaltrials.gov; Unique identifier: NCT03816865) is ongoing and ventures to assess whether programmed direct-current cardioversion increases the risk of cognitive dysfunction and the incidence of new-onset silent cerebral infarcts. In summary, the results seem to be either neutral or negative regarding the effect of cardioversion on dementia. Still, more studies are required to investigate whether converting to sinus could reduce the risk of developing dementia in the long run.

6.2. Comparison between Ablation and Oral Antiarrhythmics

Compared to oral antiarrhythmics, ablations reduce paroxysmal AF episodes and prolong the duration of sinus normorhythmia, thereby improving the quality of life [64,65]. Two recent studies demonstrated that, beyond that, catheter ablation improved cognitive function [66,67]. Using the Korean NHIS database, Kim et al. [68] compared the risk of dementia between 9119 patients undergoing ablation and 17,978 patients managed using medical therapy (antiarrhythmic or rate control drugs). During a median follow-up of 52 months, ablated patients had a significantly lower incidence and a reduced risk of dementia overall (8.1 and 5.6 per 1000 person-years, respectively; HR, 0.73; 95% CI, 0.58–0.93). The authors inferred that restoring sinus rhythm, not the ablation procedure, was the critical mechanism in reducing the risk for dementia since the ablation group had longer durations of persistent sinus rhythm when compared to the medical management arm. The ongoing DIAL-F case–control study (Cognitive Impairment in Atrial Fibrillation; Unique identifier: NCT01816308) is comparing the incidence of cognitive impairment (assessed by MoCA) in two groups of patients with AF (Patients undergoing catheter ablation for AF versus those receiving antiarrhythmic drugs) [1].

6.3. Anticoagulation

The pathophysiological mechanisms linking atrial fibrillation with cognitive impairment may be attributed to both microembolization (silent strokes) and macroembolization (overt strokes). Anticoagulation therapy is an effective measure to prevent stroke and systemic emboli in AF patients. However, anticoagulation-associated microbleeds can worsen cognitive function, concluding that intervention RCTs are needed to assess long-term outcomes. There are incoming data that suggest that OACs are lowering the risk of AF-related dementia [69–71]. Kim et al. reported that OACs had a preventive effect on dementia with an HR of 0.61 among patients with incidental AF. Friberg and Rosenqvist [70] found that patients with AF who received OAC treatment had a 29% lower risk of dementia than those with AF who were not prescribed OACs (HR, 0.71; 95% CI, 0.68–0.74) in a large-scale national cohort in Sweden. That study also showed a suggested dose-dependent effect, with patients lowering their risk the longer they are on an OAC. In another study by the same authors, it was recommended that AF patients aged >65 years, irrespective of their stroke risk score, benefit from OAC in regards to lowering the risk for dementia [71]. Several clinical trials, including GIRAF [72] and CAF [73], have investigated the role of warfarin and dabigatran in preventing cognitive decline. Both trials evaluated the participants at baseline and two years after the initiation of the study and revealed no significant differences between the two medications. Despite some studies reporting a benefit, others are suggesting that anticoagulation is associated with higher risks of developing dementia in AF, possibly due to microbleeds. In AF patients receiving warfarin, a lower time-in-therapeutic range has been associated with a higher risk of dementia [21,74]. Due to frequent suboptimal control in patients receiving warfarin, there is a significant concern for microbleeds, which can, in line with this, cause chronic cerebral injuries and inflammation and eventually lead to cognitive decline [75].

6.4. *Type of AC*

Currently, there are no randomized data on the efficacy of different OACs in preventing dementia in AF patients. In patients with an indication for AC, it is not ethically acceptable to randomize patients to OAC versus placebo to assess cognitive outcomes. On the other hand, small observational studies highlight the protective effect of DOACs compared to other pharmacological options when it comes to cognitive outcomes [70,76]. A small meta-analysis of four randomized trials that compared NOACs to warfarin demonstrated that NOACs were associated with a significant risk reduction in terms of overall stroke and systemic embolism [77,78]. Two cohort studies performed in the USA also agreed with the above outcome [76,79]. While the Danish registries showed no difference in dementia between DOACs and warfarin users [80], Kim et al. [81] enrolled 52,888 new OAC users with AF (aged ≥60 years, 31,211 NOAC users, and 21,677 warfarin users) from the Korean NHIS database, suggesting statistically significant differences in outcomes. Relative to propensity-matched warfarin users, NOAC users, regardless of whether it was apixaban, dabigatran, or rivaroxaban, tended to have a lower risk of dementia (HR, 0.78; 95% CI, 0.69–0.90). Nationwide Swedish and Danish cohort studies showed neutral results, as they concluded that the incidences of dementia when comparing warfarin and NOAC were similar [70,80].

6.5. *Anti-Inflammatory Agents*

The relationship between inflammation and cognitive decline in individuals with atrial fibrillation remains an area of active research. While some evidence suggests that inflammation plays a role in the development of cognitive impairment in patients with AF, further investigation is necessary to fully understand this relationship's nature. Specifically, additional research is required to determine the potential effectiveness of anti-inflammatory medications, such as statins, aspirin, or etanercept, in mitigating the cognitive decline associated with inflammation. Statin therapy appears to positively affect cognitive outcomes in patients with AF [70], but this will require more extensive studies to influence our current guidelines.

7. Future Directions and Conclusions

Understanding the relationship between atrial fibrillation and structural brain changes is a crucial area of investigation in cognitive neuroscience. However, the underlying structural changes that may contribute to this decline still need to be fully understood. In this review, we summarized the potential etiological link between AF and dementia, analyzed potential risk stratification or monitoring surrogates, and elaborated on therapeutic interventions that have shown benefit (Figure 2).

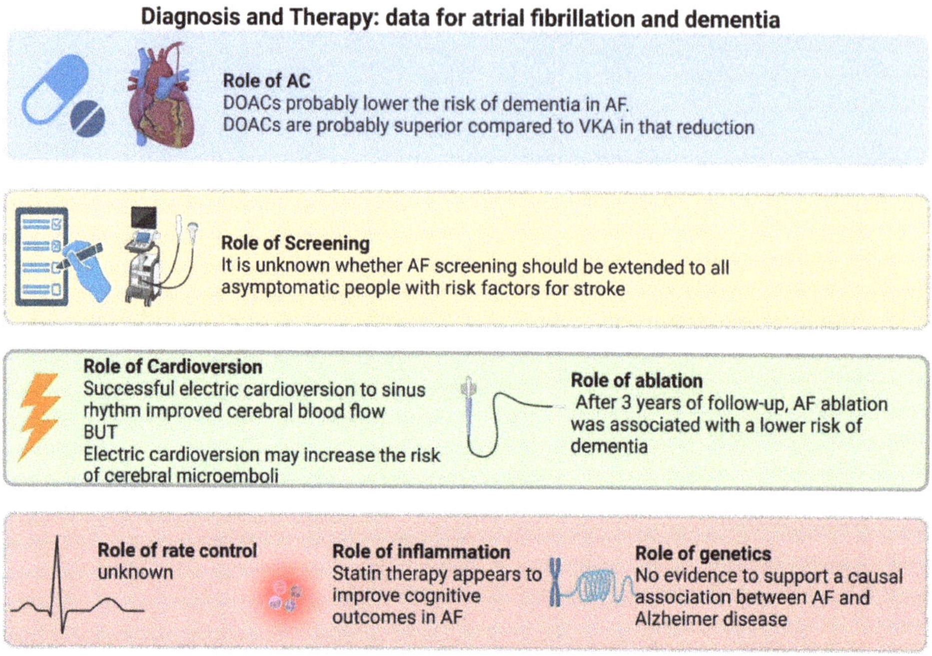

Figure 2. Summary of evidence regarding atrial fibrillation and dementia.

Hopefully, as more light is shed on the subject, more trials will be designed to solidify therapeutic guidelines. Ongoing and recently published studies, such as the SWISS-AF study [82], represent an essential step forward in clarifying the nature of this relationship. Rhythm control using pharmacological cardioversion, electrical cardioversion, or ablation to enhance cerebral perfusion and reduce the incidence of cognitive decline in patients with atrial fibrillation is of significant interest. Additionally, investigations into different anticoagulants and their effect on cognitive impairment, the potential benefits of anti-inflammatory medications such as statins in mitigating the impact of systemic inflammation associated with AF, and the role of genetics in determining susceptibility to cognitive decline are all promising areas for future research. Finally, as the neurodiagnostic tools continue to evolve, we hope to identify more and more links between neurophysiology and low blood flow states, even before atrophy begins to manifest on a macroscopic level. Continued collaboration between basic science and clinical research is critical in identifying novel approaches to preventing and managing cognitive decline in AF patients. The anticipated findings from these studies can impact early intervention strategies and ultimately improve patient outcomes.

Author Contributions: Conceptualization: D.V. and D.K.; Methodology: T.S., S.P., D.V., G.B., and P.B.; Software: D.V. and M.S.; Resources: D.V. and P.B.; Writing—Original Draft Preparation: D.V.; Writing—Review and Editing: D.K., T.S., and S.P.; Visualization: D.V.; Supervision, D.K. All authors have read and agreed to the published version of the manuscript.

Funding: This research received no external funding.

Acknowledgments: Gratitude to the Hellenic Medical Society of NY for its constant support.

Conflicts of Interest: The authors declare no conflicts of interest.

References

1. Rivard, L.; Friberg, L.; Conen, D.; Healey, J.S.; Berge, T.; Boriani, G.; Brandes, A.; Calkins, H.; Camm, A.J.; Chen, L.Y.; et al. Atrial Fibrillation and Dementia: A Report from the AF-SCREEN International Collaboration. *Circulation* **2022**, *145*, 392–409. [CrossRef] [PubMed]
2. Pistoia, F.; Sacco, S.; Tiseo, C.; Degan, D.; Ornello, R.; Carolei, A. The Epidemiology of Atrial Fibrillation and Stroke. *Cardiol. Clin.* **2016**, *34*, 255–268. [CrossRef]
3. Saglietto, A.; Ballatore, A.; Xhakupi, H.; De Ferrari, G.M.; Anselmino, M. Atrial Fibrillation and Dementia: Epidemiological Insights on an Undervalued Association. *Medicina* **2022**, *58*, 361. [CrossRef] [PubMed]
4. Liu, D.S.; Chen, J.; Jian, W.M.; Zhang, G.R.; Liu, Z.R. The association of atrial fibrillation and dementia incidence: A meta-analysis of prospective cohort studies. *J. Geriatr. Cardiol.* **2019**, *16*, 298–306. [PubMed]
5. Bunch, T.J. Atrial Fibrillation and Dementia. *Circulation* **2020**, *142*, 618–620. [CrossRef] [PubMed]
6. Pan, Y.; Wang, Y.; Wang, Y. Investigation of Causal Effect of Atrial Fibrillation on Alzheimer Disease: A Mendelian Randomization Study. *J. Am. Heart Assoc.* **2020**, *9*, e014889. [CrossRef] [PubMed]
7. Dietzel, J.; Haeusler, K.G.; Endres, M. Does atrial fibrillation cause cognitive decline and dementia? *Europace* **2018**, *20*, 408–419. [CrossRef] [PubMed]
8. Lawlor, D.A.; Harbord, R.M.; Sterne, J.A.C.; Timpson, N.; Smith, G.D. Mendelian randomization: Using genes as instruments for making causal inferences in epidemiology. *Stat. Med.* **2008**, *27*, 1133–1163. [CrossRef]
9. Diener, H.C.; Hart, R.G.; Koudstaal, P.J.; Lane, D.A.; Lip, G.Y. Atrial Fibrillation and Cognitive Function: JACC Review Topic of the Week. *J. Am. Coll. Cardiol.* **2019**, *73*, 612–619. [CrossRef] [PubMed]
10. Alonso, A.; Knopman, D.S.; Gottesman, R.F.; Soliman, E.Z.; Shah, A.J.; O'Neal, W.T.; Norby, F.L.; Mosley, T.H.; Chen, L.Y. Correlates of Dementia and Mild Cognitive Impairment in Patients with Atrial Fibrillation: The Atherosclerosis Risk in Communities Neurocognitive Study (ARIC-NCS). *J. Am. Heart Assoc.* **2017**, *6*, e006014. [CrossRef] [PubMed]
11. Liao, J.N.; Chao, T.F.; Liu, C.J.; Wang, K.L.; Chen, S.J.; Tuan, T.C.; Lin, Y.J.; Chang, S.L.; Lo, L.W.; Hu, Y.F.; et al. Risk and prediction of dementia in patients with atrial fibrillation—A nationwide population-based cohort study. *Int. J. Cardiol.* **2015**, *199*, 25–30. [CrossRef]
12. Thacker, E.L.; McKnight, B.; Psaty, B.M.; Longstreth, W.T., Jr.; Sitlani, C.M.; Dublin, S.; Arnold, A.M.; Fitzpatrick, A.L.; Gottesman, R.F.; Heckbert, S.R. Atrial fibrillation and cognitive decline: A longitudinal cohort study. *Neurology* **2013**, *81*, 119–125. [CrossRef] [PubMed]
13. Nishtala, A.; Piers, R.J.; Himali, J.J.; Beiser, A.S.; Davis-Plourde, K.L.; Saczynski, J.S.; McManus, D.D.; Benjamin, E.J.; Au, R. Atrial fibrillation and cognitive decline in the Framingham Heart Study. *Heart Rhythm.* **2018**, *15*, 166–172. [CrossRef] [PubMed]
14. Singh-Manoux, A.; Fayosse, A.; Sabia, S.; Canonico, M.; Bobak, M.; Elbaz, A.; Kivimäki, M.; Dugravot, A. Atrial fibrillation as a risk factor for cognitive decline and dementia. *Eur. Heart J.* **2017**, *38*, 2612–2618. [CrossRef] [PubMed]
15. Greenberg, S.M.; Charidimou, A. Diagnosis of Cerebral Amyloid Angiopathy: Evolution of the Boston Criteria. *Stroke* **2018**, *49*, 491–497. [CrossRef] [PubMed]
16. Whitwell, J.L.; Weigand, S.D.; Shiung, M.M.; Boeve, B.F.; Ferman, T.J.; Smith, G.E.; Knopman, D.S.; Petersen, R.C.; Benarroch, E.E.; Josephs, K.A.; et al. Focal atrophy in dementia with Lewy bodies on MRI: A distinct pattern from Alzheimer's disease. *Brain* **2007**, *130 Pt 3*, 708–719. [CrossRef]
17. Heiss, W.D.; Rosenberg, G.A.; Thiel, A.; Berlot, R.; de Reuck, J. Neuroimaging in vascular cognitive impairment: A state-of-the-art review. *BMC Med.* **2016**, *14*, 174. [CrossRef] [PubMed]
18. Zhang, M.J.; Norby, F.L.; Lutsey, P.L.; Mosley, T.H.; Cogswell, R.J.; Konety, S.H.; Chao, T.; Shah, A.M.; Solomon, S.D.; Alonso, A.; et al. Association of Left Atrial Enlargement and Atrial Fibrillation With Cognitive Function and Decline: The ARIC-NCS. *J. Am. Heart Assoc.* **2019**, *8*, e013197. [CrossRef] [PubMed]
19. Kwok, C.S.; Loke, Y.K.; Hale, R.; Potter, J.F.; Myint, P.K. Atrial fibrillation and incidence of dementia: A systematic review and meta-analysis. *Neurology* **2011**, *76*, 914–922. [CrossRef] [PubMed]
20. Santangeli, P.; Di Biase, L.; Bai, R.; Mohanty, S.; Pump, A.; Brantes, M.C.; Horton, R.; Burkhardt, J.D.; Lakkireddy, D.; Reddy, Y.M.; et al. Atrial fibrillation and the risk of incident dementia: A meta-analysis. *Heart Rhythm.* **2012**, *9*, 1761–1768. [CrossRef] [PubMed]
21. Madhavan, M.; Hu, T.Y.; Gersh, B.J.; Roger, V.L.; Killian, J.; Weston, S.A.; Graff-Radford, J.; Asirvatham, S.J.; Chamberlain, A.M. Efficacy of Warfarin Anticoagulation and Incident Dementia in a Community-Based Cohort of Atrial Fibrillation. *Mayo Clin. Proc.* **2018**, *93*, 145–154. [CrossRef] [PubMed]
22. Saito, T.; Kawamura, Y.; Sato, N.; Kano, K.; Takahashi, K.; Asanome, A.; Sawada, J.; Katayama, T.; Hasebe, N. Non-vitamin k antagonist oral anticoagulants do not increase cerebral microbleeds. *J. Stroke Cerebrovasc. Dis.* **2015**, *24*, 1373–1377. [CrossRef] [PubMed]
23. Kerr, A.J.; Simmonds, M.B.; Stewart, R.A. Influence of heart rate on stroke volume variability in atrial fibrillation in patients with normal and impaired left ventricular function. *Am. J. Cardiol.* **1998**, *82*, 1496–1500. [CrossRef] [PubMed]
24. Gardarsdottir, M.; Sigurdsson, S.; Aspelund, T.; Rokita, H.; Launer, L.J.; Gudnason, V.; Arnar, D.O. Atrial fibrillation is associated with decreased total cerebral blood flow and brain perfusion. *Europace* **2018**, *20*, 1252–1258. [CrossRef] [PubMed]

25. Shi, Y.; Thrippleton, M.J.; Makin, S.D.; Marshall, I.; I Geerlings, M.; de Craen, A.J.; A van Buchem, M.; Wardlaw, J.M. Cerebral blood flow in small vessel disease: A systematic review and meta-analysis. *J. Cereb. Blood Flow Metab.* **2016**, *36*, 1653–1667. [CrossRef] [PubMed]
26. Conway, D.S.; Lip, G.Y. Inflammation, arrhythmia burden and the thrombotic consequences of atrial fibrillation. *Eur. Heart J.* **2004**, *25*, 1761. [CrossRef] [PubMed]
27. Choudhury, A.; Chung, I.; Panja, N.; Patel, J.; Lip, G.Y. Soluble CD40 ligand, platelet surface CD40 ligand, and total platelet CD40 ligand in atrial fibrillation: Relationship to soluble P-selectin, stroke risk factors, and risk factor intervention. *Chest* **2008**, *134*, 574–581. [CrossRef] [PubMed]
28. Khan, A.A.; Lip, G.Y.H. The prothrombotic state in atrial fibrillation: Pathophysiological and management implications. *Cardiovasc. Res.* **2019**, *115*, 31–45. [CrossRef] [PubMed]
29. Kalantarian, S.; Stern, T.A.; Mansour, M.; Ruskin, J.N. Cognitive impairment associated with atrial fibrillation: A meta-analysis. *Ann. Intern. Med.* **2013**, *158 Pt 1*, 338–346. [CrossRef]
30. Kalantarian, S.; Ay, H.; Gollub, R.L.; Lee, H.; Retzepi, K.; Mansour, M.; Ruskin, J.N. Association between atrial fibrillation and silent cerebral infarctions: A systematic review and meta-analysis. *Ann. Intern. Med.* **2014**, *161*, 650–658. [CrossRef] [PubMed]
31. Gaita, F.; Corsinovi, L.; Anselmino, M.; Raimondo, C.; Pianelli, M.; Toso, E.; Bergamasco, L.; Boffano, C.; Valentini, M.C.; Cesarani, F.; et al. Prevalence of silent cerebral ischemia in paroxysmal and persistent atrial fibrillation and correlation with cognitive function. *J. Am. Coll. Cardiol.* **2013**, *62*, 1990–1997. [CrossRef] [PubMed]
32. Hahne, K.; Monnig, G.; Samol, A. Atrial fibrillation and silent stroke: Links, risks, and challenges. *Vasc. Health Risk Manag.* **2016**, *12*, 65–74. [PubMed]
33. Gupta, A.; Giambrone, A.; Gialdini, G.; Finn, C.B.; Delgado, D.; Gutierrez, J.; Wright, C.; Beiser, A.B.; Seshadri, S.; Pandya, A.; et al. Silent Brain Infarction and Risk of Future Stroke: A Systematic Review and Meta-Analysis. *Stroke* **2016**, *47*, 719–725. [CrossRef] [PubMed]
34. Healey, J.S.; Connolly, S.J.; Gold, M.R.; Israel, C.W.; Van Gelder, I.C.; Capucci, A.; Lau, C.; Fain, E.; Yang, S.; Bailleul, C.; et al. Subclinical atrial fibrillation and the risk of stroke. *N. Engl. J. Med.* **2012**, *366*, 120–129. [CrossRef] [PubMed]
35. Sigurdsson, S.; Aspelund, T.; Kjartansson, O.; Gudmundsson, E.F.; Jonsdottir, M.K.; Eiriksdottir, G.; Jonsson, P.V.; van Buchem, M.A.; Gudnason, V.; Launer, L.J. Incidence of Brain Infarcts, Cognitive Change, and Risk of Dementia in the General Population: The AGES-Reykjavik Study (Age Gene/Environment Susceptibility-Reykjavik Study). *Stroke* **2017**, *48*, 2353–2360. [CrossRef] [PubMed]
36. Azeem, F.; Durrani, R.; Zerna, C.; Smith, E.E. Silent brain infarctions and cognition decline: Systematic review and meta-analysis. *J. Neurol.* **2020**, *267*, 502–512. [CrossRef] [PubMed]
37. Blum, S.; Aeschbacher, S.; Coslovsky, M.; Meyre, P.B.; Reddiess, P.; Ammann, P.; Erne, P.; Moschovitis, G.; Di Valentino, M.; Shah, D.; et al. Long-term risk of adverse outcomes according to atrial fibrillation type. *Sci. Rep.* **2022**, *12*, 2208. [CrossRef] [PubMed]
38. Conen, D.; Rodondi, N.; Müller, A.; Beer, J.H.; Ammann, P.; Moschovitis, G.; Auricchio, A.; Hayoz, D.; Kobza, R.; Shah, D.; et al. Relationships of Overt and Silent Brain Lesions with Cognitive Function in Patients with Atrial Fibrillation. *J. Am. Coll. Cardiol.* **2019**, *73*, 989–999. [CrossRef] [PubMed]
39. Duncombe, J.; Kitamura, A.; Hase, Y.; Ihara, M.; Kalaria, R.N.; Horsburgh, K. Chronic cerebral hypoperfusion: A key mechanism leading to vascular cognitive impairment and dementia. Closing the translational gap between rodent models and human vascular cognitive impairment and dementia. *Clin. Sci.* **2017**, *131*, 2451–2468. [CrossRef] [PubMed]
40. Lavy, S.; Stern, S.; Melamed, E.; Cooper, G.; Keren, A.; Levy, P. Effect of chronic atrial fibrillation on regional cerebral blood flow. *Stroke* **1980**, *11*, 35–38. [CrossRef]
41. Cacciatore, F.; Testa, G.; Langellotto, A.; Galizia, G.; Della-Morte, D.; Gargiulo, G.; Bevilacqua, A.; Del Genio, M.T.; Canonico, V.; Rengo, F.; et al. Role of ventricular rate response on dementia in cognitively impaired elderly subjects with atrial fibrillation: A 10-year study. *Dement. Geriatr. Cogn. Disord.* **2012**, *34*, 143–148. [CrossRef] [PubMed]
42. Saglietto, A.; Scarsoglio, S.; Ridolfi, L.; Gaita, F.; Anselmino, M. Higher ventricular rate during atrial fibrillation relates to increased cerebral hypoperfusions and hypertensive events. *Sci. Rep.* **2019**, *9*, 3779. [CrossRef] [PubMed]
43. Wersching, H.; Duning, T.; Lohmann, H.; Mohammadi, S.; Stehling, C.; Fobker, M.; Conty, M.; Minnerup, J.; Ringelstein, E.; Berger, K.; et al. Serum C-reactive protein is linked to cerebral microstructural integrity and cognitive function. *Neurology* **2010**, *74*, 1022–1029. [CrossRef] [PubMed]
44. Quinn, T.J.; Gallacher, J.; Deary, I.J.; Lowe, G.D.O.; Fenton, C.; Stott, D.J. Association between circulating hemostatic measures and dementia or cognitive impairment: Systematic review and meta-analyzes. *J. Thromb. Haemost.* **2011**, *9*, 1475–1482. [CrossRef] [PubMed]
45. Cortes-Canteli, M.; Kruyer, A.; Fernandez-Nueda, I.; Marcos-Diaz, A.; Ceron, C.; Richards, A.T.; Jno-Charles, O.C.; Rodriguez, I.; Callejas, S.; Norris, E.H.; et al. Long-Term Dabigatran Treatment Delays Alzheimer's Disease Pathogenesis in the TgCRND8 Mouse Model. *J. Am. Coll. Cardiol.* **2019**, *74*, 1910–1923. [CrossRef] [PubMed]
46. Xia, C.; Vonder, M.; Sidorenkov, G.; Oudkerk, M.; de Groot, J.C.; van der Harst, P.; de Bock, G.H.; De Deyn, P.P.; Vliegenthart, R. The Relationship of Coronary Artery Calcium and Clinical Coronary Artery Disease with Cognitive Function: A Systematic Review and Meta-Analysis. *J. Atheroscler. Thromb.* **2020**, *27*, 934–958. [CrossRef] [PubMed]

47. Cortes-Canteli, M.; Gispert, J.D.; Salvadó, G.; Toribio-Fernandez, R.; Tristão Pereira, C.; Falcon, C.; Oliva, B.; Mendiguren, J.; Fernandez-Friera, L.; Sanz, J.; et al. Subclinical Atherosclerosis and Brain Metabolism in Middle-Aged Individuals: The PESA Study. *J. Am. Coll. Cardiol.* **2021**, *77*, 888–898. [CrossRef] [PubMed]
48. O'Neal, W.T.; Efird, J.T.; Dawood, F.Z.; Yeboah, J.; Alonso, A.; Heckbert, S.R.; Soliman, E.Z. Coronary artery calcium and risk of atrial fibrillation (from the multi-ethnic study of atherosclerosis). *Am. J. Cardiol.* **2014**, *114*, 1707–1712. [CrossRef] [PubMed]
49. Graves, K.G.; May, H.T.; Knowlton, K.U.; Muhlestein, J.B.; Jacobs, V.; Lappé, D.L.; Anderson, J.L.; Horne, B.D.; Bunch, T.J. Improving CHA_2DS_2-VASc stratification of non-fatal stroke and mortality risk using the Intermountain Mortality Risk Score among patients with atrial fibrillation. *Open Heart* **2018**, *5*, e000907. [CrossRef] [PubMed]
50. Esteve-Pastor, M.A.; Roldán, V.; Rivera-Caravaca, J.M.; Ramírez-Macías, I.; Lip, G.Y.H.; Marín, F. The Use of Biomarkers in Clinical Management Guidelines: A Critical Appraisal. *Thromb. Haemost.* **2019**, *119*, 1901–1919. [CrossRef] [PubMed]
51. Shin, S.Y.; Han, S.J.; Kim, J.S.; Im, S.I.; Shim, J.; Ahn, J.; Lee, E.M.; Park, Y.M.; Kim, J.H.; Lip, G.Y.; et al. Identification of Markers Associated With Development of Stroke in "Clinically Low-Risk" Atrial Fibrillation Patients. *J. Am. Heart Assoc.* **2019**, *8*, e012697. [CrossRef] [PubMed]
52. Spanos, M.; Gokulnath, P.; Chatterjee, E.; Li, G.; Varrias, D.; Das, S. Expanding the horizon of EV-RNAs: LncRNAs in EVs as biomarkers for disease pathways. *Extracell. Vesicle* **2023**, *2*, 100025. [CrossRef] [PubMed]
53. Islam, M.; Poly, T.N.; Walther, B.A.; Yang, H.-C.; Wu, C.C.; Lin, M.-C.; Chien, S.-C.; Li, Y.-C. Association between Atrial Fibrillation and Dementia: A Meta-Analysis. *Front. Aging Neurosci.* **2019**, *11*, 305. [CrossRef] [PubMed]
54. Pluta, R.; Furmaga-Jabłońska, W.; Maciejewski, R.; Ułamek-Kozioł, M.; Jabłoński, M. Brain ischemia activates beta- and gamma-secretase cleavage of amyloid precursor protein: Significance in sporadic Alzheimer's disease. *Mol. Neurobiol.* **2013**, *47*, 425–434. [CrossRef] [PubMed]
55. Hardy, J.; Allsop, D. Amyloid deposition as the central event in the aetiology of Alzheimer's disease. *Trends Pharmacol. Sci.* **1991**, *12*, 383–388. [CrossRef] [PubMed]
56. Niwa, K.; Kazama, K.; Younkin, L.; Younkin, S.G.; Carlson, G.A.; Iadecola, C. Cerebrovascular autoregulation is profoundly impaired in mice overexpressing amyloid precursor protein. *Am. J. Physiol. Heart Circ. Physiol.* **2002**, *283*, H315–H323. [CrossRef]
57. Hawkes, C.A.; Härtig, W.; Kacza, J.; Schliebs, R.; Weller, R.O.; Nicoll, J.A.; Carare, R.O. Perivascular drainage of solutes is impaired in the ageing mouse brain and in the presence of cerebral amyloid angiopathy. *Acta Neuropathol.* **2011**, *121*, 431–443. [CrossRef] [PubMed]
58. Zwimpfer, L.; Aeschbacher, S.; Krisai, P.; Coslovsky, M.; Springer, A.; Paladini, R.E.; Girod, M.; Hufschmid, J.; Knecht, S.; Badertscher, P.; et al. Neurocognitive function in patients with atrial fibrillation undergoing pulmonary vein isolation. *Front. Cardiovasc. Med.* **2022**, *9*, 1000799. [CrossRef] [PubMed]
59. Haeusler, K.G.; Kirchhof, P.; Endres, M. Left atrial catheter ablation and ischemic stroke. *Stroke* **2012**, *43*, 265–270. [CrossRef] [PubMed]
60. Nakamura, T.; Okishige, K.; Kanazawa, T.; Yamashita, M.; Kawaguchi, N.; Kato, N.; Aoyagi, H.; Yamauchi, Y.; Sasano, T.; Hirao, K. Incidence of silent cerebral infarctions after catheter ablation of atrial fibrillation utilizing the second-generation cryoballoon. *Europace* **2017**, *19*, 1681–1688. [CrossRef] [PubMed]
61. Schwarz, N.; Kuniss, M.; Nedelmann, M.; Kaps, M.; Bachmann, G.; Neumann, T.; Pitschner, H.-F.; Gerriets, T.; Lee, J.Z.; Agasthi, P.; et al. Neuropsychological decline after catheter ablation of atrial fibrillation. *Heart Rhythm.* **2010**, *7*, 1761–1767. [CrossRef]
62. Deneke, T.; Jais, P.; Scaglione, M.; Schmitt, R.; DI Biase, L.; Christopoulos, G.; Schade, A.; Mügge, A.; Bansmann, M.; Nentwich, K.; et al. Silent cerebral events/lesions related to atrial fibrillation ablation: A clinical review. *J. Cardiovasc. Electrophysiol.* **2015**, *26*, 455–463. [CrossRef] [PubMed]
63. Kirchhof, P.; Camm, A.J.; Goette, A.; Brandes, A.; Eckardt, L.; Elvan, A.; Fetsch, T.; van Gelder, I.C.; Haase, D.; Haegeli, L.M.; et al. Early Rhythm-Control Therapy in Patients with Atrial Fibrillation. *N. Engl. J. Med.* **2020**, *383*, 1305–1316. [CrossRef] [PubMed]
64. Mark, D.B.; Anstrom, K.J.; Sheng, S.; Piccini, J.P.; Baloch, K.N.; Monahan, K.H.; Daniels, M.R.; Bahnson, T.D.; Poole, J.E.; Rosenberg, Y.; et al. Effect of Catheter Ablation vs Medical Therapy on Quality of Life among Patients with Atrial Fibrillation: The CABANA Randomized Clinical Trial. *JAMA* **2019**, *321*, 1275–1285. [CrossRef] [PubMed]
65. Hindricks, G.; Potpara, T.; Dagres, N.; Arbelo, E.; Bax, J.J.; Blomström-Lundqvist, C.; Boriani, G.; Castella, M.; Dan, G.-A.; Dilaveris, P.E.; et al. 2020 ESC Guidelines for the diagnosis and management of atrial fibrillation developed in collaboration with the European Association for Cardio-Thoracic Surgery (EACTS): The Task Force for the diagnosis and management of atrial fibrillation of the European Society of Cardiology (ESC) Developed with the special contribution of the European Heart Rhythm Association (EHRA) of the ESC. *Eur. Heart J.* **2021**, *42*, 373–498. [PubMed]
66. Kirchhof, P.; Haeusler, K.G.; Blank, B.; De Bono, J.; Callans, D.; Elvan, A.; Fetsch, T.; Van Gelder, I.C.; Gentlesk, P.; Grimaldi, M.; et al. Apixaban in patients at risk of stroke undergoing atrial fibrillation ablation. *Eur. Heart J.* **2018**, *39*, 2942–2955. [CrossRef] [PubMed]
67. Jin, M.N.; Kim, T.H.; Kang, K.W.; Yu, H.T.; Uhm, J.S.; Joung, B.; Lee, M.H.; Kim, E.; Pak, H.N. Atrial Fibrillation Catheter Ablation Improves 1-Year Follow-Up Cognitive Function, Especially in Patients with Impaired Cognitive Function. *Circ. Arrhythm. Electrophysiol.* **2019**, *12*, e007197. [CrossRef] [PubMed]
68. Kim, D.; Yang, P.-S.; Sung, J.-H.; Jang, E.; Yu, H.T.; Kim, T.-H.; Uhm, J.-S.; Kim, J.-Y.; Pak, H.-N.; Lee, M.-H.; et al. Less dementia after catheter ablation for atrial fibrillation: A nationwide cohort study. *Eur. Heart J.* **2020**, *41*, 4483–4493. [CrossRef]

69. Kim, D.; Yang, P.-S.; Yu, H.T.; Kim, T.-H.; Jang, E.; Sung, J.-H.; Pak, H.-N.; Lee, M.-Y.; Lee, M.-H.; Lip, G.Y.H.; et al. Risk of dementia in stroke-free patients diagnosed with atrial fibrillation: Data from a population-based cohort. *Eur. Heart J.* **2019**, *40*, 2313–2323. [CrossRef] [PubMed]
70. Friberg, L.; Rosenqvist, M. Less dementia with oral anticoagulation in atrial fibrillation. *Eur. Heart J.* **2018**, *39*, 453–460. [CrossRef]
71. Friberg, L.; Andersson, T.; Rosenqvist, M. Less dementia and stroke in low-risk patients with atrial fibrillation taking oral anticoagulation. *Eur. Heart J.* **2019**, *40*, 2327–2335. [CrossRef] [PubMed]
72. Caramelli, B.; Yu, P.C.; Cardozo, F.A.; Magalhães, I.R.; Spera, R.R.; Amado, D.K. Effects of dabigatran versus warfarin on 2-year cognitive outcomes in old patients with atrial fibrillation: Results from the GIRAF randomized clinical trial. *BMC Med.* **2022**, *20*, 374. [CrossRef] [PubMed]
73. Bunch, T.J.; May, H.; Cutler, M.; Woller, S.C.; Jacobs, V.; Stevens, S.M. Impact of anticoagulation therapy on the cognitive decline and dementia in patients with non-valvular atrial fibrillation (cognitive decline and dementia in patients with non-valvular atrial fibrillation [CAF] trial). *J. Arrhythm.* **2022**, *38*, 997–1008. [CrossRef] [PubMed]
74. Jacobs, V.; Woller, S.C.; Stevens, S.; May, H.T.; Bair, T.L.; Anderson, J.L.; Crandall, B.G.; Day, J.D.; Johanning, K.; Long, Y.; et al. Time outside of therapeutic range in atrial fibrillation patients is associated with long-term risk of dementia. *Heart Rhythm.* **2014**, *11*, 2206–2213. [CrossRef] [PubMed]
75. van Norden, A.G.; van den Berg, H.A.; de Laat, K.F.; Gons, R.A.; van Dijk, E.J.; de Leeuw, F.E. Frontal and temporal microbleeds are related to cognitive function: The Radboud University Nijmegen Diffusion Tensor and Magnetic Resonance Cohort (RUN DMC) Study. *Stroke* **2011**, *42*, 3382–3386. [CrossRef] [PubMed]
76. Jacobs, V.; May, H.T.; Bair, T.L.; Crandall, B.G.; Cutler, M.J.; Day, J.D.; Mallender, C.; Osborn, J.S.; Stevens, S.M.; Weiss, J.P.; et al. Long-Term Population-Based Cerebral Ischemic Event and Cognitive Outcomes of Direct Oral Anticoagulants Compared with Warfarin among Long-term Anticoagulated Patients for Atrial Fibrillation. *Am. J. Cardiol.* **2016**, *118*, 210–214. [CrossRef] [PubMed]
77. Ruff, C.T.; Giugliano, R.P.; Braunwald, E.; Hoffman, E.B.; Deenadayalu, N.; Ezekowitz, M.D.; Camm, A.J.; Weitz, J.I.; Lewis, B.S.; Parkhomenko, A.; et al. Comparison of the efficacy and safety of new oral anticoagulants with warfarin in patients with atrial fibrillation: A meta-analysis of randomised trials. *Lancet* **2014**, *383*, 955–962. [CrossRef] [PubMed]
78. Wang, K.L.; Lip, G.Y.; Lin, S.J.; Chiang, C.E. Non-Vitamin K Antagonist Oral Anticoagulants for Stroke Prevention in Asian Patients with Nonvalvular Atrial Fibrillation: Meta-Analysis. *Stroke* **2015**, *46*, 2555–2561. [CrossRef] [PubMed]
79. Lutsey, P.L.; Norby, F.L.; Ensrud, K.E.; MacLehose, R.F.; Diem, S.J.; Chen, L.Y.; Alonso, A. Association of Anticoagulant Therapy with Risk of Fracture among Patients with Atrial Fibrillation. *JAMA Intern. Med.* **2020**, *180*, 245–253. [CrossRef] [PubMed]
80. Søgaard, M.; Skjøth, F.; Jensen, M.; Kjældgaard, J.N.; Lip, G.Y.H.; Larsen, T.B.; Nielsen, P.B. Nonvitamin K Antagonist Oral Anticoagulants versus Warfarin in Atrial Fibrillation Patients and Risk of Dementia: A Nationwide Propensity-Weighted Cohort Study. *J. Am. Heart Assoc.* **2019**, *8*, e011358. [CrossRef] [PubMed]
81. Kim, D.; Yang, P.-S.; Jang, E.; Yu, H.T.; Kim, T.-H.; Uhm, J.-S.; Kim, J.-Y.; Sung, J.-H.; Pak, H.-N.; Lee, M.-H.; et al. Association of anticoagulant therapy with risk of dementia among patients with atrial fibrillation. *Europace* **2021**, *23*, 184–195. [CrossRef] [PubMed]
82. Conen, D.; Rodondi, N.; Mueller, A.; Beer, J.; Auricchio, A.; Ammann, P.; Hayoz, D.; Kobza, R.; Moschovitis, G.; Shah, D.; et al. Design of the Swiss Atrial Fibrillation Cohort Study (Swiss-AF): Structural brain damage and cognitive decline among patients with atrial fibrillation. *Swiss Med. Wkly.* **2017**, *147*, w14467. [PubMed]

biomolecules

MDPI

Review

A Brief Overview of Neutrophils in Neurological Diseases

Supriya Chakraborty [1,*], Zeynab Tabrizi [1], Nairuti Nikhil Bhatt [1], Sofia Andrea Franciosa [1] and Oliver Bracko [1,2,*]

1 Department of Biology, University of Miami, Coral Gables, FL 33146, USA
2 Department of Neurology, University of Miami-Miller School of Medicine, Miami, FL 33136, USA
* Correspondence: sxc2438@miami.edu (S.C.); oliver.bracko@miami.edu (O.B.)

Abstract: Neutrophils are the most abundant leukocyte in circulation and are the first line of defense after an infection or injury. Neutrophils have a broad spectrum of functions, including phagocytosis of microorganisms, the release of pro-inflammatory cytokines and chemokines, oxidative burst, and the formation of neutrophil extracellular traps. Traditionally, neutrophils were thought to be most important for acute inflammatory responses, with a short half-life and a more static response to infections and injury. However, this view has changed in recent years showing neutrophil heterogeneity and dynamics, indicating a much more regulated and flexible response. Here we will discuss the role of neutrophils in aging and neurological disorders; specifically, we focus on recent data indicating the impact of neutrophils in chronic inflammatory processes and their contribution to neurological diseases. Lastly, we aim to conclude that reactive neutrophils directly contribute to increased vascular inflammation and age-related diseases.

Keywords: neutrophils; NETosis; vascular inflammation; neutrophil heterogeneity

1. Introduction

Neutrophils are human blood's most abundant white blood cells making up 60–70% of total leukocytes [1]. Neutrophils originate in the bone marrow from hematopoietic stem cells, and after going through successive differentiation processes, they circulate in the blood for several hours to days [2] and transmigrate out of the blood vessels into the tissue upon encountering pro-inflammatory signals [3]. Ongoing inflammation leads to up-regulation of selectins such as P-selectin on endothelial cells, which interacts with the sialo-mucin ligand P-selectin-glycoprotein–ligand-1 (PSGL-1), constitutively expressed on the surface of neutrophils. This interaction slows down the speed of free-flowing neutrophils in the blood vessel by mediating neutrophil rolling (stop and go movement) along the endothelium [4] (Figure 1a). Following the step of neutrophil rolling, a high-affinity interaction of CD18 integrins CD11a/CD18 (LFA-1) and CD11b/CD18 (Mac-1) expressed on neutrophils with inter-cellular-adhesion-molecule-1 or -2 (ICAM-1/ICAM-2) expressed on endothelium mediates the firm arrest (complete stop) of neutrophils on the vascular endothelium [5]. Neutrophils then crawl and cross the border known as the blood–brain barrier (BBB) that lies between endothelial cells and the tissue site through a process known as diapedesis. Inflammatory signals, such as complement protein C5a, Interleukin-8, and Leukotriene B4, are among the most potent chemo-attractants that guides the neutrophil within the tissue [6,7] (Figure 1a).

Neutrophils eliminate pathogens in one of four ways. First, they can phagocytose pathogens that come into cells as phagosomes, which are fused with the lysosome and degraded through various enzymes, such as acid hydrolases [8] (Figure 1b). Second, neutrophils can release granules containing cytotoxic enzymes that destroy pathogens [9] (Figure 1b). Third, they can eject their DNA chromatin with cytotoxic granules to trap the pathogen in a process known as NETosis [10] (Figure 1b). Fourth, neutrophils also undergo oxidative bursts [11]. After the elimination of pathogens, most neutrophils undergo apoptosis or other methods of cell death, and cell debris is phagocytosed by macrophages [12]. The remaining neutrophils reverse migrate into the blood and go to the bone marrow, spleen, and liver for degradation [3]. Despite playing vital roles in the body's defense, neutrophils are responsible for initiating or progressing various diseases such as psoriasis, rheumatoid arthritis, systemic lupus erythematosus, coronary heart disease, and asthma [13]. Moreover, the involvement of neutrophils in low-level chronic inflammatory diseases such as Alzheimer's is being investigated [14,15].

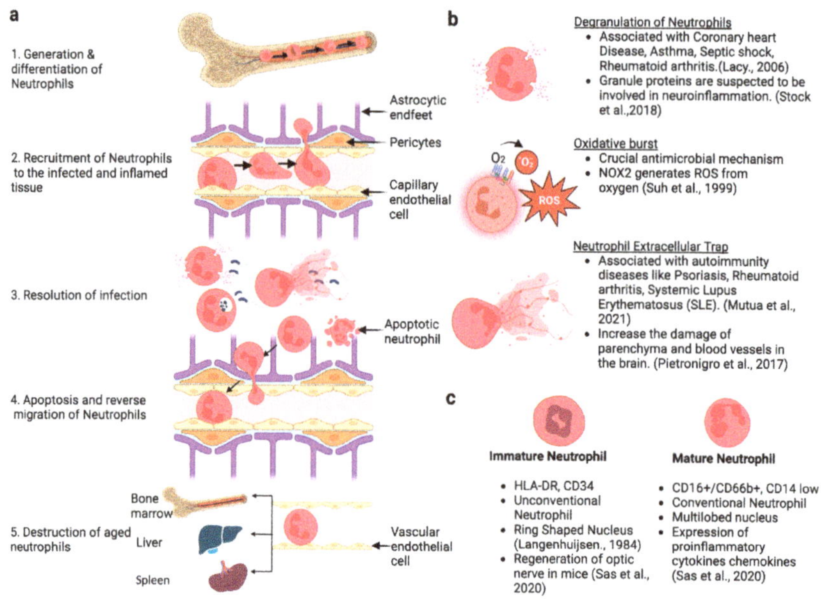

Figure 1. A brief overview of the neutrophil life cycle. (**a**) Generation and maturation of the neutrophil in the bone marrow and their entry into the blood after attaining maturity, functional responses upon encountering pathogens, and apoptosis or clearance after pathogen clearance; (**b**) mechanism of effector actions of neutrophil; (**c**) distinctions between immature and mature neutrophils based on cellular morphology and molecular markers [13,16–23].

With the implementation of single-cell sequencing, sophisticated programs, and software to analyze the data [24], the utilization of bone marrow chimera mice [25] is challenging the idea of a single homogeneous 'type' of neutrophils. Instead, it is becoming increasingly clear that the circulating neutrophil population is heterogeneous [24,26] (Figure 1c). Therefore, this review will discuss the overall development and differentiation of neutrophils, their heterogeneity, and their involvement in neurological diseases.

2. Generation, Development, and Life Cycle of Neutrophils

Like all leukocytes, neutrophils begin their life cycle in the bone marrow from hematopoietic stem cells, which differentiate into multipotent precursor (MPP) cells. MPPs then divide and generate lymphoid and myeloid progenitor (LMP) cells. LMPs undergo subsequent differentiation to give rise to granulocyte–macrophage precursors (GMP), which divide to generate neutrophil precursors [27,28]. Neutrophil precursors differentiate into myeloblast, promyelocyte, myelocyte, metamyelocyte, and band cell stages, which ultimately give rise to segmented neutrophils [16,17].

Myeloblasts are considered the earliest neutrophil progenitors, with a large nucleus and cytoplasm devoid of granules. Promyelocytes are larger than myeloblasts, the shape of their nucleus is rounded, and it contains azurophilic granules with myeloperoxidase as the primary compound [17,28]. In the myelocyte stage, the rounded nucleus of the promyelocyte begins to indent, and the cell size also decreases. At this stage of neutrophil development, secondary granules start forming, made up of compounds such as lactoferrin, lysozyme, and collagenase [3]. In the metamyelocyte stage, the size of the cell decreases further, and the nucleus becomes more densely packed with chromatin. It becomes kidney-shaped, and both azurophilic and secondary granules are observed. In the band cell stage, the cell size decreases even further, and the relative shape of the nucleus becomes more elongated and lobular [17].

Behind the differentiation of neutrophils from the hematopoietic stem cell, several molecules and factors have a crucial role. Granulocyte-colony stimulating factor is one of them. It has vital roles in the neutrophil life cycle, such as controlling the differentiation of neutrophils from the hematopoietic stem cells, the production and survival of neutrophils [29], and their release from the bone marrow into the blood [30,31]. Apart from this, chemokine receptors CXCR2 and CXCR4 control the decision of retention or release of the neutrophils from mouse bone marrow; CXCR4 (which is the receptor for the chemokine CXCL12) is mainly associated with the retention of neutrophils in the marrow and CXCR2 (which is the receptor of the CXCL1 and CXCL2) is related to the event of a neutrophil release from the bone marrow [32]. In humans, IL8 (CXCL8) is the major ligand for CXCR1 and CXCR2. On top of these, several other transcription factors, such as the CCAAT enhancer binding protein α gene [33] and PU.1 [34], are essential in the later stages of neutrophil differentiation.

3. Common Techniques and Methodologies Used to Study Neutrophil Biology

Proteomics: There are three main types of proteomics: expression proteomics, structural proteomics, and functional proteomics [35,36]. Expression proteomics quantitatively identifies and characterizes proteins in various sample groups (Control, treatment). Proteins are separated using two-dimensional gel electrophoresis, and then the separated proteins are identified using mass spectrometry (MALDI TOF MS, LC-ESI MS/MS) [37]. On the other hand, the primary objective of structural proteomics is to elucidate the structure and composition of proteins and study the protein–protein interactions [38]. In this proteomics, target proteins are first immunoprecipitated, then separated by gel electrophoresis. Finally, they are identified using mass spectrometry [37,39]. In contrast, functional proteomics is about studying the function of proteins [40], whereas mass spectrometry is widely used with other techniques [37].

Single-cell sequencing: This is a widely used state-of-the-art technique for studying neutrophils. The main benefit of this technique is that it can identify the expression of various genes and transcripts at a single-cell resolution [41]. For example, using single-cell RNA sequencing, it has been shown that the circulating population of neutrophils is heterogenous, contrary to the popular belief of one homogenous population of neutrophils [42]. On top of that, the data obtained from the single-cell sequencing of neutrophil are processed through sophisticated algorithms such as diffusion mapping, which lead to further advancement of understanding about the development of neutrophils [43].

Single-cell Metabolomics: Studying immune cells such as neutrophils with a short life span is challenging but essential. One of the ways to understand cellular metabolism is to study the whole metabolome, which is the collection of all metabolites at a particular time point [44]. The composition of metabolites is measured at the single-cell level to understand cellular metabolism processes in individual cells and populations. It has been recently used to discover the role of glutathione depletion in the spontaneous apoptosis of neutrophils [45].

Multiphoton Imaging: Another way to obtain real-time information about highly dynamic immune cells such as neutrophils is to use two-photon and three-photon imaging, which, unlike confocal microscopy, uses two photons for better penetration into the sample and less photodamage [46,47]. Using this technique, researchers have shown the extravasation of neutrophils. For example, micro emboli formation in sickle-cell disease and NET formation occurs in the lungs upon receiving inflammatory cues [48–52]. Similarly, another group of researchers utilized multiphoton imaging to study neutrophil-extracellular-trap formation in vivo [53].

4. Contribution of Neutrophils in Neurological Diseases

4.1. Ischemic Stroke

Ischemic stroke is defined by the occlusion of blood vessels in the brain due to thrombus formation. Within minutes, neutrophils in mouse blood can be detected at the ischemic site, guided by inflammatory cytokines such as Il-1, Il-6, CXCL1, and TNF and chemokines such as CCL-2, CCL-3, and CCL-5 [54,55]. In addition, damage-associated molecular patterns such as high mobility group box protein, heat shock protein, and DNA released by the death of the cells help in their recruitment and activation in mice and patients (Figure 2) [55,56]. Neutrophils then adhere to the endothelium and transmigrate inside the tissue space. The pro-inflammatory cytokines, proteases, and reactive oxygen species secreted by neutrophils exacerbate cerebral inflammation, cause blood–brain barrier breakdown, and recruit neutrophil reinforcement to the thrombus site [55]. Neutrophils can also exaggerate the pro-thrombogenic pathways by direct interaction with platelets, proteolytic cleavage of clotting factors, and releasing a neutrophil extracellular trap that binds and entraps platelets [57,58].

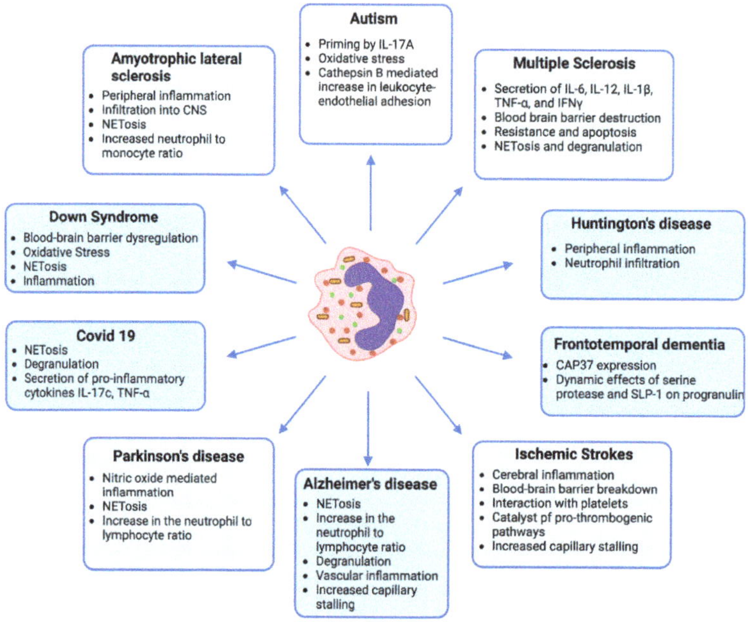

Figure 2. An overview of neutrophils' functions in various neurological diseases.

Further research has shown that the neutrophil population has functional heterogeneity. The pro-inflammatory 'N1' neutrophils have an anti-inflammatory 'N2' counterpart. The N2 neutrophils secrete TGFβ and IL-10 and have a neuroprotective and immune-suppressive function [59], but further research is needed to establish whether N1 and N2 neutrophils are ontogenically different. Recent studies in mice have shown that reducing the adhesion and neutrophil number using an antibody against the neutrophil protein Ly6G increased reperfusion by reducing the number of capillary obstructions in the capillaries [60,61]. Overall, the ischemic stroke outcome was less severe when neutrophils were depleted.

4.2. Parkinson's Disease

Parkinson's disease (PD) is a progressive neurodegenerative disorder that causes dopaminergic neuron loss in the brain, primarily in the substantia nigra, which has essential roles in controlling movement. Besides the accumulation of alpha-synuclein and loss of dopaminergic neurons in the substation nigra, PD encompasses both central and peripheral inflammation. Many studies uncover the involvement of immune cells, such as neutrophils, in PD [62]. In mice, one of the effector mechanisms of neutrophils is the use of large concentrations of nitric oxide, which is generated by nitric oxide synthase [63]. Patients with PD have elevated expression of neuronal nitric oxide synthase in their circulating neutrophils, indicating a role in increased oxidative stress in PD patients [64]. Nitric oxide reacts with superoxide anion and can give rise to peroxynitrite, which is shown to trigger the release of the neutrophil extracellular trap ex vivo, exacerbating inflammation [65]. Since neutrophils are one of the most abundant white blood cells in the blood, they are often utilized to assess peripheral inflammation. For example, leucine-rich repeat kinase two has essential roles in Parkinson's disease. Its kinase activity is measured by the phosphorylation status of Rab10, which can be extracted from peripheral blood neutrophils [66]. The neutrophil-to-lymphocyte ratio is a way to estimate inflammation. Although its role in Parkinson's is still under investigation, some studies show that the patients generally have a higher neutrophil-to-lymphocyte ratio than the healthy controls [62,67].

4.3. COVID-19

Severe acute respiratory syndrome coronavirus-2 causes COVID-19 disease, primarily damaging the lungs, and causes mild to moderate respiratory difficulty in many patients [68]. Inflammation, which occurs during the acute phase of infection, recruits neutrophils to the lungs, which then undergo degranulation with the secretion of NETs and pro-inflammatory cytokines. This results from animal models in an increase in inflammation and associated tissue damage [69]. Almost half the people with COVID-19 are now developing long-term symptoms [70]. Neutrophils, among the other leukocytes, play a vital role in the post-COVID-19 changes in the lungs and other organs, including the brain. Compared to control patients, people with post-COVID-19 interstitial lung changes have shown increased levels of pro-inflammatory cytokines such as TNF and IL-17C in the plasma and reported upregulation of proteins known for neutrophil chemotaxis such as CCL20 and CCL25 (Figure 2) [71]. This group also showed more circulating neutrophils (indicating peripheral inflammation), higher concentration of myeloperoxidase (peroxidase expressed predominantly by neutrophils), and increased NETosis compared to controls [71]. The upregulation of NETs during long COVID-19 increases the expression of fibrogenic mediators, which can lead to thrombosis in the lung, kidney, or other organs [71,72]. It is proposed that neutrophils specifically harm the microvasculature by causing endothelial cell swelling and small clots [73,74]. Patients with long COVID-19 also reported varying degrees of cognitive impairment, commonly called 'brain fog.' It is reported that SARS-CoV-2 can enter the central nervous system via dysregulation of the BBB [75,76]. Once inside the brain, it can constrict the capillaries and reduce the blood flow via binding ACE2 on the pericytes [77]. SARS-CoV-2 also increases the viscosity of the blood, which causes the reduced blood flow in the cerebral capillaries [78]. As seen in other dementia or stroke models, this reduction in the blood flow might be one of the mechanisms behind cognitive impairment in the long COVID-19.

4.4. Huntington's Disease

Huntington's disease (HD) is a rare neurodegenerative disease that causes neurodegeneration due to a mutation in the Huntington gene [79]. For a long time, the role of the immune system in HD has remained underappreciated, but the recent focus on this area has started to uncover some fascinating insights. For example, compared to the healthy controls, the plasma of HD patients has been shown to contain increased IL-6, MMP-9, VEGF, TGF-β1, and decreased IL-18 levels [80]. Interestingly, Il-6 deficiency in a mouse model has been shown to exacerbate the HD symptoms and thus may point towards the importance of optimal IL-6 in the body [81]. In HD cases, it is thought that there is only a marginal influx of peripheral immune cells, including neutrophils [79]. In another study, however, the effect on neutrophils went in the other direction. Using BACHD mice, a well-characterized mouse model of HD, the authors observed that the BACHD mice had 60% fewer infiltrating neutrophils in the peritoneum than wild-type mice [81]. More data need to be reported on the effect of neutrophils on mouse models and patients with HD.

4.5. *Amyotrophic Lateral Sclerosis*

Amyotrophic lateral sclerosis (ALS) is a neurodegenerative disease caused by neurogenic amyotrophy and degeneration of upper and lower motor neurons [82]. Dysregulated inflammatory pathways are prevalent in patients with amyotrophic lateral sclerosis, characterized by increased levels of pro-inflammatory cytokines and subsequent immune cell infiltration into the CNS, such as neutrophils [83]. Relatively high levels of inflammatory cytokines, such as IFN-β, TNF-α, IL-6, and IL-8, were identified in the plasma of ALS participants. These pro-inflammatory cytokines have an imperative role in ALS mouse models in neutrophil activity and trafficking [83–85]. High levels of IL-6 and sIL-6R have been demonstrated in several chronic inflammatory and autoimmune diseases [86,87]. A study has shown the effect of systemic IL-6-mediated inflammation on endothelial cell (EC) death damaging the CNS barrier in ALS individuals (Figure 2) [88]. In patients, IL-6R can be released by proteolytic shedding from neutrophils or by secretion from monocytes of an alternatively spliced messenger RNA (mRNA) species as a soluble form (sIL-6R). Increased shedding of sIL-6R promotes IL-6/sIL-6R complex formation in the blood [89,90]. This complex activates the trans-signaling pathway upon binding to glycoprotein 130 on the target cell membrane, activating JAK and other signal transduction molecules, including MAPK, ERK, P13K, and STAT. This results in a pro-inflammatory response in the endothelial cells and a de novo synthesis of monocyte-attracting chemokines and vascular cell adhesion molecules, leading to the extravasation of inflammatory cells into the CNS [88]. It is believed that such a swamp of cells will cause EC degeneration, the principal cause of damage to the brain–CNS barrier in ALS.

Apart from the role of inflammatory cytokines, other immune factors are associated with ALS that can enhance the activity, trafficking, and survival of neutrophils, including Leukotriene B4, platelet-activating factor, and C5a [91]. Several studies have shown a correlation between the increased percentage of neutrophils within the total leukocyte population and disease progression [92–94]. Accordingly, the increased neutrophil-to-monocyte ratio in patients with ALS suggests progressive alterations in peripheral myeloid populations, which may contribute to functional changes associated with the disease [92]. In the skeletal muscle, neutrophils are recognized for their role in mediating myofiber damage and atrophy. It was found that ALS patients have significant infiltrations of mast cells and phagocytic neutrophils into their skeletal muscles. These cells orchestrate interactions with each other, myofibers, and motor endplates, leading to neuromuscular junction denervation and muscular atrophy. In mouse models, an influx of endomysial elastase-expressing neutrophils may cause such atrophy in muscles, which is reportedly upregulated in muscular dystrophy, impairing myogenesis [95,96]. The data suggest that neutrophil hyperactivation with a high potential for cytotoxicity is involved in neutrophil-mediated inflammation due to NET formation, indicating that the uncontrolled activation of neutrophils likely contributes to muscular pathology ALS.

Studies have shown a link between immune cell features, such as population frequency and expression markers, and disease features. Accordingly, CD16 (FcgRIIIb) expression on mouse neutrophils is associated with disease severity and disease progression rate. As neutrophils age in culture, they lose expression of CD16 in parallel with declining functionality, increasing phagocytosis and oxidative stress [96,97]. This suggests the possibility of reactive oxygen species production during chronic neutrophil activation that exacerbates motor neuron degeneration [91]. A study by Murdock et al. showed that the immune system affects ALS patients differently depending on their gender. Using two separate survival models, the study examined the potential role of neutrophils in ALS patients and determined whether sex affects immunity in those individuals [98]. First, they showed the association of increasing peripheral neutrophils with increased mortality in ALS [99,100]. Second, they revealed that the role of neutrophils is complicated by sex, as low peripheral neutrophil levels were associated with more prolonged median survival in females. Low neutrophil levels, however, did not correlate with increased survival in male participants. This discrepancy could be explained by various mechanisms, including the direct effects of sex hormones on neutrophil activity due to the immunomodulatory effects of male and female hormones and the alternation of the immune environment by sex within the CNS [92,94]. The data suggest a profound impact of neutrophils in the disease progression, and sex-specific neutrophil responses could contribute to sex differences seen in ALS patients. Much more work is needed to untangle the role of neutrophils and their sex-specific roles.

4.6. Multiple Sclerosis

Multiple sclerosis (MS) is a chronic inflammatory disorder of the central nervous system (CNS), characterized by the deterioration of the myelin sheath that protects the nerve fibers. Neutrophil infiltration into the CNS, followed by cytokines secretion such as TNF-α, IL-6, IL-12, IL-1β, and IFN-γ, triggers the inflammatory cascade, causing BBB injury in MS patients [101].

IL-17, known to be produced by Th17 cells and neutrophils, is also reported to be upregulated in the brain of experimental autoimmune encephalomyelitis (EAE) mice during the early stages of the disease. IL-17 is known to disrupt the BBB, which facilitates the migration of inflammatory cells into the brain and also affects the production of other inflammatory chemokines and cytokines associated with EAE, including CXCL1 and CXCL2 [102,103]. Observations suggest neutrophils may damage neurons directly through CXCR2 signaling, a key regulator of neutrophil neurotoxicity. A study found that neutrophils isolated from control EAE mice induced severe neuronal cell death via CXCR2-mediated ROS generation in neutrophils. Neutrophil-specific CXCR2 deletion is sufficient to rescue this effect by preventing ROS production. We can conclude that MS's pathogenesis requires CXCR2 signaling [104,105].

Increased neutrophil numbers are primarily attributed to a decrease in spontaneous apoptosis. The existence of a pro-inflammatory environment appears to modify neutrophil lifespan in patients with relapsing-remitting MS (RRMS), as it was shown that RRMS neutrophils have greater apoptosis resistance than neutrophils from healthy controls (HC) [106], stating that an inflammatory environment can influence neutrophil survival and apoptosis. Such an environment can result from inflammatory priming agents such as IL-1β, IL-6, IFN-y, G-CSF, GM-CSF, or IL-8, which can delay neutrophil apoptosis. In addition, elevated levels of TLR2, CD43, FPR1, and CXCR1 in RRMS patients' neutrophils further suggest that the chronic inflammatory milieu primes neutrophils in RRMS [107]. Primed neutrophils from RRMS patients showed enhanced effector mechanisms, including degranulation, oxidative burst, and release of NETs [108]. Increased levels of NETs are linked to pathophysiological conditions, which are believed to be induced by inflammatory mediators, including IL-8. MS patients show higher levels of IL-8 [109], which also prolongs neutrophil survival [110]. Interestingly, a study found that the subset of RRMS patients with high NETs in the serum was significantly enriched in male patients compared to females, suggesting that NETosis may be gender-specific in this disease and could be involved in certain aspects of MS pathogenesis, such as the opening of the BBB [111].

4.7. Autism

Autism spectrum disorder (ASD) is a pediatric heterogeneous neuropsychiatric disorder characterized by social and communication deficits, language impairment, and ritualistic or repetitive behaviors [112]. Dysregulations in innate and adaptive immune cells have been implicated in the pathogenesis of ASD [113]. Thereby, oxidants and pro-inflammatory cytokines generate important effects.

Past investigations found high levels of pro-inflammatory cytokines, including IL-6 and IL-17A, in ASD patients and autistic mice [114,115]. In ASD subjects, IL-17A produced by immune cells such as Th17 and monocytes/macrophages cells can activate neutrophils. Primed neutrophils might worsen overall inflammation, indicating they play a crucial role in developing peripheral inflammation. A study has found the upregulation of IL-17A/IL-17R in neutrophils of ASD patients, whose interaction is required in oxidative stress regulation in neutrophils. Activation of IL-17A/IL-17R in neutrophils correlates with elevated NOX2/ROS signaling [113].

Excessive oxidative stress due to the activation of NFκB and iNOS is also reported in autistic subjects and BTBR T + Itpr3tf/J (BTBR) mice. This study showed that sulforaphane, an Nrf2 activation mediator, inhibited NFκB-iNOS signaling in neutrophils and cerebellum [116], given the importance of Nrf2 in controlling immune system response and cellular oxidative stress. It was suggested that sulforaphane would attenuate the effects of oxidative stress on neutrophils and brain tissue in autism via activating Nrf2 in microglial cells and macrophages.

ASD neutrophils may also cause increased oxidative stress in response to environmental pollutants such as i-2-ethylhexyl phthalate (DEHP). Neutrophils in such patients are unable to mount an antioxidant response, which was shown by lower expression levels of Nrf2 [117].

Epithelial cell-derived neutrophil-activating peptide-78 (ENA-78/CXCL5) is a CXC chemokine that attracts and activates neutrophils [118]. Accompanied by IL-8, it can induce neutrophil chemotaxis and activation by increasing intracellular calcium levels and elastase release [119], stating that the dysfunction of the immune system can act as a crucial factor in the development of autism.

Highly reactive neutrophil cathepsin B can initiate leukocyte–endothelial cell adhesion by promoting leukocyte Mac-1 activation and its interaction with endothelial ICAM-1. Accordingly, Mac-1 in neutrophils and ICAM-1 in endothelial cells is found to be significantly higher in autistic mice, underlying the role of neutrophil cathepsin B in inducing neurovascular inflammation during autism [120].

4.8. Down Syndrome

Down syndrome (DS) is one of the most common genetic disorders caused by complete or partial triplication of chromosome 21 [101]. Individuals with DS have a higher risk of Alzheimer's disease because patients have an extra copy of the amyloid precursor protein (APP) gene located on chromosome 21, which is trivalent in this population [121]. Due to mutations in APP and genetic risk genes, such as PSEN1 and PSEN2 which alter amyloid concentrations, patients with DS are more prone to develop Alzheimer's disease at an earlier age (30–60 years) than those with late-onset Alzheimer's disease (≥65 years) [122].

It is generally accepted that patients with DS are susceptible to developing infections and chronic inflammatory conditions [122,123], which might also play a critical role in the onset and progression of dementia in these individuals. A meta-analysis study found high expression levels of IL-1β, TNF-α, IFN-γ, and IL-6 cytokines in DS patients compared to healthy control [124]. These cytokines stimulate neutrophil activation and their trafficking into the CNS, underlining the importance of primed neutrophils in neuroinflammation in DS patients. Elevated levels of inflammation can lead to Toll-like receptor (TLR) signaling dysregulation. A study reported the increased expression levels of TLR2 on neutrophils and altered gene expression of key regulators proteins involved in its signal propagation [125]. It is possible that excessive cytokine levels in DS may reflect abnormal signaling of TLR pathways. Since the end point of these pathways results in the release of inflammatory mediators, DS subjects may display a hyperresponsive immune reaction [124]. Furthermore, TLR2 and TLR4 regulate neutrophil function in DS patients, including apoptosis, adhesion, and activating pro-inflammatory signaling pathways, including the NF-κB pathway. Stimulation of the NF-κB pathway induces transcriptional machinery of pro-inflammatory factors, such as iNOS, NOX2, IL-6, and other chemokines/cytokines [10,113]. It has also been found that platelet TLR4 activation can lead to robust neutrophil activation and NET formation, which can cause BBB damage in DS subjects. The same symptoms were observed in Alzheimer's disease when NETs develop intravascularly and intraparenchymally [10,14,126]. These observations raise the possibility that the same pathologic mechanisms might contribute to the progression of both diseases.

Although many of the inflammatory pathways described for AD would apply directly to DS, there are some particular inflammatory genes on chromosome 21 that affect inflammatory responses in the DS brain, including a highly expressed CXADR gene, which functions as an adhesion molecule and is associated with endothelial tight junctions [127,128]. It has been shown that CXADR not only can induce stress-activated mitogen-activated protein kinase (MAPK) pathways in the heart leading to increased production of IFNγ, IL-12, IL-1β, TNFα, and IL-6 but also has a significant role in BBB permeability facilitating trans-endothelial migration of neutrophils [127]. The increased migration of neutrophils damages the tight junction and increases the permeability of the BBB. This damage occurs when neutrophil elastase (NE) is released, disrupting cadherin–cadherin binding, thus increasing BBB permeability [129,130]. We can conclude that altered expression of the CXADR gene on the endothelial cells of the cerebral vasculature in DS subjects can affect inflammatory cells' infiltration into the brain, such as neutrophils, and influence the inflammatory response.

DS patients are generally believed to be exposed to excessive amounts of oxidative stress due to dosage imbalances of the cytoplasmic enzyme Cu^{2+}/Zn^{2+} superoxide dismutase (SOD-1) gene, which is also found in Chr 21 [131,132]. SOD plays a fundamental role in regulating reactive oxygen species (ROS) levels, a major contributor to oxidative stress, neuronal death, and disease progression [133]. Chronic neutrophil activation may exacerbate motor neuron degeneration in DS by producing reactive oxygen species. Indeed, neutrophils have been implicated in other neurodegenerative diseases such as AD [134] and produce significantly higher levels of reactive oxygen species in these patients [135], which is plausible since the functionally deleterious role of neutrophils in AD could be mirrored in DS. We can therefore suggest inhibiting NADPH oxidase enzymes (NOX) as a potential target to reduce ROS, leading to decreased capillary stalling and increased CBF [136].

4.9. Frontotemporal Dementia (FTD)

FTD is the second most common form of dementia and is a collection of neurodegenerative disorders affecting mainly the frontal and temporal lobes of the brain. Its symptom can vary widely between persons, but some include behavioral changes such as a lack of empathy, increased inappropriate social behavior, speech and language problems, and motor disorders [137]. Neuroinflammation, macrophage infiltration, and microglial activation are reported in some patients with FTD [138,139]. Although microglia are thought to be one of the critical immune cells in the progression of FTD [140], recent data indicate the involvement of other immune cells in the disease progression. Employing a mouse model with frontotemporal lobular degeneration, the authors observed the differential expression of genes associated with neutrophils and monocytes [141]. Moreover, the presence of the neutrophilic protein, 37 kDa cationic antimicrobial protein (CAP37), is detected in the brain of a patient with FTD, indicating the involvement of neutrophils in this disease [21].

Another critically important protein in FTD is progranulin, whose mutation is a major cause of the onset and progression of the disease [142]. Progranulin involves multiple pathways, from wound healing after injury to inflammation [143]. Single-nucleus RNA sequencing of FTD patients' cerebral cortex has shown profound neurovascular dysfunction [144]. Interestingly, there is also a dynamic interaction between neutrophils and progranulin; neutrophils release serine protease, and elastase can cleave progranulin into smaller fragments of granulin peptides [145]. At least, one of the granulin fragments, granulin-B, has been shown to elicit an inflammatory response [142]. In contrast, another neutrophil-secreted protein, leukocyte protease inhibitor-1 (SLP-1), can inhibit the generation of granulin fragments from progranulin [146]. Therefore, understanding the role of innate immune cells such as neutrophils in the progression or resolution of FTD is crucial and might lead to the development of therapeutic interventions.

5. Contribution of Neutrophils in Alzheimer's Disease (AD)

An increasing number of studies provide evidence indicating that inflammation and the innate immune system play a pivotal role in the development of AD [147]. Therefore, neutrophil activation and alteration may contribute to increased vascular inflammation associated with AD [148]. A study conducted with a transgenic mouse model of AD revealed that neutrophil depletion significantly reduced Alzheimer-like neuropathology and improved memory in mice with cognitive deficits [14]. One of the pathological hallmarks of Alzheimer's disease is the generation of senile plaques that make up beta-amyloid peptides (Aβ) [149]. Aβ plays an essential role in recruiting neutrophils to inflamed tissues by triggering a shift in lymphocyte function-associated antigen-1 (LFA-1) from a low to high-affinity state, resulting in increased LFA1-dependent adhesion of neutrophils in AD mouse models [14]. High-affinity LFA-1 may also signal the arrest of neutrophils within the parenchyma, causing neutrophil accumulation and leading to the widespread neutrophil-dependent central nervous system (CNS) damage [14]. The data suggest an interaction between the innate immune system and senile plaques linking the periphery with the CNS [150]. Furthermore, the adhesion molecule, Mac-1 (CD11b/CD18), is expressed at higher levels in sporadic AD patients than in control patients. This increase may be correlated with disease severity and the progression of dementia [151]. It should also be noted that the α subunit of Mac-1, CD11b, is thought to be expressed exclusively on leukocytes, where it interacts with the adhesion molecule-1 (ICAM-1, CD54) and other endothelial surface ligands [152]. In this regard, the significant elevation in blood neutrophils, Mac-1, in AD patients suggests a state of neutrophil activation. This activation is likely a consequence of increased levels of TNF-α, IL-6, ICAM-1, CRP, and other inflammatory-immune markers observed in AD patients' serum (Figure 3) [151,153,154].

Neutrophil-associated inflammatory proteins may affect memory and executive function in patients with AD. Lipocalin-2 (LCN2), a pro-inflammatory molecule secreted into the peripheral circulation by neutrophils, is believed to cause cognitive decline [148,155]. Accordingly, several studies have reported the upregulation of LCN2 in AD mouse models and patients' serum and brain tissue [156–158]. How LCN2 contributes to cognitive decline in AD patients is not fully understood. LNC2 may affect neutrophil adhesion by altering the expression of the neutrophil adhesion factors CD62L (L-selectin), Mac-1 (CD11b/CD18.), and CD51/CD61 (αvβ3) [159], and the increased adhesion causes an over activation of the inflammatory cascade. The overactivation, in turn, may reduce proper cellular function contributing to neuronal dysfunction. It is unclear how LCN2 contributes to cognitive decline in AD patients. It may affect neutrophil adhesion by altering the expression of neutrophil adhesion factors CD62L (L-selectin), Mac-1 (CD11b/CD18.), and CD51/CD61 (αvβ3). It is also possible that LCN2 induces chemokine production and enhances neutrophil trafficking into the brain. In addition to LCN2, four different neutrophil adhesive and activation promoter proteins have been associated with executive function deficits in mild AD, including MPO, IL-8, TNF -α, and MIP-1β (CCL4) [148]. We can therefore conclude that neutrophil-associated inflammatory proteins and their related pathways potentially contribute to AD progression and severity.

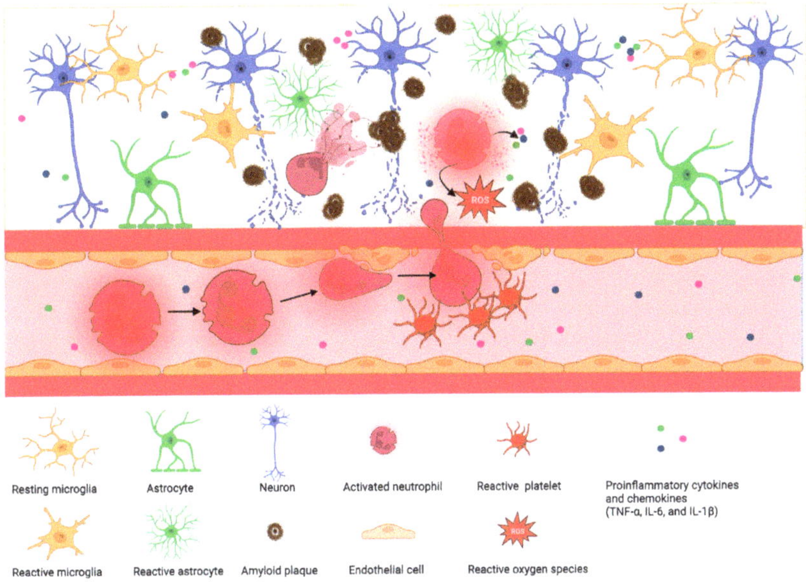

Figure 3. Diagrammatic representation of cellular events in the Alzheimer's vasculature. Increased deposition of amyloid plaques in the brain leads to inflammatory processes; reactive microglia and astrocytes release pro-inflammatory cytokines IL-1β, TNF-α, and IL-6 are released to the bloodstream and attract and activate innate immune cells such as neutrophils and platelets. Peripheral inflammation contributes to reactive neutrophils and platelets. Those, in turn, are more prone to react with blood vessels, likely at sites of reduced blood–brain barrier permeability. Neutrophils are now more likely to stall blood vessels and degranulation inside the vessels. Furthermore, neutrophils interact with reactive platelets, transmigrate into the brain parenchyma, undergo degranulation and NETosis, and secrete reactive oxygen species and pro-inflammatory chemokines and cytokines. This vicious cycle contributes to low-level chronic inflammation in patients with Alzheimer's.

The tight cohesion of leukocytes causes capillary stalling to the cortical capillary endothelium. This adhesion contributes to cerebral blood flow (CBF) reduction in mouse models of AD. Studies have shown the contribution of pro-inflammatory mediators such as IL-1β, IL-6, and TNF-α increased vascular inflammation and likely increased capillary stalling [160]. Elevated levels of inflammatory cytokines can enhance neutrophil infiltration, which may also contribute to a decrease in CBF. In this regard, Cruz Hernández et al. investigated the role of neutrophil adhesion in capillary segments, which exacerbates AD symptoms due to CBF reduction. By administrating a fluorescently tagged antibody against the neutrophil marker Ly6G, they could detect immediate improvement in CBF in amyloid precursor proteins (APP) mice [134]. In patients and mouse models of AD, elevated NADPH-oxidase 2 (NOX2) increases oxidative stress in the microvasculature, and oxidative stress contributes to capillary stalling and vascular inflammation. A recent preprint has demonstrated in an AD mouse model, that inhibiting NOX2 improves short-term memory [135,161]. These improvements were associated with increases in CBF by reducing capillary stalling in the APP/PS1 mouse model [135]. Although neutrophils impact AD progression and severity, very little is known about the mechanism affecting neutrophil changes.

Neutrophils tend to act in different forms when pathogens invade. These changes include ROS production causing an oxidative burst, phagocytosis, and degranulation. Neutrophils have elevated amounts of NADPH-oxidase, which has the essential function of generating reactive oxygen species [162]. Elevated levels of ROS lead to neutrophil hyperactivation and increase NET formation. As neutrophils are a crucial medium for inflammation-activated vascular and tissue damage, the inappropriate activation of neutrophils may cause oxidative stress and magnifies inflammatory responses. Neutrophil activation can also trigger apoptosis and hamper their ability to function full-fledged against the impulse [134]. The experimental analysis noted that the flow of aged neutrophils increases ROS production, which is directly related to Aβ deposition in patients [163,164].

6. Conclusions and Future Perspectives

It has become evident that neutrophils are not only important as first responders after infection or injury following a static response. Neutrophils can react dynamically and highly sophisticatedly to orchestrate a response fighting the threat and attract other immune cells. In aging and neurological diseases, neutrophils appear more activated, likely driving a vicious cycle, further increasing the low-level chronic vascular inflammation associated with neurological diseases. These reactive neutrophils will likely release pro-inflammatory cytokines and chemokines, forming more NETs in the vessels and the parenchyma. Moreover, the inflammasome was shown to be more active in neurological diseases such as Alzheimer's, driving further NET formation [165,166]. Altogether, these factors drive an inflammatory environment in the periphery and the brain even further.

It is imminent, despite their short lifespan, that neutrophils are contributing to the increased inflammation associated with neurological disease development, so that inhibiting reactive neutrophils could contribute to slower disease progression. Therefore, a better understanding of neutrophil behaviors in neurological diseases will be crucial to better understanding the transition to more reactive neutrophils. The question of neutrophil heterogeneity is still very much open and needs further investigation using singe-cell and other approaches. Overall, due to the nature of neutrophils, in vivo, studies are critical to investigate neutrophils in the tissue and their interaction with the vasculature or sites of injury.

Neutrophils are significantly different between rodent models and humans, and this needs to be considered when interpreting results. Furthermore, sterile housing conditions also significantly impact neutrophil maturation and the immune system in model organisms [167,168]. For example, wild mice can be used or co-housed with pet-shop mice to make the immune system more mature [169]. Another example to address this problem is the recently humanized neutrophil mouse model [170]. More studies are needed to overcome these shortcomings.

There are many open questions which mean more research in model organisms, and patient is needed to understand the specific contributions of neutrophils in neurological diseases. However, neutrophils eventually drive inflammation in the vasculature and contribute to aging and age-related disease.

Author Contributions: S.C. wrote portions of the initial draft of this paper and modified the initial draft. S.C. prepared the figures. Z.T., N.N.B., S.A.F. and O.B. wrote sections of the manuscript. S.C. and O.B. conceived the original idea. O.B. supervised the project. All authors have read and agreed to the published version of the manuscript.

Funding: Supported by grants from the NIH (3R21AG075798-01A1S) to SF, NIH (R21AG075798 and R21AG082193), Alzheimer's Association (AARG 22-974437), and The Florida Department of Health (23A12) to OB.

Institutional Review Board Statement: Not applicable.

Conflicts of Interest: The authors declare no conflict of interest.

References

1. Dancey, J.T.; Deubelbeiss, K.A.; Harker, L.A.; Finch, C.A. Neutrophil kinetics in man. *J. Clin. Investig.* **1976**, *58*, 705–715. [CrossRef] [PubMed]
2. Pillay, J.; Braber, I.D.; Vrisekoop, N.; Kwast, L.M.; de Boer, R.J.; Borghans, J.A.M.; Tesselaar, K.; Koenderman, L. In vivo labeling with $2H_2O$ reveals a human neutrophil lifespan of 5.4 days. *Blood* **2010**, *116*, 625–627. [CrossRef] [PubMed]

3. Hidalgo, A.; Chilvers, E.R.; Summers, C.; Koenderman, L. The Neutrophil Life Cycle. *Trends Immunol.* **2019**, *40*, 584–597. [CrossRef] [PubMed]
4. Sundd, P.; Pospieszalska, M.K.; Cheung, L.S.-L.; Konstantopoulos, K.; Ley, K. Biomechanics of leukocyte rolling. *Biorheology* **2011**, *48*, 1–35. [CrossRef] [PubMed]
5. Chesnutt, B.C.; Smith, D.F.; Raffler, N.A.; Smith, M.L.; White, E.J.; Ley, K. Induction of LFA-1-Dependent Neutrophil Rolling on ICAM-1 by Engagement of E-Selectin. *Microcirculation* **2006**, *13*, 99–109. [CrossRef]
6. De Oliveira, S.; Rosowski, E.E.; Huttenlocher, A. Neutrophil migration in infection and wound repair: Going forward in reverse. *Nat. Rev. Immunol.* **2016**, *16*, 378–391. [CrossRef]
7. Filippi, M.-D. Neutrophil transendothelial migration: Updates and new perspectives. *Blood* **2019**, *133*, 2149–2158. [CrossRef]
8. Rosales, C. Neutrophil: A Cell with Many Roles in Inflammation or Several Cell Types? *Front. Physiol.* **2018**, *9*, 113. [CrossRef]
9. Sheshachalam, A.; Srivastava, N.; Mitchell, T.; Lacy, P.; Eitzen, G. Granule Protein Processing and Regulated Secretion in Neutrophils. *Front. Immunol.* **2014**, *5*, 448. [CrossRef]
10. Papayannopoulos, V. Neutrophil extracellular traps in immunity and disease. *Nat. Rev. Immunol.* **2018**, *18*, 134–147. [CrossRef]
11. Chen, Y.; Junger, W.G. Measurement of Oxidative Burst in Neutrophils. *Methods Mol. Biol.* **2012**, *844*, 115–124. [CrossRef] [PubMed]
12. Savill, J.S.; Wyllie, A.H.; Henson, J.E.; Walport, M.J.; Henson, P.M.; Haslett, C. Macrophage phagocytosis of aging neutrophils in inflammation. Programmed cell death in the neutrophil leads to its recognition by macrophages. *J. Clin. Investig.* **1989**, *83*, 865–875. [CrossRef] [PubMed]
13. Lacy, Mechanisms of Degranulation in Neutrophils. *Allergy Asthma Clin. Immunol.* **2006**, *2*, 98–108. [CrossRef]
14. Zenaro, E.; Pietronigro, E.; Bianca, V.D.; Piacentino, G.; Marongiu, L.; Budui, S.; Turano, E.; Rossi, B.; Angiari, S.; Dusi, S.; et al. Neutrophils promote Alzheimer's disease-like pathology and cognitive decline via LFA-1 integrin. *Nat. Med.* **2015**, *21*, 880–886. [CrossRef] [PubMed]
15. Kinney, J.W.; Bemiller, S.M.; Murtishaw, A.S.; Leisgang, A.M.; Salazar, A.M.; Lamb, B.T. Inflammation as a central mechanism in Alzheimer's disease. *Alzheimers Dement.* **2018**, *4*, 575–590. [CrossRef] [PubMed]
16. La Cour, L.F. X.—Mitosis and Cell Differentiation in the Blood. *Proc. R. Soc. Edinburgh. Sect. B Boil.* **1944**, *62*, 73–85. [CrossRef]
17. Bainton, D.F.; Ullyot, J.L.; Farquhar, M.G. The development of neutrophilic polymorphonuclear leukocytes in human bone marrow. *J. Exp. Med.* **1971**, *134*, 907–934. [CrossRef]
18. Pietronigro, E.C.; Della Bianca, V.; Zenaro, E.; Constantin, G. NETosis in Alzheimer's Disease. *Front. Immunol.* **2017**, *8*, 211. [CrossRef]
19. McKenna, E.; Mhaonaigh, A.U.; Wubben, R.; Dwivedi, A.; Hurley, T.; Kelly, L.A.; Stevenson, N.J.; Little, M.A.; Molloy, E.J. Neutrophils: Need for Standardized Nomenclature. *Front. Immunol.* **2021**, *12*, 602963. [CrossRef]
20. Mutua, V.; Gershwin, L.J. A Review of Neutrophil Extracellular Traps (NETs) in Disease: Potential Anti-NETs Therapeutics. *Clin. Rev. Allergy Immunol.* **2021**, *61*, 194–211. [CrossRef]
21. Stock, A.J.; Kasus-Jacobi, A.; Pereira, H.A. The role of neutrophil granule proteins in neuroinflammation and Alzheimer's disease. *J. Neuroinflamm.* **2018**, *15*, 240. [CrossRef] [PubMed]
22. Suh, Y.-A.; Arnold, R.S.; Lassegue, B.; Shi, J.; Xu, X.X.; Sorescu, D.; Chung, A.B.; Griendling, K.K.; Lambeth, J.D. Cell transformation by the superoxide-generating oxidase Mox1. *Nat. Cell Biol.* **1999**, *401*, 79–82. [CrossRef]
23. Langenhuijsen, M.M.A.C. Neutrophils with ring-shaped nuclei in myeloproliferative disease. *Br. J. Haematol.* **1984**, *58*, 227–230. [CrossRef] [PubMed]
24. Wigerblad, G.; Cao, Q.; Brooks, S.; Naz, F.; Gadkari, M.; Jiang, K.; Gupta, S.; O'neil, L.; Dell'orso, S.; Kaplan, M.J.; et al. Single-Cell Analysis Reveals the Range of Transcriptional States of Circulating Human Neutrophils. *J. Immunol.* **2022**, *209*, 772–782. [CrossRef]
25. Coulibaly, A.P. Neutrophil modulation of behavior and cognition in health and disease: The unexplored role of an innate immune cell. *Immunol. Rev.* **2022**, *311*, 177–186. [CrossRef]
26. Sas, A.R.; Carbajal, K.S.; Jerome, A.D.; Menon, R.; Yoon, C.; Kalinski, A.L.; Giger, R.J.; Segal, B.M. A new neutrophil subset promotes CNS neuron survival and axon regeneration. *Nat. Immunol.* **2020**, *21*, 1496–1505. [CrossRef] [PubMed]
27. Borregaard, N. Neutrophils, from Marrow to Microbes. *Immunity* **2010**, *33*, 657–670. [CrossRef]
28. Bartels, M.; Murphy, K.; Rieter, E.; Bruin, M. Understanding chronic neutropenia: Life is short. *Br. J. Haematol.* **2016**, *172*, 157–169. [CrossRef]
29. Shirakawa, K.; Sano, M. Neutrophils and Neutrophil Extracellular Traps in Cardiovascular Disease: An Overview and Potential Therapeutic Approaches. *Biomedicines* **2022**, *10*, 1850. [CrossRef]
30. Metcalf, D.; Nicola, N.A. Proliferative effects of purified granulocyte colony-stimulating factor (G-CSF) on normal mouse hemopoietic cells. *J. Cell. Physiol.* **1983**, *116*, 198–206. [CrossRef]
31. Roberts, A.W. G-CSF: A key regulator of neutrophil production, but that's not all! *Growth Factors* **2005**, *23*, 33–41. [CrossRef] [PubMed]
32. Eash, K.J.; Greenbaum, A.; Gopalan, P.K.; Link, D.C. CXCR2 and CXCR4 antagonistically regulate neutrophil trafficking from murine bone marrow. *J. Clin. Investig.* **2010**, *120*, 2423–2431. [CrossRef] [PubMed]
33. Zhang, D.-E.; Zhang, P.; Wang, N.-D.; Hetherington, C.J.; Darlington, G.J.; Tenen, D.G. Absence of granulocyte colony-stimulating factor signaling and neutrophil development in CCAAT enhancer binding protein α-deficient mice. *Proc. Natl. Acad. Sci. USA* **1997**, *94*, 569–574. [CrossRef] [PubMed]

34. Nerlov, C.; Graf, T. PU. 1 induces myeloid lineage commitment in multipotent hematopoietic progenitors. *Genes Dev.* **1998**, *12*, 2403–2412. [CrossRef]
35. Hanash, S. Disease proteomics. *Nature* **2003**, *422*, 226–232. [CrossRef]
36. Al-Amrani, S.; Al-Jabri, Z.; Al-Zaabi, A.; Alshekaili, J.; Al-Khabori, M. Proteomics: Concepts and applications in human medicine. *World J. Biol. Chem.* **2021**, *12*, 57–69. [CrossRef]
37. Luerman, G.C.; Uriarte, S.M.; Rane, M.J.; McLeish, K.R. Application of proteomics to neutrophil biology. *J. Proteom.* **2010**, *73*, 552–561. [CrossRef]
38. Yee, C.S.; Seyedsayamdost, M.R.; Chang, M.C.Y.; Nocera, D.G.; Stubbe, J. Generation of the R2 Subunit of Ribonucleotide Reductase by Intein Chemistry: Insertion of 3-Nitrotyrosine at Residue 356 as a Probe of the Radical Initiation Process. *Biochemistry* **2003**, *42*, 14541–14552. [CrossRef]
39. Mikesh, L.M.; Ueberheide, B.; Chi, A.; Coon, J.J.; Syka, J.E.; Shabanowitz, J.; Hunt, D.F. The utility of ETD mass spectrometry in proteomic analysis. *Biochim. Biophys. Acta* **2006**, *1764*, 1811–1822. [CrossRef]
40. Monti, M.; Orru, S.; Pagnozzi, D.; Pucci, P. Functional proteomics. *Clin. Chim. Acta* **2005**, *357*, 140–150. [CrossRef]
41. Chen, H.; Albergante, L.; Hsu, J.Y.; Lareau, C.A.; Bosco, G.L.; Guan, J.; Zhou, S.; Gorban, A.N.; Bauer, D.E.; Aryee, M.J.; et al. Single-cell trajectories reconstruction, exploration and mapping of omics data with STREAM. *Nat. Commun.* **2019**, *10*, 1903. [CrossRef] [PubMed]
42. Xie, X.; Shi, Q.; Wu, P.; Zhang, X.; Kambara, H.; Su, J.; Yu, H.; Park, S.-Y.; Guo, R.; Ren, Q.; et al. Single-cell transcriptome profiling reveals neutrophil heterogeneity in homeostasis and infection. *Nat. Immunol.* **2020**, *21*, 1119–1133. [CrossRef] [PubMed]
43. Grieshaber-Bouyer, R.; Radtke, F.A.; Cunin, P.; Stifano, G.; Levescot, A.; Vijaykumar, B.; Nelson-Maney, N.; Blaustein, R.B.; Monach, P.A.; Nigrovic, P.A.; et al. The neutrotime transcriptional signature defines a single continuum of neutrophils across biological compartments. *Nat. Commun.* **2021**, *12*, 2856. [CrossRef] [PubMed]
44. Zenobi, R. Single-Cell Metabolomics: Analytical and Biological Perspectives. *Science* **2013**, *342*, 1243259. [CrossRef]
45. Yuyun, X.; Fan, Y.; Weiping, W.; Qing, Y.; Bingwei, S. Metabolomic analysis of spontaneous neutrophil apoptosis reveals the potential involvement of glutathione depletion. *Innate Immun.* **2021**, *27*, 31–40. [CrossRef]
46. Helmchen, F.; Denk, W. Deep tissue two-photon microscopy. *Nat. Methods* **2005**, *2*, 932–940. [CrossRef]
47. Horton, N.G.; Wang, K.; Kobat, D.; Clark, C.G.; Wise, F.W.; Schaffer, C.B.; Xu, C. In vivo three-photon microscopy of subcortical structures within an intact mouse brain. *Nat. Photonics* **2013**, *7*, 205–209. [CrossRef]
48. Kreisel, D.; Nava, R.G.; Li, W.; Zinselmeyer, B.H.; Wang, B.; Lai, J.; Pless, R.; Gelman, A.E.; Krupnick, A.S.; Miller, M.J. In vivo two-photon imaging reveals monocyte-dependent neutrophil extravasation during pulmonary inflammation. *Proc. Natl. Acad. Sci. USA* **2010**, *107*, 18073–18078. [CrossRef]
49. Bennewitz, M.; Jimenez, M.; Vats, R.; Tutuncuoglu, E.; Jonassaint, J.; Kato, G.; Gladwin, M.T.; Sundd, P. Lung vaso-occlusion in sickle cell disease mediated by arteriolar neutrophil-platelet microemboli. *J. Clin. Investig.* **2017**, *2*, e89761. [CrossRef]
50. Looney, M.R.; Thornton, E.E.; Sen, D.; Lamm, W.J.; Glenny, R.W.; Krummel, M.F. Stabilized imaging of immune surveillance in the mouse lung. *Nat. Methods* **2011**, *8*, 91–96. [CrossRef]
51. Vats, R.; Kaminski, T.W.; Brzoska, T.; Leech, J.A.; Tutuncuoglu, E.; Katoch, O.; Jonassaint, J.C.; Tejero, J.; Novelli, E.M.; Pradhan-Sundd, T.; et al. Liver-to-lung microembolic NETs promote gasdermin D–dependent inflammatory lung injury in sickle cell disease. *Blood* **2022**, *140*, 1020–1037. [CrossRef] [PubMed]
52. Brzoska, T.; Kaminski, T.W.; Bennewitz, M.F.; Sundd, P. Live Imaging of the Lung. *Curr. Protoc. Cytom.* **2020**, *95*, e80. [CrossRef] [PubMed]
53. Byun, D.J.; Kim, Y.M.; Hyun, Y.-M. Real-time observation of neutrophil extracellular trap formation in the inflamed mouse brain via two-photon intravital imaging. *Lab. Anim. Res.* **2022**, *38*, 16. [CrossRef] [PubMed]
54. Segel, G.B.; Halterman, M.W.; Lichtman, M.A. The paradox of the neutrophil's role in tissue injury. *J. Leukoc. Biol.* **2011**, *89*, 359–372. [CrossRef] [PubMed]
55. Kolaczkowska, E.; Kubes, P. Neutrophil recruitment and function in health and inflammation. *Nat. Rev. Immunol.* **2013**, *13*, 159–175. [CrossRef]
56. Zhang, Q.; Raoof, M.; Chen, Y.; Sumi, Y.; Sursal, T.; Junger, W.; Brohi, K.; Itagaki, K.; Hauser, C.J. Circulating mitochondrial DAMPs cause inflammatory responses to injury. *Nature* **2010**, *464*, 104–107. [CrossRef]
57. Jickling, G.C.; Liu, D.; Ander, B.P.; Stamova, B.; Zhan, X.; Sharp, F.R. Targeting Neutrophils in Ischemic Stroke: Translational Insights from Experimental Studies. *J. Cereb. Blood Flow Metab.* **2015**, *35*, 888–901. [CrossRef]
58. Ruhnau, J.; Schulze, J.; Dressel, A.; Vogelgesang, A. Thrombosis, Neuroinflammation, and Poststroke Infection: The Multifaceted Role of Neutrophils in Stroke. *J. Immunol. Res.* **2017**, *2017*, 5140679. [CrossRef]
59. Wanrooy, B.J.; Wen, S.W.; Wong, C.H. Dynamic roles of neutrophils in post-stroke neuroinflammation. *Immunol. Cell Biol.* **2021**, *99*, 924–935. [CrossRef]
60. El Amki, M.; Glück, C.; Binder, N.; Middleham, W.; Wyss, M.T.; Weiss, T.; Meister, H.; Luft, A.; Weller, M.; Weber, B.; et al. Neutrophils Obstructing Brain Capillaries Are a Major Cause of No-Reflow in Ischemic Stroke. *Cell Rep.* **2020**, *33*, 108260. [CrossRef]
61. Erdener, Ş.E.; Tang, J.; Kılıç, K.; Postnov, D.; Giblin, J.T.; Kura, S.; Chen, I.C.A.; Vayisoğlu, T.; Sakadžić, S.; Schaffer, C.B.; et al. Dynamic capillary stalls in reperfused ischemic penumbra contribute to injury: A hyperacute role for neutrophils in persistent traffic jams. *J. Cereb. Blood Flow Metab.* **2021**, *41*, 236–252. [CrossRef] [PubMed]

62. Uçar, C.A.; Çokal, B.G.; Artık, H.A.; Inan, L.E.; Yoldaş, T.K. Comparison of neutrophil–lymphocyte ratio (NLR) in Parkinson's disease subtypes. *Neurol. Sci.* **2017**, *38*, 287–293. [CrossRef] [PubMed]
63. Bogdan, C. Nitric oxide and the immune response. *Nat. Immunol.* **2001**, *2*, 907–916. [CrossRef] [PubMed]
64. Gatto, E.M.; Riobó, N.A.; Carreras, M.C.; Cherñavsky, A.; Rubio, A.; Satz, M.L.; Poderoso, J.J. Overexpression of neutrophil neuronal nitric oxide synthase in Parkinson's disease. *Nitric Oxide* **2000**, *4*, 534–539. [CrossRef]
65. Manda-Handzlik, A.; Bystrzycka, W.; Cieloch, A.; Glodkowska-Mrowka, E.; Jankowska-Steifer, E.; Heropolitanska-Pliszka, E.; Skrobot, A.; Muchowicz, A.; Ciepiela, O.; Wachowska, M.; et al. Nitric oxide and peroxynitrite trigger and enhance release of neutrophil extracellular traps. *Cell. Mol. Life Sci.* **2020**, *77*, 3059–3075. [CrossRef] [PubMed]
66. Fan, Y.; Howden, A.J.; Sarhan, A.R.; Lis, P.; Ito, G.; Martinez, T.N.; Brockmann, K.; Gasser, T.; Alessi, D.R.; Sammler, E.M. Interrogating Parkinson's disease LRRK2 kinase pathway activity by assessing Rab10 phosphorylation in human neutrophils. *Biochem. J.* **2018**, *475*, 23–44. [CrossRef] [PubMed]
67. Muñoz-Delgado, L.; Macías-García, D.; Jesús, S.; Martín-Rodríguez, J.F.; Labrador-Espinosa, M.Á.; Jiménez-Jaraba, M.V.; Adarmes-Gómez, A.; Carrillo, F.; Mir, P. Peripheral Immune Profile and Neutrophil-to-Lymphocyte Ratio in Parkinson's Disease. *Mov. Disord.* **2021**, *36*, 2426–2430. [CrossRef] [PubMed]
68. Hu, B.; Guo, H.; Zhou, P.; Shi, Z.L. Characteristics of SARS-CoV-2 and COVID-19. *Nat. Rev. Microbiol.* **2021**, *19*, 141–154. [CrossRef]
69. McKenna, E.; Wubben, R.; Isaza-Correa, J.M.; Melo, A.M.; Mhaonaigh, A.U.; Conlon, N.; O'donnell, J.S.; Cheallaigh, C.N.; Hurley, T.; Stevenson, N.J.; et al. Neutrophils in COVID-19: Not Innocent Bystanders. *Front. Immunol.* **2022**, *13*, 864387. [CrossRef]
70. Groff, D.; Sun, A.; Ssentongo, A.E.; Ba, D.M.; Parsons, N.; Poudel, G.R.; Lekoubou, A.; Oh, J.S.; Ericson, J.E.; Ssentongo, P.; et al. Short-term and Long-term Rates of Postacute Sequelae of SARS-CoV-2 Infection: A Systematic Review. *JAMA Netw. Open* **2021**, *4*, e2128568. [CrossRef]
71. George, P.M.; Reed, A.; Desai, S.R.; Devaraj, A.; Faiez, T.S.; Laverty, S.; Kanwal, A.; Esneau, C.; Liu, M.K.; Kamal, F.; et al. A persistent neutrophil-associated immune signature characterizes post–COVID-19 pulmonary sequelae. *Sci. Transl. Med.* **2022**, *14*, eabo5795. [CrossRef] [PubMed]
72. Wu, F.; Zhao, S.; Yu, B.; Chen, Y.-M.; Wang, W.; Song, Z.-G.; Hu, Y.; Tao, Z.-W.; Tian, J.-H.; Pei, Y.-Y.; et al. A new coronavirus associated with human respiratory disease in China. *Nature* **2020**, *579*, 265–269. [CrossRef]
73. Østergaard, L. SARS-CoV-2 related microvascular damage and symptoms during and after COVID-19: Consequences of capillary transit-time changes, tissue hypoxia and inflammation. *Physiol. Rep.* **2021**, *9*, e14726. [CrossRef] [PubMed]
74. Lowenstein, C.J.; Solomon, S.D. Severe COVID-19 is a Microvascular Disease. *Circulation* **2020**, *142*, 1609–1611. [CrossRef] [PubMed]
75. Buzhdygan, T.P.; DeOre, B.J.; Baldwin-Leclair, A.; Bullock, T.A.; McGary, H.M.; Khan, J.A.; Razmpour, R.; Hale, J.F.; Galie, P.A.; Potula, R.; et al. The SARS-CoV-2 spike protein alters barrier function in 2D static and 3D microfluidic in-vitro models of the human blood-brain barrier. *Neurobiol. Dis.* **2020**, *146*, 105131. [CrossRef]
76. Krasemann, S.; Glatzel, M.; Pless, O. Response to: SARS-CoV-2 and type I interferon signaling in brain endothelial cells: Blurring the lines between friend or foe. *Stem Cell Rep.* **2022**, *17*, 1014–1015. [CrossRef]
77. Hirunpattarasilp, C.; James, G.; Kwanthongdee, J.; Freitas, F.; Huo, J.; Sethi, H.; Kittler, J.T.; Owens, R.J.; McCoy, L.E.; Attwell, D. SARS-CoV-2 triggers pericyte-mediated cerebral capillary constriction. *Brain* **2023**, *146*, 727–738. [CrossRef]
78. Sloop, G.D.; Pop, G.; Weidman, J.J.; Cyr, J.A.S. COVID-19 Demonstrates That Inflammation Is a Hyperviscous State. *Cureus* **2022**, *14*, e30603. [CrossRef]
79. Machiela, E.; Jeloka, R.; Caron, N.S.; Mehta, S.; Schmidt, M.E.; Baddeley, H.J.E.; Tom, C.M.; Polturi, N.; Xie, Y.; Mattis, V.B.; et al. The Interaction of Aging and Cellular Stress Contributes to Pathogenesis in Mouse and Human Huntington Disease Neurons. *Front. Aging Neurosci.* **2020**, *12*, 524369. [CrossRef]
80. Chang, K.-H.; Wu, Y.-R.; Chen, Y.-C.; Chen, C.-M. Plasma inflammatory biomarkers for Huntington's disease patients and mouse model. *Brain Behav. Immun.* **2015**, *44*, 121–127. [CrossRef]
81. Kwan, W.; Träger, U.; Davalos, D.; Chou, A.; Bouchard, J.; Andre, R.; Miller, A.; Weiss, A.; Giorgini, F.; Cheah, C.; et al. Mutant huntingtin impairs immune cell migration in Huntington disease. *J. Clin. Investig.* **2012**, *122*, 4737–4747. [CrossRef] [PubMed]
82. Desport, J.C.; Preux, P.M.; Magy, L.; Boirie, Y.; Vallat, J.M.; Beaufrère, B.; Couratier, P. Factors correlated with hypermetabolism in patients with amyotrophic lateral sclerosis. *Am. J. Clin. Nutr.* **2001**, *74*, 328–334. [CrossRef] [PubMed]
83. Beers, D.R.; Appel, S.H. Immune dysregulation in amyotrophic lateral sclerosis: Mechanisms and emerging therapies. *Lancet Neurol.* **2019**, *18*, 211–220. [CrossRef]
84. Ehrhart, J.; Smith, A.J.; Kuzmin-Nichols, N.; Zesiewicz, T.A.; Jahan, I.; Shytle, R.D.; Kim, S.-H.; Sanberg, C.D.; Vu, T.H.; Gooch, C.L.; et al. Humoral factors in ALS patients during disease progression. *J. Neuroinflamm.* **2015**, *12*, 127. [CrossRef]
85. Aebischer, J.; Moumen, A.; Sazdovitch, V.; Seilhean, D.; Meininger, V.; Raoul, C. Elevated levels of IFNγ and LIGHT in the spinal cord of patients with sporadic amyotrophic lateral sclerosis. *Eur. J. Neurol.* **2012**, *19*, 752–759. [CrossRef] [PubMed]
86. Jones, S.A.; Richards, P.J.; Scheller, J.; Rose-John, S. IL-6 transsignaling: The in vivo consequences. *J. Interferon Cytokine Res.* **2005**, *25*, 241–253. [CrossRef]
87. Rose-John, S.; Waetzig, G.H.; Scheller, J.; Grötzinger, J.; Seegert, D. The IL-6/sIL-6R complex as a novel target for therapeutic approaches. *Expert Opin. Ther. Targets* **2007**, *11*, 613–624. [CrossRef]

88. Garbuzova-Davis, S.; Ehrhart, J.; Sanberg, P.R.; Borlongan, C.V. Potential Role of Humoral IL-6 Cytokine in Mediating Pro-Inflammatory Endothelial Cell Response in Amyotrophic Lateral Sclerosis. *Int. J. Mol. Sci.* **2018**, *19*, 423. [CrossRef]
89. Leggate, M.; Nowell, M.A.; Jones, S.A.; Nimmo, M.A. The response of interleukin-6 and soluble interleukin-6 receptor isoforms following intermittent high intensity and continuous moderate intensity cycling. *Cell Stress Chaperones* **2010**, *15*, 827–833. [CrossRef]
90. Müllberg, J.; Dittrich, E.; Graeve, L.; Gerhartz, C.; Yasukawa, K.; Taga, T.; Kishimoto, T.; Heinrich, P.C.; Rose-John, S. Differential shedding of the two subunits of the interleukin-6 receptor. *FEBS Lett.* **1993**, *332*, 174–178. [CrossRef]
91. Murdock, B.J.; Goutman, S.A.; Boss, J.; Kim, S.; Feldman, E.L. Amyotrophic Lateral Sclerosis Survival Associates with Neutrophils in a Sex-specific Manner. *Neurol. Neuroimmunol. Neuroinflamm.* **2021**, *8*, e953. [CrossRef] [PubMed]
92. Murdock, B.J.; Bender, D.E.; Kashlan, S.R.; Figueroa-Romero, C.; Backus, C.; Callaghan, B.C.; Goutman, S.A.; Feldman, E.L. Increased ratio of circulating neutrophils to monocytes in amyotrophic lateral sclerosis. *Neurol. Neuroimmunol. Neuroinflamm.* **2016**, *3*, e242. [CrossRef] [PubMed]
93. Shang, H.-F.; Wei, Q.-Q.; Hou, Y.-B.; Zhang, L.-Y.; Ou, R.-W.; Cao, B.; Chen, Y.-P. Neutrophil-to-lymphocyte ratio in sporadic amyotrophic lateral sclerosis. *Neural Regen. Res.* **2022**, *17*, 875. [CrossRef] [PubMed]
94. Leone, M.A.; Mandrioli, J.; Russo, S.; Cucovici, A.; Gianferrari, G.; Lisnic, V.; Muresanu, D.F.; Giuliani, F.; Copetti, M.; The Pooled Resource Open-Access ALS Clinical Trials Consortium; et al. Neutrophils-to-Lymphocyte Ratio Is Associated with Progression and Overall Survival in Amyotrophic Lateral Sclerosis. *Biomedicines* **2022**, *10*, 354. [CrossRef] [PubMed]
95. Arecco, N.; Clarke, C.J.; Jones, F.K.; Simpson, D.M.; Mason, D.; Beynon, R.J.; Pisconti, A. Elastase levels and activity are increased in dystrophic muscle and impair myoblast cell survival, proliferation and differentiation. *Sci. Rep.* **2016**, *6*, 24708. [CrossRef] [PubMed]
96. Moulding, D.A.; Hart, C.A.; Edwards, S.W. Regulation of neutrophil FcγRIIIb (CD16) surface expression following delayed apoptosis in response to GM-CSF and sodium butyrate. *J. Leukoc. Biol.* **1999**, *65*, 875–882. [CrossRef]
97. McGill, R.B.; Steyn, F.J.; Ngo, S.T.; Thorpe, K.A.; Heggie, S.; Ruitenberg, M.J.; Henderson, R.D.; McCombe, P.A.; Woodruff, T.M. Monocytes and neutrophils are associated with clinical features in amyotrophic lateral sclerosis. *Brain Commun.* **2020**, *2*, fcaa013. [CrossRef]
98. Murdock, B.J.; Zhou, T.; Kashlan, S.R.; Little, R.J.; Goutman, S.A.; Feldman, E.L. Correlation of Peripheral Immunity with Rapid Amyotrophic Lateral Sclerosis Progression. *JAMA Neurol.* **2017**, *74*, 1446–1454. [CrossRef]
99. Kim, C.F.; Moalem-Taylor, G. Detailed characterization of neuro-immune responses following neuropathic injury in mice. *Brain Res.* **2011**, *1405*, 95–108. [CrossRef]
100. Klein, S.L.; Flanagan, K.L. Sex differences in immune responses. *Nat. Rev. Immunol.* **2016**, *16*, 626–638. [CrossRef]
101. Zhou, Q.; Jia, R.; Dang, J. Correlation between the Neutrophil-to-Lymphocyte Ratio and Multiple Sclerosis: Recent Understanding and Potential Application Perspectives. *Neurol. Res. Int.* **2022**, *2022*, 3265029. [CrossRef] [PubMed]
102. Wojkowska, D.; Szpakowski, P.; Ksiazek-Winiarek, D.; Leszczynski, M.; Glabinski, A. Interactions between Neutrophils, Th17 Cells, and Chemokines during the Initiation of Experimental Model of Multiple Sclerosis. *Mediat. Inflamm.* **2014**, *2014*, 590409. [CrossRef] [PubMed]
103. Huppert, J.; Closhen, D.; Croxford, A.; White, R.; Kulig, P.; Pietrowski, E.; Bechmann, I.; Becher, B.; Luhmann, H.J.; Waisman, A.; et al. Cellular mechanisms of IL-17-induced blood-brain barrier disruption. *FASEB J.* **2010**, *24*, 1023–1034. [CrossRef] [PubMed]
104. Khaw, Y.M.; Cunningham, C.; Tierney, A.; Sivaguru, M.; Inoue, M. Neutrophil-selective deletion of Cxcr2 protects against CNS neurodegeneration in a mouse model of multiple sclerosis. *J. Neuroinflamm.* **2020**, *17*, 49. [CrossRef]
105. Liu, L.; Belkadi, A.; Darnall, L.; Hu, T.; Drescher, C.; Cotleur, A.C.; Padovani-Claudio, D.; He, T.; Choi, K.; Lane, T.E.; et al. CXCR2-positive neutrophils are essential for cuprizone-induced demyelination: Relevance to multiple sclerosis. *Nat. Neurosci.* **2010**, *13*, 319–326. [CrossRef] [PubMed]
106. Naegele, M.; Tillack, K.; Reinhardt, S.; Schippling, S.; Martin, R.; Sospedra, M. Neutrophils in multiple sclerosis are characterized by a primed phenotype. *J. Neuroimmunol.* **2012**, *242*, 60–71. [CrossRef]
107. Hertwig, L.; Pache, F.; Romero-Suarez, S.; Stürner, K.H.; Borisow, N.; Behrens, J.; Bellmann-Strobl, J.; Seeger, B.; Asselborn, N.; Ruprecht, K.; et al. Distinct functionality of neutrophils in multiple sclerosis and neuromyelitis optica. *Mult. Scler. J.* **2016**, *22*, 160–173. [CrossRef]
108. De Bondt, M.; Hellings, N.; Opdenakker, G.; Struyf, S. Neutrophils: Underestimated Players in the Pathogenesis of Multiple Sclerosis (MS). *Int. J. Mol. Sci.* **2020**, *21*, 4558. [CrossRef]
109. Lund, B.T.; Ashikian, N.; Ta, H.Q.; Chakryan, Y.; Manoukian, K.; Groshen, S.; Gilmore, W.; Cheema, G.S.; Stohl, W.; Burnett, M.E.; et al. Increased CXCL8 (IL-8) expression in Multiple Sclerosis. *J. Neuroimmunol.* **2004**, *155*, 161–171. [CrossRef]
110. Costantini, C.; Micheletti, A.; Calzetti, F.; Perbellini, O.; Pizzolo, G.; Cassatella, M.A. Neutrophil activation and survival are modulated by interaction with NK cells. *Int. Immunol.* **2010**, *22*, 827–838. [CrossRef]
111. Tillack, K.; Naegele, M.; Haueis, C.; Schippling, S.; Wandinger, K.-P.; Martin, R.; Sospedra, M. Gender differences in circulating levels of neutrophil extracellular traps in serum of multiple sclerosis patients. *J. Neuroimmunol.* **2013**, *261*, 108–119. [CrossRef] [PubMed]

112. Haghighatfard, A.; Asl, E.Y.; Bahadori, R.A.; Aliabadian, R.; Farhadi, M.; Mohammadpour, F.; Tabrizi, Z. FOXP2 down expression is associated with executive dysfunctions and electrophysiological abnormalities of brain in Autism spectrum disorder; a neuroimaging genetic study. *Autism Dev. Lang. Impair.* **2022**, *7*, 23969415221126391. [CrossRef] [PubMed]
113. Nadeem, A.; Ahmad, S.F.; Attia, S.M.; Al-Ayadhi, L.Y.; Bakheet, S.A.; Al-Harbi, N.O. Oxidative and inflammatory mediators are upregulated in neutrophils of autistic children: Role of IL-17A receptor signaling. *Prog. Neuro Psychopharmacol. Biol. Psychiatry* **2019**, *90*, 204–211. [CrossRef] [PubMed]
114. He, X.; Zhou, M.; Yang, T.; Ren, J.-K.; Sun, L.; Liu, T.-Y.; Sun, J.-B.; Ma, P.-J.; Liu, H.-T.; Fang, J.-Q.; et al. Early postnatal activation of A2ARs alleviates social deficits by attenuating the abnormal infiltration of peripheral neutrophils in the BTBR T + Itpr3 tf/J mouse model of autism. *Res. Sq.* **2022**, *preprint*. [CrossRef]
115. Akintunde, M.E.; Rose, M.; Krakowiak, P.; Heuer, L.; Ashwood, P.; Hansen, R.; Hertz-Picciotto, I.; Van de Water, J. Increased production of IL-17 in children with autism spectrum disorders and co-morbid asthma. *J. Neuroimmunol.* **2015**, *286*, 33–41. [CrossRef] [PubMed]
116. Nadeem, A.; Ahmad, S.F.; Al-Harbi, N.O.; Attia, S.M.; Bakheet, S.A.; Ibrahim, K.E.; Alqahtani, F.; Alqinyah, M. Nrf2 activator, sulforaphane ameliorates autism-like symptoms through suppression of Th17 related signaling and rectification of oxidant-antioxidant imbalance in periphery and brain of BTBR T+tf/J mice. *Behav. Brain Res.* **2019**, *364*, 213–224. [CrossRef]
117. Nadeem, A.; Ahmad, S.F.; Al-Harbi, N.O.; Al-Ayadhi, L.Y.; Alanazi, M.M.; Alfardan, A.S.; Attia, S.M.; Algahtani, M.; Bakheet, S.A. Dysregulated Nrf2 signaling in response to di(2-ethylhexyl) phthalate in neutrophils of children with autism. *Int. Immunopharmacol.* **2022**, *106*, 108619. [CrossRef]
118. Bisset, L.R.; Schmid-Grendelmeier, P. Chemokines and their receptors in the pathogenesis of allergic asthma: Progress and perspective. *Curr. Opin. Pulm. Med.* **2005**, *11*, 35–42. [CrossRef]
119. Mostafa, G.A.; Al-Ayadhi, L.Y. The possible link between elevated serum levels of epithelial cell-derived neutrophil-activating peptide-78 (ENA-78/CXCL5) and autoimmunity in autistic children. *Behav. Brain Funct.* **2015**, *11*, 11. [CrossRef]
120. Wang, H.; Yin, Y.-X.; Gong, D.-M.; Hong, L.-J.; Wu, G.; Jiang, Q.; Wang, C.-K.; Blinder, P.; Long, S.; Han, F.; et al. Cathepsin B inhibition ameliorates leukocyte-endothelial adhesion in the BTBR mouse model of autism. *CNS Neurosci. Ther.* **2019**, *25*, 476–485. [CrossRef]
121. Fortea, J.; Quiroz, Y.T.; Ryan, N.S. Lessons from Down syndrome and autosomal dominant Alzheimer's disease. *Lancet Neurol.* **2023**, *22*, 5–6. [CrossRef]
122. Boerwinkle, A.H.; Gordon, B.A.; Wisch, J.; Flores, S.; Henson, R.L.; Butt, O.H.; McKay, N.; Chen, C.D.; Benzinger, T.L.; Fagan, A.M.; et al. Comparison of amyloid burden in individuals with Down syndrome versus autosomal dominant Alzheimer's disease: A cross-sectional study. *Lancet Neurol.* **2023**, *22*, 55–65. [CrossRef] [PubMed]
123. Ram, G.; Chinen, J. Infections and immunodeficiency in Down syndrome. *Clin. Exp. Immunol.* **2011**, *164*, 9–16. [CrossRef] [PubMed]
124. Huggard, D.; Kelly, L.; Ryan, E.; McGrane, F.; Lagan, N.; Roche, E.; Balfe, J.; Leahy, T.R.; Franklin, O.; Doherty, D.; et al. Increased systemic inflammation in children with Down syndrome. *Cytokine* **2020**, *127*, 154938. [CrossRef] [PubMed]
125. Huggard, D.; Koay, W.J.; Kelly, L.; McGrane, F.; Ryan, E.; Lagan, N.; Roche, E.; Balfe, J.; Leahy, T.R.; Franklin, O.; et al. Altered Toll-like Receptor Signalling in Children with Down Syndrome. *Mediat. Inflamm.* **2019**, *2019*, 4068734. [CrossRef]
126. Clark, S.R.; Ma, A.C.; Tavener, S.A.; McDonald, B.; Goodarzi, Z.; Kelly, M.M.; Patel, K.D.; Chakrabarti, S.; McAvoy, E.; Sinclair, G.D.; et al. Platelet TLR4 activates neutrophil extracellular traps to ensnare bacteria in septic blood. *Nat. Med.* **2007**, *13*, 463–469. [CrossRef]
127. Wilcock, D.M. Neuroinflammation in the aging down syndrome brain; lessons from Alzheimer's disease. *Curr. Gerontol. Geriatr. Res.* **2012**, *2012*, 170276. [CrossRef]
128. Tomko, R.P.; Xu, R.; Philipson, L. HCAR and MCAR: The human and mouse cellular receptors for subgroup C adenoviruses and group B coxsackieviruses. *Proc. Natl. Acad. Sci. USA* **1997**, *94*, 3352–3356. [CrossRef]
129. Allport, J.R.; Ding, H.; Collins, T.; Gerritsen, M.E.; Luscinskas, F.W. Endothelial-dependent Mechanisms Regulate Leukocyte Transmigration: A Process Involving the Proteasome and Disruption of the Vascular Endothelial–Cadherin Complex at Endothelial Cell-to-Cell Junctions. *J. Exp. Med.* **1997**, *186*, 517–527. [CrossRef]
130. Bolton, S.; Anthony, D.; Perry, V. Loss of the tight junction proteins occludin and zonula occludens-1 from cerebral vascular endothelium during neutrophil-induced blood–brain barrier breakdown in vivo. *Neuroscience* **1998**, *86*, 1245–1257. [CrossRef]
131. Akinci, O.; Mihci, E.; Tacoy, S.; Kardelen, F.; Keser, I.; Aslan, M. Neutrophil oxidative metabolism in Down syndrome patients with congenital heart defects. *Environ. Mol. Mutagen.* **2010**, *51*, 57–63. [CrossRef]
132. Zemlan, F.P.; Thienhaus, O.J.; Bosmann, H.B. Superoxide dismutase activity in Alzheimer's disease: Possible mechanism for paired helical filament formation. *Brain Res.* **1989**, *476*, 160–162. [CrossRef] [PubMed]
133. Ferrari, M.; Stagi, S. Oxidative Stress in Down and Williams-Beuren Syndromes: An Overview. *Molecules* **2021**, *26*, 3139. [CrossRef] [PubMed]
134. Dong, Y.; Lagarde, J.; Xicota, L.; Corne, H.; Chantran, Y.; Chaigneau, T.; Crestani, B.; Bottlaender, M.; Potier, M.C.; Aucouturier, P.; et al. Neutrophil hyperactivation correlates with Alzheimer's disease progression. *Ann. Neurol.* **2018**, *83*, 387–405. [CrossRef] [PubMed]

135. Hernández, J.C.C.; Bracko, O.; Kersbergen, C.J.; Muse, V.; Haft-Javaherian, M.; Berg, M.; Park, L.; Vinarcsik, L.K.; Ivasyk, I.; Rivera, D.A.; et al. Neutrophil adhesion in brain capillaries reduces cortical blood flow and impairs memory function in Alzheimer's disease mouse models. *Nat. Neurosci.* **2019**, *22*, 413–420. [CrossRef]
136. Ruiz-Uribe, N.E.; Bracko, O.; Swallow, M.; Omurzakov, A.; Dash, S.; Uchida, H.; Xiang, D.; Haft-Javaherian, M.; Falkenhain, K.; Lamont, M.E.; et al. Vascular oxidative stress causes neutrophil arrest in brain capillaries, leading to decreased cerebral blood flow and contributing to memory impairment in a mouse model of Alzheimer's disease. *bioRxiv* **2023**. [CrossRef]
137. Bright, F.; Werry, E.L.; Dobson-Stone, C.; Piguet, O.; Ittner, L.M.; Halliday, G.M.; Hodges, J.R.; Kiernan, M.C.; Loy, C.T.; Kassiou, M.; et al. Neuroinflammation in frontotemporal dementia. *Nat. Rev. Neurol.* **2019**, *15*, 540–555. [CrossRef]
138. Bellucci, A.; Bugiani, O.; Ghetti, B.; Spillantini, M.G. Presence of Reactive Microglia and Neuroinflammatory Mediators in a Case of Frontotemporal Dementia with P301S Mutation. *Neurodegener. Dis.* **2011**, *8*, 221–229. [CrossRef]
139. Raz, L.; Knoefel, J.E.; Bhaskar, K. The neuropathology and cerebrovascular mechanisms of dementia. *J. Cereb. Blood Flow Metab.* **2016**, *36*, 172–186. [CrossRef]
140. Broe, M.; Hodges, J.; Schofield, E.; Shepherd, C.; Kril, J.; Halliday, G. Staging disease severity in pathologically confirmed cases of frontotemporal dementia. *Neurology* **2003**, *60*, 1005–1011. [CrossRef]
141. Sirkis, D.W.; Geier, E.G.; Bonham, L.W.; Karch, C.M.; Yokoyama, J.S. Recent Advances in the Genetics of Frontotemporal Dementia. *Curr. Genet. Med. Rep.* **2019**, *7*, 41–52. [CrossRef]
142. Ahmed, Z.; Mackenzie, I.R.; Hutton, M.L.; Dickson, D.W. Progranulin in frontotemporal lobar degeneration and neuroinflammation. *J. Neuroinflamm.* **2007**, *4*, 7. [CrossRef] [PubMed]
143. He, Z.; Ong, C.H.P.; Halper, J.; Bateman, A. Progranulin is a mediator of the wound response. *Nat. Med.* **2003**, *9*, 225–229. [CrossRef] [PubMed]
144. Gerrits, E.; Giannini, L.A.A.; Brouwer, N.; Melhem, S.; Seilhean, D.; Le Ber, I.; Kamermans, A.; Kooij, G.; de Vries, H.E.; Boddeke, E.W.G.M.; et al. Neurovascular dysfunction in GRN-associated frontotemporal dementia identified by single-nucleus RNA sequencing of human cerebral cortex. *Nat. Neurosci.* **2022**, *25*, 1034–1048. [CrossRef] [PubMed]
145. Cenik, B.; Sephton, C.F.; Cenik, B.K.; Herz, J.; Yu, G. Progranulin: A proteolytically processed protein at the crossroads of inflammation and neurodegeneration. *J. Biol. Chem.* **2012**, *287*, 32298–32306. [CrossRef]
146. Suh, H.-S.; Choi, N.; Tarassishin, L.; Lee, S.C. Regulation of Progranulin Expression in Human Microglia and Proteolysis of Progranulin by Matrix Metalloproteinase-12 (MMP-12). *PLoS ONE* **2012**, *7*, e35115. [CrossRef]
147. Bagyinszky, E.; Van Giau, V.; Shim, K.; Suk, K.; An, S.S.A.; Kim, S. Role of inflammatory molecules in the Alzheimer's disease progression and diagnosis. *J. Neurol. Sci.* **2017**, *376*, 242–254. [CrossRef]
148. Bawa, K.K.; Initiative, F.T.A.D.N.; Krance, S.; Herrmann, N.; Cogo-Moreira, H.; Ouk, M.; Yu, D.; Wu, C.-Y.; Black, S.E.; Lanctôt, K.L.; et al. A peripheral neutrophil-related inflammatory factor predicts a decline in executive function in mild Alzheimer's disease. *J. Neuroinflamm.* **2020**, *17*, 84. [CrossRef]
149. Fu, H.; Li, J.; Du, P.; Jin, W.; Gao, G.; Cui, D. Senile plaques in Alzheimer's disease arise from Aβ-and Cathepsin D-enriched mixtures leaking out during intravascular haemolysis and microaneurysm rupture. *FEBS Lett.* **2022**, *597*, 1007–1040. [CrossRef]
150. Baik, S.H.; Cha, M.Y.; Hyun, Y.M.; Cho, H.; Hamza, B.; Kim, D.K.; Han, S.H.; Choi, H.; Kim, K.H.; Moon, M.; et al. Migration of neutrophils targeting amyloid plaques in Alzheimer's disease mouse model. *Neurobiol. Aging* **2014**, *35*, 1286–1292. [CrossRef]
151. Scali, C.; Prosperi, C.; Bracco, L.; Piccini, C.; Baronti, R.; Ginestroni, A.; Sorbi, S.; Pepeu, G.; Casamenti, F. Neutrophils CD11b and fibroblasts PGE2 are elevated in Alzheimer's disease. *Neurobiol. Aging* **2002**, *23*, 523–530. [CrossRef]
152. Smith, C.W.; Burns, A.R.; Simon, S.I. Co-operative signaling between leukocytes and endothelium mediating firm attachment. *Vasc. Adhes. Mol. Inflamm.* **1999**, 39–64. [CrossRef]
153. Singh, V.K. Neuroautoimmunity: Pathogenic Implications for Alzheimer's Disease. *Gerontology* **1997**, *43*, 79–94. [CrossRef] [PubMed]
154. Nogueira-Neto, J.; Cardoso, A.S.C.; Monteiro, H.P.; Fonseca, F.L.A.; Ramos, L.R.; Junqueira, V.B.C.; Simon, K.A. Basal neutrophil function in human aging: Implications in endothelial cell adhesion. *Cell Biol. Int.* **2016**, *40*, 796–802. [CrossRef] [PubMed]
155. Song, J.; Kim, O.Y. Perspectives in Lipocalin-2: Emerging biomarker for medical diagnosis and prognosis for Alzheimer's disease. *Clin. Nutr. Res.* **2018**, *7*, 1–10. [CrossRef] [PubMed]
156. Ferreira, A.C.; Pinto, V.; Mesquita, S.D.; Novais, A.; Sousa, J.C.; Correia-Neves, M.; Sousa, N.; Palha, J.A.; Marques, F. Lipocalin 2 is involved in emotional behaviors and cognitive function. *Front. Cell. Neurosci.* **2013**, *7*, 122. [CrossRef] [PubMed]
157. Mesquita, S.D.; Ferreira, A.C.; Falcao, A.M.; Sousa, J.C.; Oliveira, T.G.; Correia-Neves, M.; Sousa, N.; Marques, F.; Palha, J.A. Lipocalin 2 modulates the cellular response to amyloid beta. *Cell Death Differ.* **2014**, *21*, 1588–1599. [CrossRef] [PubMed]
158. Kang, H.; Shin, H.J.; An, H.S.; Jin, Z.; Lee, J.Y.; Lee, J.; Kim, K.E.; Jeong, E.A.; Choi, K.Y.; McLean, C.; et al. Role of Lipocalin-2 in Amyloid-Beta Oligomer-Induced Mouse Model of Alzheimer's Disease. *Antioxidants* **2021**, *10*, 1657. [CrossRef] [PubMed]
159. Schroll, A.; Eller, K.; Feistritzer, C.; Nairz, M.; Sonnweber, T.; Moser, P.A.; Rosenkranz, A.R.; Theurl, I.; Weiss, G. Lipocalin-2 ameliorates granulocyte functionality. *Eur. J. Immunol.* **2012**, *42*, 3346–3357. [CrossRef] [PubMed]
160. Crumpler, R.; Roman, R.J.; Fan, F. Capillary Stalling: A Mechanism of Decreased Cerebral Blood Flow in AD/ADRD. *J. Exp. Neurol.* **2021**, *2*, 149. [CrossRef] [PubMed]
161. Park, L.; Zhou, P.; Pitstick, R.; Capone, C.; Anrather, J.; Norris, E.H.; Younkin, L.; Younkin, S.; Carlson, G.; McEwen, B.S.; et al. Nox2-derived radicals contribute to neurovascular and behavioral dysfunction in mice overexpressing the amyloid precursor protein. *Proc. Natl. Acad. Sci. USA* **2008**, *105*, 1347–1352. [CrossRef] [PubMed]

162. Barua, S.; Kim, J.Y.; Yenari, M.A.; Lee, J.E. The role of NOX inhibitors in neurodegenerative diseases. *IBRO Rep.* **2019**, *7*, 59–69. [CrossRef] [PubMed]
163. Matamalas, A.; Bagó, J.; D'agata, E.; Pellisé, F. Validity and reliability of photographic measures to evaluate waistline asymmetry in idiopathic scoliosis. *Eur. Spine J.* **2016**, *25*, 3170–3179. [CrossRef] [PubMed]
164. Uhl, B.; Vadlau, Y.; Zuchtriegel, G.; Nekolla, K.; Sharaf, K.; Gaertner, F.; Massberg, S.; Krombach, F.; Reichel, C.A. Aged neutrophils contribute to the first line of defense in the acute inflammatory response. *Blood J. Am. Soc. Hematol.* **2016**, *128*, 2327–2337. [CrossRef] [PubMed]
165. White, C.S.; Lawrence, C.B.; Brough, D.; Rivers-Auty, J. Inflammasomes as therapeutic targets for Alzheimer's disease. *Brain Pathol.* **2017**, *27*, 223–234. [CrossRef]
166. Münzer, P.; Negro, R.; Fukui, S.; di Meglio, L.; Aymonnier, K.; Chu, L.; Cherpokova, D.; Gutch, S.; Sorvillo, N.; Shi, L.; et al. NLRP3 Inflammasome Assembly in Neutrophils Is Supported by PAD4 and Promotes NETosis Under Sterile Conditions. *Front. Immunol.* **2021**, *12*, 683803. [CrossRef]
167. Nauseef, W.M. How human neutrophils kill and degrade microbes: An integrated view. *Immunol. Rev.* **2007**, *219*, 88–102. [CrossRef]
168. Zschaler, J.; Schlorke, D.; Arnhold, J. Differences in Innate Immune Response between Man and Mouse. *Crit. Rev. Immunol.* **2014**, *34*, 433–454. [CrossRef]
169. Graham, A.L. Naturalizing mouse models for immunology. *Nat. Immunol.* **2021**, *22*, 111–117. [CrossRef]
170. Zheng, Y.; Sefik, E.; Astle, J.; Karatepe, K.; Öz, H.H.; Solis, A.G.; Jackson, R.; Luo, H.R.; Bruscia, E.M.; Halene, S.; et al. Human neutrophil development and functionality are enabled in a humanized mouse model. *Proc. Natl. Acad. Sci. USA* **2022**, *119*, e2121077119. [CrossRef]

biomolecules

MDPI

Review

Vascular Effects on Cerebrovascular Permeability and Neurodegeneration

Nurul Sulimai [1], Jason Brown [1] and David Lominadze [1,2,*]

[1] Department of Surgery, College of Medicine, University of South Florida Morsani, Tampa, FL 33612, USA; nurulsulimai@usf.edu (N.S.); jasonb3@usf.edu (J.B.)
[2] Department of Molecular Pharmacology and Physiology, College of Medicine, University of South Florida Morsani, Tampa, FL 33612, USA
* Correspondence: dlominadze@usf.edu

Abstract: Neurons and glial cells in the brain are protected by the blood brain barrier (BBB). The local regulation of blood flow is determined by neurons and signal conducting cells called astrocytes. Although alterations in neurons and glial cells affect the function of neurons, the majority of effects are coming from other cells and organs of the body. Although it seems obvious that effects beginning in brain vasculature would play an important role in the development of various neuroinflammatory and neurodegenerative pathologies, significant interest has only been directed to the possible mechanisms involved in the development of vascular cognitive impairment and dementia (VCID) for the last decade. Presently, the National Institute of Neurological Disorders and Stroke applies considerable attention toward research related to VCID and vascular impairments during Alzheimer's disease. Thus, any changes in cerebral vessels, such as in blood flow, thrombogenesis, permeability, or others, which affect the proper vasculo-neuronal connection and interaction and result in neuronal degeneration that leads to memory decline should be considered as a subject of investigation under the VCID category. Out of several vascular effects that can trigger neurodegeneration, changes in cerebrovascular permeability seem to result in the most devastating effects. The present review emphasizes the importance of changes in the BBB and possible mechanisms primarily involving fibrinogen in the development and/or progression of neuroinflammatory and neurodegenerative diseases resulting in memory decline.

Keywords: blood-brain-barrier; blood proteins; fibrinogen and cognitive impairment

1. Introduction

During this past decade, greater emphasis in neuroscience has been given to problems associated with vascular cognitive impairment and dementia (VCID) [1]. As a result, out of 132 Research Priorities that have been used as a guide by the National Health, Lung, and Blood Institute for this last decade, two compelling questions relate to vascular effects on neurodegeneration. They are "What interdependencies between the brain/peripheral nervous system are important to the development, progression, manifestations, and treatment of cardiac and vascular disease?" and "What pathobiology underlies vascular causes of cognitive decline?" [2]. Inflammation can be one of the main causes in the development of VCID.

Research has shown a strong link between cardiovascular diseases, along with cerebrovascular diseases, and subsequent cognitive impairment and dementia [3]. During some neuroinflammatory diseases, VCID is often presented as a co-morbidity. For example, Alzheimer's disease (AD) is the leading cause of dementia, and it is often accompanied by VCID. It is estimated that 40% of AD patients also have some form of VCID. Vascular dementia accounts for about 15–30% of dementia cases worldwide [3,4]. Other diseases that are known to cause a substantial cognitive impairment are traumatic brain injury

(TBI) [5–8] and multiple sclerosis (MS) [9]. Although dementia is not the primary clinical sign associated with stroke, stroke almost doubles the risk of developing dementia later in life [10]. The risk of dementia in stroke patients after the incident depends on the lesion size and location, but stroke survivors also suffer worsened cognition years later for reasons not well understood [10]. While incidence of dementia can be close to 5% after transient ischemic stroke, its occurrence can reach 34% after severe stroke [3,4]. Ischemic stroke is the most common type of stroke, making up 87% of all strokes where brain ischemia causes substantial neuronal damage. It has been shown that aggravated peripheral inflammatory response to stroke caused by preceding systemic inflammation has deleterious actions on components of the neurovascular unit (NVU) that may affect BBB integrity [11]. Some common mechanisms associated with cerebrovascular-driven cognitive impairment are associated with accumulation of abnormal proteins, oxidative stress, early synaptic disconnection, and apoptosis leading to cell death. All these abnormalities can be positively accompanied by alterations in the blood brain barrier (BBB). There are many studies that link BBB dysfunction with dementia in humans [12–15] and in animal models [14–17]. The objective of this review is to underline the importance of changes in the cerebrovascular permeability resulting in accumulation of blood plasma proteins and particularly of fibrinogen (Fg) in the extravascular space of the brain, as well as to discuss some possible mechanisms involved in the development and/or progression of neuroinflammatory and neurodegenerative diseases resulting in memory decline.

2. Inflammation and Thrombogenesis

Inflammation is one the most important factors that cause changes in normal homeostasis in the body. Many neurodegenerative diseases are associated with inflammation and considered neuroinflammatory diseases. For example, TBI [18–20], AD [21], MS [22,23], and stroke [24] are considered neuroinflammatory diseases. The main components of the circulatory system that can be affected by inflammation and may result in neurodegeneration are blood cells, such as platelets, leukocytes, and erythrocytes, and vascular wall components, such as endothelial cells (ECs), smooth muscle cells, and pericytes. For example, blood samples from AD patients showed an increased number of activated platelets compared to that in samples from the control group [25]. Platelets, major players in hemostasis and thrombosis [21], have also been known to have a significant effect during inflammation [26]. Activation of platelets and their increased aggregation have been documented during neurodegenerative diseases such as TBI [27], AD [25], and MS [28]. It has been shown that AD mutations result in a significantly hyperactivated state of circulating platelets where the platelets from 3XTg-AD mice adhere more avidly on matrices and have an increased ability to form thrombi during normal flow condition [29]. Therefore, it is well-accepted that platelets are not only activated as a result of inflammation during inflammatory diseases but also can cause or exacerbate pathological processes and result in further neurodegeneration.

Increased thrombosis in the circulatory system would reasonably be expected to affect blood flow in small cerebral vessels, impairing exchange between the vessels and neurons and thus leading to vasculo-neuronal uncoupling. For example, it is known that TBI is associated with an almost immediate reduction in cerebral blood flow (CBF). It was reported that in the period of 2 h after a controlled cortical impact causing mild-to-moderate TBI, microthrombi occluded up to 70% of venules and 33% of arterioles [30].The reduced CBF seen in the traumatic penumbra caused by the formation of thrombi in the microcirculation [30] leads to secondary damages causing impairment in neuronal function and neurodegeneration, ultimately resulting in memory reduction [31].

The role of platelets in MS pathology has been speculated due to their interaction with leukocytes during their penetration of the BBB and the release of platelet-EC adhesion molecule-1 (PECAM-1) to the circulation [32]. In normal conditions, the basal expression of adhesion molecules is low; however, the expression of adhesion molecules on ECs and leucocytes is upregulated during inflammation [33]. Increased soluble PECAM-1 (sPECAM-1) is detected in the sera of MS patients [34]. The increased levels of sPECAM-1 may be a result of its increased release from microvessels and leukocytes during inflammation [34]. Platelet extravasation has been described in inflammatory reactions as a consequence of vascular rupture or increased permeability of undamaged venular endothelium by a transcellular route [35]. More commonly, platelets are described as "pathfinders" to direct leukocyte recruitment to the sites of their extravasation [36], and the significant role platelets play in the leukocyte recruitment into the inflamed brain microvessels was validated [37].

3. BBB Breakdown and Extravasation of Blood Cells

Leukocyte extravasation occurs primarily in post-capillary venules where shear stress is low [38]. The process of leukocyte migration from the blood stream to the extravascular space involves multiple steps. It begins with flowing leukocytes decelerating and slowly rolling on the activated endothelium [38,39], followed by adhesion strengthening and spreading, intravascular crawling, and finally, transcellular and/or paracellular transmigration [39]. Specific interactions of Fg with leukocytes and with ECs through its respective receptors on these cells, such as integrin αMβ2 and intercellular adhesion molecule-1 (ICAM-1), results in migration of leukocytes through ECs [40,41]. It has been shown that Fg dose dependently increases the adhesion of leukocytes to human umbilical vein ECs [42]. In an inflammatory condition, where Fg is elevated, it can be assumed that there is an increased adhesion of leukocytes on the luminal surface of ECs, leading to intraintimal accumulation and then extravasation of leukocytes. In fact, the finding of large depositions of Fg on the luminal surface of ECs in vivo represents a hallmark of certain inflammatory conditions, such as atherothrombosis [43]. Thus, it is possible that inflammation-induced elevation of the blood content of Fg potentially exacerbates the neuroinflammatory pathology.

Besides leukocytes, erythrocytes could also be found being extravasated during a breakdown of the BBB. However, if the vascular wall is not damaged significantly enough to allow penetration of red blood cells (RBCs), they do not cross the BBB even if the vessels are permeable to other cells or plasma components [44]. Extravasation of erythrocytes has been shown to cause oxidative injury to the brain [45]. It leads to the deposition of hemoglobin-derived neurotoxic products, including free iron. Decompartmentalization of iron from erythrocytes can cause brain edema and lipid peroxidation, leading to oxidative damages and neuronal death [45,46].

Oxidative Stress in Cerebrovascular Disease

Under normal conditions, there is a balance between oxidant and antioxidant systems preventing oxidative damage. Oxidative stress develops when generation of reactive oxygen species (ROSs) is enhanced and/or ROS scavenging is impaired. Iron-derived ROSs are implicated in the pathogenesis of various vascular disorders, including vasculitis and reperfusion injury [47]. In the brain, ROSs have been shown to significantly alter BBB permeability and promote monocyte transmigration across the BBB. Therefore, any molecules that potentially generate enhanced ROSs potentially exacerbate neuroinflammation [48].

4. Increased Cerebrovascular Permeability and Neurodegeneration

4.1. Paracellular

Overall, changes in cerebrovascular permeability play the main role in the development of VCID. There are two modes of vascular permeability: paracellular (between the cells) [49] and transcellular (through the cells) [50–52]. Depending on the size of a substance, one or the other pathway can be used in extravasation. For example, a moderately large virus may not fit the gaps formed for a paracellular transport and would likely use a transcellular pathway involving vesicular transport [53]. In normal conditions, brain vessels are characterized with higher transendothelial electrical resistance than in peripheral circulation, indicating tighter junctions and therefore suggesting lesser paracellular transport than in skeletal muscle [54]. Similarly, much less transcellular transport (caveolar transport) occurs in brain vessels than in peripheral circulation [55]. During pathologies (inflammation), slight increases in these transport mechanisms can result in devastating consequences. These effects may result in enhanced water transport through paracellular and transcellular (via aquaporins) pathways and result in edema formation [56]. It is noteworthy that the physical breakdown of the BBB (rupture of vessels) that can occur during stroke or moderate to severe TBI and result in the accumulation of blood cells in the brain can result in changes of neuronal function and thus neurodegeneration [46]. Damage of vessels that leads to vascular rupture results in bleeding and the accumulation of blood components in the brain tissue. This process may not be considered a result of "vascular permeability" changes. Vascular permeability changes may occur in non-ruptured vessels that can be a result of alterations in the function of paracellular and/or transcellular transports. Altogether, changes in BBB integrity (vascular rupture) and/or permeability (enhanced paracellular and/or transcellular transports) inevitably lead to neurodegeneration and can result in memory reduction.

Chronic inflammation may also be one of the causes of enhanced cerebrovascular permeability. One of the indications of inflammation is a microvascular leakage of plasma substances and proteins and their deposition in the subendothelial matrix and interstitium [49]. We have shown a reduction in some endothelial junction proteins, along with increased endothelial layer permeability to albumin, caused by an elevated level of Fg, which is known to be associated with inflammation [57,58]. One of the possible mechanisms of increased paracellular transport can be explained by the findings that activity of inflammatory matrix metalloproteinases (MMPs) increases in many neurodegenerative diseases and after ischemic central nervous system (CNS) injury [59,60]. MMPs directly affect junction proteins and basement membrane extracellular matrix proteins [59]. Involvement of MMP-9 in increased extravascular deposition of Fg and an accompanied reduction in short-term memory has been found during TBI [61]. The levels of many adherence junctions and tight junction proteins are reduced in various neurodegenerative diseases, such as in AD and other diseases associated with dementia [44], amyotrophic lateral sclerosis [62], MS [63], and some animal models of neurodegeneration, such as aging [64].

4.2. Transcellular

Another pathway of cerebrovascular permeability, transcellular transport, includes caveolar transcytosis. While small molecules mainly use the paracellular pathway, high molecular weight proteins (e.g., Fg) cross the vessel wall mainly via caveolar transcytosis [52,65]. The main effect of this transport as opposed to the paracellular transport is that it can move relatively large proteins across the BBB. Crossing of the vascular wall for large proteins of the plasma via paracellular transport requires physically opening endothelial junctions wide enough to allow extravasation of these proteins.

On the other hand, caveolae that can be ~30–80 nm in diameter [66–68], with the neck diameter reaching ~56 nm in ECs [69], can accommodate large proteins. For example, albumin and Fg, with their Stokes–Einstein radiuses of about 3.5 nm [70,71] and 8.4 nm [72], respectively, can easily fit into a caveola and be transported across the BBB.

Among the continuous endothelium found in many types of tissue, including lung, muscle, and brain, the ECs in the brain have more restricted permeability [73]. The critical characteristic of brain endothelium is that it establishes the barrier limits for the diffusion of blood-borne solutes and restricts molecular exchange [73]. These features include specialized tight junctions that restrict diffusion of molecules, a small number of endocytic vesicles, and lowered rates of transcytosis relative to peripheral vasculature [44]. Thus, in normal conditions, caveolar transcytosis is quite low and very little macromolecular cargo crosses the cerebral capillary endothelium [55]. However, during inflammation, activation of ECs results in enhanced caveolar transcytosis that can have devastating effects on brain cells [74–76].

Alterations in endothelial layer integrity, caveolar transcytosis, and the basal membrane result in the accumulation of high molecular weight proteins, normally found in plasma, in the extravascular space [77]. One such protein is Fg [77]. It is evident that deposition of Fg in the extravascular space of brain tissue during inflammatory diseases such as AD [78] and TBI [8,79] is associated with a decline in memory. Enhanced deposition of Fg results in favorable conditions for the formation of Fg-containing protein complexes such as Fg-amyloid beta (Aβ) [75,78,80] and Fg-cellular prion protein (PrP^C) [81]. It has been shown that PrP^C can be endocytosed via caveolae [82]. We found that at an elevated level (e.g., during inflammation), Fg is transcytosed [74], is extravasated by caveolae [75], and then can directly interact with PrP^C [83]. It is known that, during neurodegenerative diseases, endogenous PrP^C undergoes a transformation to a conformationally altered scrapie prion protein (PrP^{Sc}) that accumulates in the brain as insoluble aggregates [84]. The binding of Fg to PrP^{Sc} has also been documented [85]. It has also been found that PrP^C can bind readily to Aβ, indicating that it may act as a receptor that initiates a chain of events leading to neuronal destruction [86]. In addition, it has been shown that the specific interaction of Fg with Aβ [75,80] modifies Fg's structure, leading to an abnormal fibrin clot formation more resistant to degradation [80,87]. Combined, these results suggest that a possible interaction of extravasated Fg with PrP^C and Aβ may result in the formation of aggregates highly resistant to degradation and lead to the neurodegeneration seen during neuroinflammatory diseases. In addition, the deposition of Fg and the formation of Fg-containing protein complexes in the extravascular space of the brain results in increased water transport and its accumulation in the interstitium, leading to the formation of edema and the resultant neurodegeneration [88].

5. Acute Phase Proteins in VCID

As a result of the development of systemic inflammation, blood plasma proteins such as albumin, Fg, C-reactive protein (CRP), and possibly some other high molecular weight acute reactant proteins may contribute to VCID. In response to injury and infection, at the expense of albumin synthesis, the liver enhances the synthesis of certain plasma proteins collectively known as acute phase proteins (APPs) [89]. The magnitude of the increase in the levels of these proteins varies. While CRP and serum amyloid A (SAA) can reach plasma levels of several hundred to a thousand-fold following acute inflammation, levels of haptoglobin and Fg do not increase more than two to tenfold [89,90]. Moreover, while the levels of CRP and SAA rapidly return to their normal range after the inflammation subsides, the levels of haptoglobin and Fg stay elevated for more than 21 days [90].

CRP is a 21 kD protein that has a similar structure to SAA [91]. Measuring the levels of CRP has been a standard of care in the clinic that can be a useful objective index to monitor the effectiveness of a therapy for a disease (inflammatory). At a normal level (0.8–9 μg/mL), CRP does not affect the BBB permeability [91]. On the other hand, it has been shown that when the level of CRP exceeds 2.5 μg/mL, it increases paracellular permeability of the BBB, affecting function but not the level of expression of tight junction proteins [91].

SAA is a small protein with a molecular weight of 12.5 kD that can be found in the blood of healthy individuals at the level of 20–50 μg/mL [92]. In subclinical inflammation, and for patients receiving glucocorticoid or immunosuppressive therapy, it has been suggested that SAA is a more sensitive biomarker than CRP [93]. It has been shown that Apo-SAA dose dependently increased the rat brain ECs permeability, shown by a significant reduction in transendothelial electrical resistance [94]. Interestingly, it has been shown that the circulating lipid free form of the SAA in the human plasma is 100 times lesser than that of SAA associated with high-density lipoprotein (HDL) [92]. Furthermore, the SAA-mediated impairment of the BBB was shown to be inhibited by the addition of HDL related to SAA in plasma [94]. Whether the free form of native SAA impairs the integrity of the BBB in pathological conditions remains unclear [94]. Given the SAA characteristic of high lipophilicity and the fact that most of the circulating SAA is associated with HDL, it is suggested that only a small amount of lipid-free SAA plays a major role in the BBB permeability changes compared to the other APPs.

Haptoglobin is an acute phase glycoprotein that can be found in the serum of all mammals [95]. As it binds to free hemoglobin (Hb) with a high affinity, haptoglobin's primary function is to facilitate Hb clearance. Hb is the prominent blood protein involved in transporting oxygen in the circulation. When unbound to haptoglobin and in the absence of other clearance mechanisms, free Hb can catalyze the formation of free radicals and mediate oxidative damages [95]. Haptoglobin is primarily synthesized in the liver. However, it has been shown that oligodendroglia can also synthesize haptoglobin, releasing it into the extracellular space, where it shows protective effects on brain cells from damages mediated by hemolytic product during intracerebral hemorrhage [96]. Haptoglobin production has also been described in other tissues, such as lung, skin, and kidney, during inflammatory conditions [97]. Although zonulin, a pre-haptoglobin precursor protein, has been shown to enhance small intestinal permeability, its direct effect in mediating BBB permeability has been questioned [98]. There are contrasting findings where preclinical in vitro data showed that zonulin potentially impaired BBB permeability [99], while other studies did not find evidence of its significant contribution in BBB permeability changes [98]. Taking into consideration that haptoglobin is found at a very low level in the normal brain [96] and that it has been shown to protect against Hb-induced toxicity [100], the prevailing role of haptoglobin can be considered to be neuroprotection.

Fg is an acute phase reactant protein that is increased during inflammation [90]. The blood content of Fg increases not only during neuroinflammatory diseases such as AD [101], MS [102], TBI [103], or stroke) [104], but also during other inflammatory diseases such as cardiovascular diseases [105,106] and cancer [107]. It has been widely shown that Fg and its derivative fibrin are not only markers of inflammation [90], but also cause inflammatory responses [57,108–111]. The cause-and-effect relationship between elevated blood levels of Fg (HFg) and cardiovascular disease has been shown, and HFg is presumed to be more than just a byproduct of an inflammatory cardiovascular disease. It may independently or interactively modulate the severity and/or the progression of cardiovascular disease [105]. Fg deposition in brain parenchyma has been documented during conditions with an impaired BBB, such as MS [112]. In fact, extravascular deposition of Fg in the brain parenchyma seen in autopsy tissue samples of patients who suffered from MS is indicative of BBB impairment [113].

Changes in blood rheological properties that are caused by changes in blood viscosity, blood flow, RBC aggregation, leukocyte activity, and platelet thrombogenesis are associated with HFg. A significant role of Fg in blood viscosity changes has long been known [114–117]. In addition, it is established that Fg is directly involved in platelet thrombogenesis [118]. We have shown that HFg that occurs during hypertension (an inflammatory disease) enhanced development of platelet thrombogenesis [119,120]. Interaction of Fg and leukocytes during their activation has been well defined [42,121]. In addition, the role of Fg in increased RBC aggregation has been well established [122]. We have shown that a direct interaction of Fg with erythrocytes has a significant effect on RBC aggregation [123]. Furthermore, a direct correlation between blood viscosity and RBC aggregation during hypertension is well established [124,125]. Combined, these effects of Fg can easily result in development of hemostasis during inflammatory diseases and cause vasculo-neuronal uncoupling in the brain.

Specific interaction of Fg with the microvascular endothelium and the resultant vasoconstriction has been shown [126]. We have also shown that at elevated levels, Fg can activate ECs [111] and enhance caveolar transcytosis of proteins [127], resulting in the increased BBB permeability seen during TBI [128]. While out of vasculature, Fg can interact with and activate astrocytes [129,130] and, by directly interacting with neurons, generate (ROS), NO, and mitochondrial superoxide in these cells [83]. Moreover, as it is converted to fibrin in the extravascular space, it induces perivascular microglial clustering, promotes demyelination, and promotes dendrite and spine elimination in neurons, which has been shown to be associated with neurodegeneration and reduced neuronal density [131,132].

Although Fg is mainly synthesized and generated in the liver [133], in addition to being situated in plasma, it is accumulated in α-granules of platelets [134,135]. The release of Fg from α-granules [134] occurs slower than secretion from dense granules [136] and possibly as a second phase of platelet activation after content of the dense granules is released. Fg deposited on activated endothelial cells can become a binding site for even non-activated platelets via their surface receptor $\alpha_{IIb}\beta_3$ [137]. It has been known that RBC aggregation can be promoted by several plasma proteins, such as Fg, α2-macroglobulin, and immunoglobulins M and G [138]. However, it was found that the only protein that had an effect on RBC aggregation on a biologically relevant level was Fg [138]. We have found that Fg had a specific interaction with RBCs via, most likely, integrin-like receptors on the surface of erythrocytes and promoted RBC aggregation [123]. Combined, these results suggest that during inflammation, when its blood content is elevated, Fg can be involved in platelet thrombogenesis and RBC aggregation, leading to blood flow reduction and decreasing supply to neural tissue with necessary nutrients, resulting in neurodegeneration. These findings suggest that Fg can be one of the most prominent agents in the circulatory system involved in vascular effects of neurodegeneration and memory reduction, i.e., in VCID. Some inflammation-induced and VCI-mediated mechanisms involved in memory decline are presented in Figure 1.

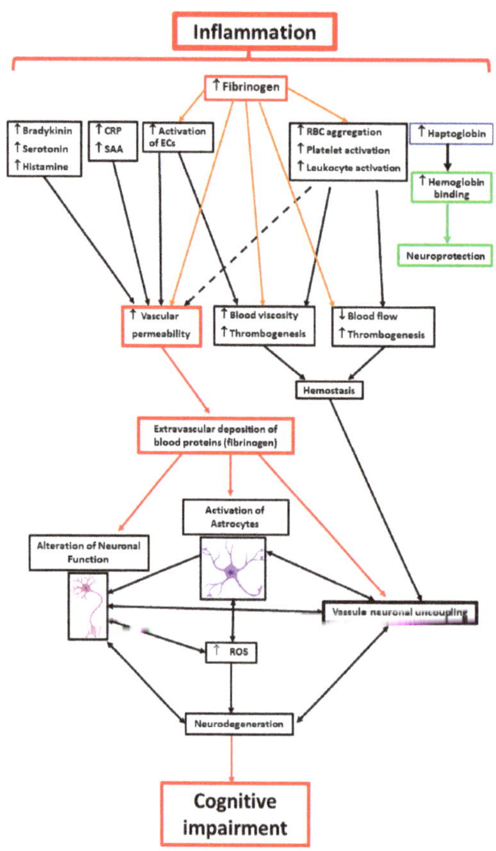

Figure 1. Schematic representation of the hypothesis of inflammation-induced vascular cognitive impairment. Among other factors that can affect vasculo-neuronal uncoupling, the factor that stands

out is fibrinogen (Fg). It can be involved in changes of vascular permeability in various ways. Changes in vascular permeability result in direct activation of astrocytes and neurons through increased deposition of blood proteins (particularly of Fg) in the extravascular space, leading to neurodegeneration and a reduction in memory. Orange arrows indicate direct effects of Fg. Red arrows indicate direct effects. Dotted arrow indicates indirect effect. Green boxes and the arrow define anti-inflammatory pathway, while red boxes emphasize effects with strongest effects in VCID. Abbreviations: RBC—red blood cells, ECs—endothelial cells, CRP—C reactive protein, SAA—serum amyloid A, ROS—reactive oxygen species.

6. Other Proteins Involved in VCID

It has been shown that Aβ is generated in both brain and peripheral tissues and is released into the circulatory system [139], where its level is correlated with increased risk of AD development [140–142]. Blood-derived Aβ can enter the brain tissue and cause neuronal dysfunction [143,144]. Strong association of Aβ peptide with Fg was linked to severity of AD [80]. A correlation of Aβ pathology and impairment in memory during TBI has been suggested [145]. In addition, there is evidence that links the occurrence of TBI to the onset and progression of AD and cognitive impairment [146]. Repetitive mild TBI has been shown to accelerate Aβ deposition, lipid peroxidation, and cognitive impairment in a transgenic mouse model of AD [146].

The extravasation of Aβ has a major role in the accumulation of Aβ in the CNS [147]. There is also evidence that Aβ accumulation itself affects brain vasculature and changes the function of the NVU [144]. It has been shown that the association of Fg and Aβ alters thrombosis [78] and results in the formation of clots with an abnormal structure and resistance to fibrinolysis [87]. Although the formation of complexes containing Fg/fibrin [148] and Aβ [149,150] is the hallmark of AD [78,80,148], some evidence indicates that the content of Aβ alone has a limited effect on memory [151,152]. These results suggest that formation of Fg–Aβ complexes can have a greater effect on memory reduction than the extravascular deposition of Fg or Aβ alone.

Although Aβ is strongly associated with AD [80], there is evidence of the greater role of cellular prion protein (PrP^C) in memory reduction [151,152]. In addition, the role of PrP^C in TBI-associated memory reduction has been shown [153]. PrP^C is a cell surface, glycosylphosphatidylinositol anchored glycoprotein, abundantly expressed in neurons, glial cells [154], and endothelial cells [155]. It was shown that PrP^C participates in Aβ transcytosis through the BBB [156] and Aβ-mediated memory reduction during TBI [151]. We have recently shown that Fg can specifically interact with PrP^C on the surface of astrocytes [81,130]. Moreover, our data showed that Fg can form a complex with PrP^C in the extravascular space of mouse brains during mild-to-moderate TBI and was accompanied by short-term memory reduction [61]. TBI causes Aβ-PrP^C-Fyn kinase activation, which also induces tau phosphorylation [153]. Fyn kinase is localized in the postsynaptic density of the brain, which is the primary site of signal transduction and processing, and its activity is linked to synaptic function [157,158].

It is well known that the deposition of Aβ plaques and tau-associated neurofibrillary tangles are a hallmark of AD. Both the deposition of amyloid and tau proteins have been implicated in the memory decline present during AD [159]. In fact, the direct interaction between the Aβ and specific regions of tau has recently been defined, suggesting that targeting only Aβ or only tau may not be the best treatment strategy during AD [160]. However, there are some conflicting data regarding effects of Aβ and tau on memory. Some data indicate that the content of tau, but not the levels of Aβ, in cerebrospinal fluid is associated with the severity of short-term memory impairment present in AD patients [161]. Other studies indicate that hyperphosphorylation of tau is not directly responsible for Aβ-induced neurodegeneration in vitro [162], and amyloid deposition has a greater association with microglial activation and memory reduction than tau pathology does [163]. Similarly, it has been shown that tau has a limited role in Aβ-induced memory impairment [164]. These results suggest that Aβ may have a greater effect than tau in memory reduction. This point can be substantiated by the fact that tau is exclusively present in nonvascular brain cells while Aβ, in addition to its presence in brain cells, can also be extravasated from the blood stream to further increase its overall content in the brain during neuroinflammation.

The neuropeptide substance P (SP) was first identified in the brain and gut in the early 1930s by Euler and Gaddum [165]. It is widely distributed in the central, peripheral, and enteric nervous systems and acts as a neurotransmitter and a neuromodulator that has a potent hypotensive property. It has been shown that during acute brain injury, SP was found perivascularly and linked to vasogenic edema formation [166]. It is suggested that SP plays a major role in secondary injury during neuroinflammatory diseases such as TBI. It has been found that SP mediates an increase in vascular permeability leading to the formation of edema [167]. Furthermore, SP itself may have a direct role in learning and memory, as it has been shown that blocking SP receptor expression in the hippocampus in the neostriatum impairs learning and memory in tested rats [168].

7. Some Other Inflammatory Agents Commonly Associated with BBB Disruption

Several other inflammatory mediators have been involved in modulation of BBB permeability. These are bradykinin, which increases BBB permeability by acting on B_2 receptors, serotonin, which affects BBB permeability in some but not all studies, and histamine, one of the few CNS neurotransmitters consistently associated with BBB impairment [169]. Increased BBB permeability leads to the neurodegeneration and reduction in memory seen during diseases such as AD [144] and TBI [7,8,61,75].

8. Conclusions

In conclusion, we would like to emphasize that most of the systemic effects that are conveyed to neurons originate in the circulation. These effects exclude genetic, epigenetic, and some sensory effects that could directly affect neuronal function, which are not considered in the present review. Systemic alterations undeniably affect the composition of blood and properties of blood cells, plasma proteins, vascular cells, and/or vessels affecting blood flow. All these changes can positively influence the BBB integrity. As a result, any pathological alteration of the BBB property results in abnormal effects in glial and neuronal functions, eventually leading to possible neuroinflammation and neurodegeneration with high incidence of memory reduction. All these emphasize an imperative importance to study mechanisms of cerebrovascular permeability during various neurodegenerative diseases. Studying the link between BBB dysfunction and dementia might be key in finding the right window for intervention. To help accelerate the development of new and existing biomarkers for VCID, the MarkVCID consortium was formed under cooperative agreements with the NINDS and the National Institute on Aging in 2016 [170], involving multicenter studies whose mission is to identify and validate biomarkers for VCID. The overall goal of the consortium is to deliver high-quality biomarkers ready for use in clinical trials aimed at generating scientific breakthroughs in deeper understanding and treatment of VCID. New discoveries will not only open the possibility of restoring and/or maintaining properties of the intact BBB, but also will exploit ways for safe delivery of drugs through the BBB during various pathologies. Currently diagnosis for VCID is limited to clinical signs of dementia and/or magnetic resonance imaging, which may take place later in the course of the disease as a part of intervention or prevention. Therefore, developing biomarkers, preferably noninvasive markers, for early detection and prevention of VCID is an imperative goal for the future.

Author Contributions: Conceptualization, D.L. and N.S.; writing—original draft preparation, D.L. and N.S.; writing—review and editing, D.L., N.S. and J.B.; funding acquisition, D.L. All authors have read and agreed to the published version of the manuscript.

Funding: This work was supported by the NIH grant # HL146832 and USF COM HSC-18330 Funds.

Institutional Review Board Statement: Not applicable.

Informed Consent Statement: Not applicable.

Data Availability Statement: Not applicable.

Conflicts of Interest: The authors declare no conflict of interest.

Abbreviations

RBCs—red blood cells, ECs—endothelial cells, CRP—C reactive protein, SAA—serum amyloid A, ROSs—reactive oxygen species.

References

1. Corriveau, R.A.; Bosetti, F.; Emr, M.; Gladman, J.T.; Koenig, J.I.; Moy, C.S.; Pahigiannis, K.; Waddy, S.P.; Koroshetz, W. The Science of Vascular Contributions to Cognitive Impairment and Dementia (VCID): A Framework for Advancing Research Priorities in the Cerebrovascular Biology of Cognitive Decline. *Cell. Mol. Neurobiol.* **2016**, *36*, 281–288. [CrossRef] [PubMed]
2. Ochocinska, M.J.; Zlokovic, B.V.; Searson, P.C.; Crowder, A.T.; Kraig, R.P.; Ljubimova, J.Y.; Mainprize, T.G.; Banks, W.A.; Warren, R.Q.; Kindzelski, A.; et al. NIH workshop report on the trans-agency blood–brain interface workshop 2016: Exploring key challenges and opportunities associated with the blood, brain and their interface. *Fluids Barriers CNS* **2017**, *14*, 12. [CrossRef] [PubMed]
3. Gorelick Philip, B.; Scuteri, A.; Black, S.; Decarli, C.; Greenberg, S.; Iadecola, C.; Launer, L.; Laurent, S.; Lopez, O.; Nyenhuis, D.; et al. Vascular contributions to cognitive impairment and dementia: A statement for healthcare professionals from the american heart association/american stroke association. *Stroke* **2011**, *42*, 2672–2713. [CrossRef] [PubMed]
4. Wolters, F.J.; Ikram, M.A. Epidemiology of Vascular Dementia. *Arterioscler. Thromb. Vasc. Biol.* **2019**, *39*, 1542–1549. [CrossRef]
5. Yang, S.H.; Gustafson, J.; Gangidine, M.; Stepien, D.; Schuster, R.; Pritts, T.A.; Goodman, M.D.; Remick, D.G.; Lentsch, A.B. A murine model of mild traumatic brain injury exhibiting cognitive and motor deficits. *J. Surg. Res.* **2013**, *184*, 981–988. [CrossRef]
6. Gorman, L.; Shook, B.; Becker, D. Traumatic brain injury produces impairments in long-term and recent memory. *Brain Res.* **1993**, *614*, 29–36. [CrossRef]
7. Muradashvili, N.; Charkviani, M.; Sulimai, N.; Tyagi, N.; Crosby, J.; Lominadze, D. Effects of fibrinogen synthesis inhibition on vascular cognitive impairment during traumatic brain injury in mice. *Brain Res.* **2020**, *1751*, 147208. [CrossRef]
8. Muradashvili, N.; Tyagi, S.C.; Lominadze, D. Localization of fibrinogen in the vasculo-astrocyte interface after cortical contusion injury in mice. *Brain Sci.* **2017**, *7*, 77. [CrossRef]
9. Benedict, R.H.B.; Amato, M.P.; DeLuca, J.; Geurts, J.J.G. Cognitive impairment in multiple sclerosis: Clinical management, MRI, and therapeutic avenues. *Lancet Neurol.* **2020**, *19*, 860–871. [CrossRef]
10. Doyle, K.P.; Buckwalter, M.S. Immunological mechanisms in poststroke dementia. *Curr. Opin. Neurol.* **2020**, *33*, 30–36. [CrossRef]
11. McColl, B.W.; Allan, S.M.; Rothwell, N.J. Systemic infection, inflammation and acute ischemic stroke. *Neuroscience* **2009**, *158*, 1049–1061. [CrossRef]
12. Fiala, M.; Liu, Q.N.; Sayre, J.; Pop, V.; Brahmandam, V.; Graves, M.C.; Vinters, H.V. Cyclooxygenase-2-positive macrophages infiltrate the Alzheimer's disease brain and damage the blood–brain barrier. *Eur. J. Clin. Investig.* **2002**, *32*, 360–371. [CrossRef]
13. Hay, J.R.; Johnson, V.E.; Young, A.M.; Smith, D.H.; Stewart, W. Blood-brain barrier disruption is an early event that may persist for many years after traumatic brain injury in humans. *J. Neuropathol. Exp. Neurol.* **2015**, *74*, 1147–1157.
14. van de Haar, H.J.; Burgmans, S.; Hofman, P.A.M.; Verhey, F.R.J.; Jansen, J.F.A.; Backes, W.H. Blood–brain barrier impairment in dementia: Current and future in vivo assessments. *Neurosci. Biobehav. Rev.* **2015**, *49*, 71–81. [CrossRef] [PubMed]
15. Gorelick, P.B.; Counts, S.E.; Nyenhuis, D. Vascular cognitive impairment and dementia. *Biochim. Biophys. Acta (BBA)-Mol. Basis Dis.* **2016**, *1862*, 860–868. [CrossRef] [PubMed]
16. Montagne, A.; Zhao, Z.; Zlokovic, B.V. Alzheimer's disease: A matter of blood-brain barrier dysfunction? *J. Exp. Med.* **2017**, *214*, 3151–3169. [CrossRef]
17. Yang, L.; Song, J.; Nan, D.; Wan, Y.; Guo, H. Cognitive Impairments and blood-brain Barrier Damage in a Mouse Model of Chronic Cerebral Hypoperfusion. *Neurochem. Res.* **2022**, *47*, 3817–3828. [CrossRef] [PubMed]
18. Ghirnikar, R.S.; Lee, Y.L.; Eng, L.F. Inflammation in Traumatic Brain Injury: Role of Cytokines and Chemokines. *Neurochem. Res.* **1998**, *23*, 329–340. [CrossRef]
19. Morganti-Kossmann, M.C.; Rancan, M.; Stahel, P.F.; Kossmann, T. Inflammatory response in acute traumatic brain injury: A double-edged sword. *Curr. Opin. Crit. Care* **2002**, *8*, 101–105. [CrossRef]
20. Cederberg, D.; Siesjö, P. What has inflammation to do with traumatic brain injury? *Child's Nerv. Syst.* **2010**, *26*, 221–226. [CrossRef]
21. Paul, J.; Strickland, S.; Melchor, J.P. Fibrin deposition accelerates neurovascular damage and neuroinflammation in mouse models of Alzheimer's disease. *J. Exp. Med.* **2007**, *204*, 1999–2008. [CrossRef]
22. Frohman, E.M.; Racke, M.K.; Raine, C.S. Multiple Sclerosis—The Plaque and Its Pathogenesis. *N. Engl. J. Med.* **2006**, *354*, 942–955. [CrossRef]

23. Hauser, S.L.; Oksenberg, J.R. The Neurobiology of Multiple Sclerosis: Genes, Inflammation, and Neurodegeneration. *Neuron* **2006**, *52*, 61–76. [CrossRef] [PubMed]
24. Sandoval, K.E.; Witt, K.A. Blood-brain barrier tight junction permeability and ischemic stroke. *Neurobiol. Dis.* **2008**, *32*, 200–219 [CrossRef] [PubMed]
25. Sevush, S.; Jy, W.; Horstman, L.L.; Mao, W.-W.; Kolodny, L.; Ahn, Y.S. Platelet Activation in Alzheimer Disease. *Arch. Neurol.* **1998**, *55*, 530–536. [CrossRef] [PubMed]
26. Behari, M.; Shrivastava, M. Role of platelets in neurodegenerative diseases: A universal pathophysiology. *Int. J. Neurosci.* **2013**, *123*, 287–299. [CrossRef]
27. Sillesen, M.; Johansson, P.I.; Rasmussen, L.S.; Jin, G.; Jepsen, C.H.; Imam, A.M.; Hwabejire, J.; Lu, J.; Duggan, M.; Velmahos, G.; et al. Platelet activation and dysfunction in a large-animal model of traumatic brain injury and hemorrhage. *J. Trauma Acute Care Surg.* **2013**, *74*, 1252–1259. [CrossRef]
28. Sheremata, W.A.; Jy, W.; Horstman, L.L.; Ahn, Y.S.; Alexander, J.S.; Minagar, A. Evidence of platelet activation in multiple sclerosis. *J. Neuroinflamm.* **2008**, *5*, 27. [CrossRef]
29. Canobbio, I.; Visconte, C.; Oliviero, B.; Guidetti, G.; Zarà, M.; Pula, G.; Torti, M. Increased platelet adhesion and thrombus formation in a mouse model of Alzheimer's disease. *Cell. Signal.* **2016**, *28*, 1863–1871. [CrossRef]
30. Schwarzmaier, S.M.; Kim, S.W.; Trabold, R.; Plesnila, N. Temporal Profile of Thrombogenesis in the Cerebral Microcirculation after Traumatic Brain Injury in Mice. *J. Neurotrauma* **2010**, *27*, 121–130. [CrossRef]
31. Birdsill, A.C.; Carlsson, C.M.; Willette, A.A.; Okonkwo, O.C.; Johnson, S.C.; Xu, G.; Oh, J.M.; Gallagher, C.L.; Koscik, R.L.; Jonaitis, E.M.; et al. Low cerebral blood flow is associated with lower memory function in metabolic syndrome. *Obesity* **2013**, *21*, 1313–1320. [CrossRef]
32. Horstman, L.L.; Jy, W.; Ahn, Y.S.; Zivadinov, R.; Maghzi, A.H.; Etemadifar, M.; Steven Alexander, J.; Minagar, A. Role of platelets in neuroinflammation: A wide-angle perspective. *J. Neuroinflamm.* **2010**, *7*, 10. [CrossRef]
33. Lassmann, H.; Rössler, K.; Zimprich, F.; Vass, K. Expression of Adhesion Molecules and Histocompatibility Antigens at the Blood-Brain Barrier. *Brain Pathol.* **1991**, *1*, 115–123. [CrossRef]
34. Losy, J.; Niezgoda, A.; Wender, M. Increased serum levels of soluble PECAM-1 in multiple sclerosis patients with brain gadolinium-enhancing lesions. *J. Neuroimmunol.* **1999**, *99*, 169–172. [CrossRef] [PubMed]
35. Feng, D.; Nagy, J.A.; Pyne, K.; Dvorak, H.F.; Dvorak, A.M. Platelets Exit Venules by a Transcellular Pathway at Sites of F–Met Peptide–Induced Acute Inflammation in Guinea Pigs. *Int. Arch. Allergy Immunol.* **1998**, *116*, 188–195. [CrossRef]
36. Zuchtriegel, G.; Uhl, B.; Puhr-Westerheide, D.; Pörnbacher, M.; Lauber, K.; Krombach, F.; Reichel, C.A. Platelets Guide Leukocytes to Their Sites of Extravasation. *PLoS Biol.* **2016**, *14*, e1002459. [CrossRef]
37. Carvalho-Tavares, J.; Hickey, M.J.; Hutchison, J.; Michaud, J.; Sutcliffe, I.T.; Kubes, P. A Role for Platelets and Endothelial Selectins in Tumor Necrosis Factor-α–Induced Leukocyte Recruitment in the Brain Microvasculature. *Circ. Res.* **2000**, *87*, 1141–1148. [CrossRef]
38. Weirather, J.; Frantz, S. Chapter 2—Role of the Innate Immune System in Ischemic Heart Failure. In *Inflammation in Heart Failure*; Blankesteijn, W.M., Altara, R., Eds.; Academic Press: Boston, MA, USA, 2015; pp. 19–38.
39. Ley, K.; Laudanna, C.; Cybulsky, M.I.; Nourshargh, S. Getting to the site of inflammation: The leukocyte adhesion cascade updated. *Nat. Rev. Immunol.* **2007**, *7*, 678–689. [CrossRef] [PubMed]
40. Forsyth, C.B.; Solovjov, D.A.; Ugarova, T.P.; Plow, E.F. Integrin αMβ2-Mediated Cell Migration to Fibrinogen and Its Recognition Peptides. *J. Exp. Med.* **2001**, *193*, 1123–1134. [CrossRef] [PubMed]
41. Roche, Y.; Pasquier, D.; Rambeaud, J.-J.; Seigneurin, D.; Duperray, A. Fibrinogen mediates bladder cancer cell migration in an ICAM-1-dependent pathway. *Thromb. Haemost.* **2017**, *89*, 1089–1097. [CrossRef]
42. Languino, L.R.; Plescia, J.; Duperrray, A.; Brian, A.A.; Plow, E.F.; Geltosky, J.E.; Alteri, D.C. Fibrinogen mediates leukocyte adhesion to vascular endothelium through an ICAM-1-dependent pathway. *Cell* **1993**, *73*, 1423–1434. [CrossRef] [PubMed]
43. Stary, H.C.; Chandler, A.B.; Dinsmore, R.E.; Fuster, V.; Glagov, S.; Insull, W.; Rosenfeld, M.E.; Schwartz, C.J.; Wagner, W.D.; Wissler, R.W. A Definition of Advanced Types of Atherosclerotic Lesions and a Histological Classification of Atherosclerosis. *Circulation* **1995**, *92*, 1355–1374. [CrossRef] [PubMed]
44. Zlokovic, B.V. The Blood-Brain Barrier in Health and Chronic Neurodegenerative Disorders. *Neuron* **2008**, *57*, 178–201. [CrossRef]
45. Wu, J.; Hua, Y.; Keep, R.F.; Schallert, T.; Hoff, J.T.; Xi, G. Oxidative brain injury from extravasated erythrocytes after intracerebral hemorrhage. *Brain Res.* **2002**, *953*, 45–52. [CrossRef] [PubMed]
46. Zlokovic, B.V. Neurovascular pathways to neurodegeneration in Alzheimer's disease and other disorders. *Nat. Rev. Neurosci.* **2011**, *12*, 723–738. [CrossRef]
47. Balla, J.; Jacob, H.S.; Balla, G.; Nath, K.; Eaton, J.W.; Vercellotti, G.M. Endothelial-cell heme uptake from heme proteins: Induction of sensitization and desensitization to oxidant damage. *Proc. Natl. Acad. Sci. USA* **1993**, *90*, 9285–9289. [CrossRef]
48. Schreibelt, G.; Kooij, G.; Reijerkerk, A.; van Doorn, R.; Gringhuis, S.I.; van der Pol, S.; Weksler, B.B.; Romero, I.A.; Couraud, P.-O.; Piontek, J.; et al. Reactive oxygen species alter brain endothelial tight junction dynamics via RhoA, PI3 kinase, and PKB signaling. *FASEB J.* **2007**, *21*, 3666–3676. [CrossRef]
49. Mehta, D.; Malik, A.B. Signaling mechanisms regulating endothelial permeability. *Physiol. Rev.* **2006**, *86*, 279–367. [CrossRef]
50. Stan, R.-V.; Marion, K.; Palade, G.E. PV-1 is a component of the fenestral and stomatal diaphragms in fenestrated endothelia. *Proc. Natl. Acad. Sci. USA* **1999**, *96*, 13203–13207. [CrossRef]

51. Stan, R.V. Endothelial stomatal and fenestral diaphragms in normal vessels and angiogenesis. *J. Cell. Mol. Med.* **2007**, *11*, 621–643. [CrossRef]
52. Komarova, Y.; Malik, A.B. Regulation of endothelial permeability via paracellular and transcellular transport pathways. *Annu. Rev. Physiol.* **2010**, *72*, 463–493. [CrossRef]
53. Erickson, M.A.; Banks, W.A. Transcellular routes of blood–brain barrier disruption. *Exp. Biol. Med.* **2022**, *247*, 788–796. [CrossRef]
54. Lochhead, J.J.; Yang, J.; Ronaldson, P.T.; Davis, T.P. Structure, Function, and Regulation of the Blood-Brain Barrier Tight Junction in Central Nervous System Disorders. *Front. Physiol.* **2020**, *11*, 914. [CrossRef]
55. Tuma, P.L.; Hubbard, A.L. Transcytosis: Crossing cellular barriers. *Physiol. Rev.* **2003**, *83*, 871–932. [CrossRef] [PubMed]
56. Unterberg, A.W.; Stover, J.; Kress, B.; Kiening, K.L. Edema and brain trauma. *Neuroscience* **2004**, *129*, 1019–1027. [CrossRef] [PubMed]
57. Tyagi, N.; Roberts, A.M.; Dean, W.L.; Tyagi, S.C.; Lominadze, D. Fibrinogen induces endothelial cell permeability. *Mol. Cell. Biochem.* **2008**, *307*, 13–22. [CrossRef]
58. Patibandla, P.K.; Tyagi, N.; Dean, W.L.; Tyagi, S.C.; Roberts, A.M.; Lominadze, D. Fibrinogen induces alterations of endothelial cell tight junction proteins. *J. Cell. Physiol.* **2009**, *221*, 195–203. [CrossRef] [PubMed]
59. Rosenberg, G.A. Matrix metalloproteinases and their multiple roles in neurodegenerative diseases. *Lancet Neurol.* **2009**, *8*, 205–216. [CrossRef]
60. Cheng, T.; Petraglia, A.L.; Li, Z.; Thiyagarajan, M.; Zhong, Z.; Wu, Z.; Liu, D.; Maggirwar, S.B.; Deane, R.; Fernández, J.A.; et al. Activated protein C inhibits tissue plasminogen activator–induced brain hemorrhage. *Nat. Med.* **2006**, *12*, 1278–1285. [CrossRef]
61. Muradashvili, N.; Benton, R.L.; Saatman, K.E.; Tyagi, S.C.; Lominadze, D. Ablation of matrix metalloproteinase-9 gene decreases cerebrovascular permeability and fibrinogen deposition post traumatic brain injury in mice. *Metab. Brain Dis.* **2015**, *30*, 411–426. [CrossRef]
62. Henkel, J.S.; Beers, D.R.; Wen, S.; Bowser, R.; Appel, S.H. Decreased mRNA expression of tight junction proteins in lumbar spinal cords of patients with ALS. *Neurology* **2009**, *72*, 1614–1616. [CrossRef]
63. Huber, J.D.; Egleton, R.D.; Davis, T.P. Molecular physiology and pathophysiology of tight junctions in the blood–brain barrier. *Trends Neurosci.* **2001**, *24*, 719–725. [CrossRef] [PubMed]
64. Bell, R.D.; Winkler, E.A.; Sagare, A.P.; Singh, I.; LaRue, B.; Deane, R.; Zlokovic, B.V. Pericytes control key neurovascular functions and neuronal phenotype in the adult brain and during brain aging. *Neuron* **2010**, *68*, 409–427. [CrossRef]
65. Muradashvili, N.; Khundmiri, S.J.; Tyagi, R.; Gartung, A.; Dean, W.L.; Lee, M.-J.; Lominadze, D. Sphingolipids affect fibrinogen-induced caveolar transcytosis and cerebrovascular permeability. *Am. J. Physiol. Cell Physiol.* **2014**, *307*, C169–C179. [CrossRef]
66. Massey, K.A.; Schnitzer, J.E. Targeting and imaging signature caveolar molecules in lungs. *Proc. Am. Thorac. Soc.* **2009**, *6*, 419–430. [CrossRef]
67. Parton, R.G.; del Pozo, M.A. Caveolae as plasma membrane sensors, protectors and organizers. *Nat. Rev. Mol. Cell Biol.* **2013**, *14*, 98–112. [CrossRef]
68. Schlörmann, W.; Steiniger, F.; Richter, W.; Kaufmann, R.; Hause, G.; Lemke, C.; Westermann, M. The shape of caveolae is omega-like after glutaraldehyde fixation and cup-like after cryofixation. *Histochem. Cell Biol.* **2010**, *133*, 223–228. [CrossRef] [PubMed]
69. Richter, T.; Floetenmeyer, M.; Ferguson, C.; Galea, J.; Goh, J.; Lindsay, M.R.; Morgan, G.P.; Marsh, B.J.; Parton, R.G. High-resolution 3D quantitative analysis of caveolar ultrastructure and caveola-cytoskeleton interactions. *Traffic* **2008**, *9*, 893–909. [CrossRef]
70. Marchant, R.E.; Kang, I.; Sit, P.S.; Zhou, Y.; Todd, B.A.; Eppell, S.J.; Lee, I. Molecular views and measurements of hemostatic processes using atomic force microscopy. *Curr. Protein Pept. Sci.* **2002**, *3*, 249–274. [CrossRef] [PubMed]
71. Fu, B.M.; Shen, S. Structural mechanisms of acute VEGF effect on microvessel permeability. *Am. J. Physiol.-Heart Circ. Physiol.* **2003**, *284*, H2124–H2135. [CrossRef]
72. Potschka, M. Universal calibration of gel permeation chromatography and determination of molecular shape in solution. *Anal. Biochem.* **1987**, *162*, 47–64. [CrossRef]
73. Ayloo, S.; Gu, C. Transcytosis at the blood–brain barrier. *Curr. Opin. Neurobiol.* **2019**, *57*, 32–38. [CrossRef] [PubMed]
74. Muradashvili, N.; Tyagi, R.; Lominadze, D. A dual-tracer method for differentiating transendothelial transport from paracellular leakage in vivo and in vitro. *Front. Physiol.* **2012**, *3*, 166–172. [CrossRef] [PubMed]
75. Muradashvili, N.; Tyagi, R.; Tyagi, N.; Tyagi, S.C.; Lominadze, D. Cerebrovascular disorders caused by hyperfibrinogenemia. *J. Physiol.* **2016**, *594*, 5941–5957. [CrossRef]
76. Sulimai, N.; Lominadze, D. Fibrinogen and Neuroinflammation During Traumatic Brain Injury. *Mol. Neurobiol.* **2020**, *57*, 4692–4703. [CrossRef] [PubMed]
77. Chodobski, A.; Zink, B.J.; Szmydynger-Chodobska, J. Blood-brain barrier pathophysiology in traumatic brain injury. *Transl. Stroke Res.* **2011**, *2*, 492–516. [CrossRef]
78. Cortes-Canteli, M.; Paul, J.; Norris, E.H.; Bronstein, R.; Ahn, H.J.; Zamolodchikov, D.; Bhuvanendran, S.; Fenz, K.M.; Strickland, S. Fibrinogen and β-Amyloid association alters thrombosis and fibrinolysis: A possible contributing factor to Alzheimer's Disease. *Neuron* **2010**, *66*, 695–709. [CrossRef]
79. Jenkins, D.R.; Craner, M.J.; Esiri, M.M.; DeLuca, G.C. The contribution of fibrinogen to inflammation and neuronal density in human traumatic brain injury. *J. Neurotrauma* **2018**, *35*, 2259–2271. [CrossRef]

80. Ahn, H.J.; Zamolodchikov, D.; Cortes-Canteli, M.; Norris, E.H.; Glickman, J.F.; Strickland, S. Alzheimer's disease peptide β-amyloid interacts with fibrinogen and induces its oligomerization. *Proc. Natl. Acad. Sci. USA* **2010**, *107*, 21812–21817. [CrossRef]
81. Charkviani, M.; Muradashvili, N.; Sulimai, N.H.; Lominadze, D. Fibrinogen—Cellular Prion Protein Complex Formation on Astrocytes. *J. Neurophysiol.* **2020**, *124*, 536–543. [CrossRef]
82. Peters, P.J.; Mironov, A.; Peretz, D.; van Donselaar, E.; Leclerc, E.; Erpel, S.; DeArmond, S.J.; Burton, D.R.; Williamson, R.A.; Vey, M.; et al. Trafficking of prion proteins through a caveolae-mediated endosomal pathway. *J. Cell Biol.* **2003**, *162*, 703–717. [CrossRef]
83. Sulimai, N.; Brown, J.; Lominadze, D. The Effects of Fibrinogen's Interactions with Its Neuronal Receptors, Intercellular Adhesion Molecule-1 and Cellular Prion Protein. *Biomolecules* **2021**, *11*, 1381. [CrossRef] [PubMed]
84. Prusiner, S.B. Prions. *Proc. Natl. Acad. Sci. USA* **1998**, *95*, 13363–13383. [CrossRef] [PubMed]
85. Fischer, M.B.; Roeckl, C.; Parizek, P.; Schwarz, H.P.; Aguzzi, A. Binding of disease-associated prion protein to plasminogen. *Nature* **2000**, *408*, 479–483. [CrossRef] [PubMed]
86. Lauren, J.; Gimbel, D.A.; Nygaard, H.B.; Gilbert, J.W.; Strittmatter, S.M. Cellular prion protein mediates impairment of synaptic plasticity by amyloid-[bgr] oligomers. *Nature* **2009**, *457*, 1128–1132. [CrossRef]
87. Zamolodchikov, D.; Strickland, S. Aβ delays fibrin clot lysis by altering fibrin structure and attenuating plasminogen binding to fibrin. *Blood* **2012**, *119*, 3342–3351. [CrossRef]
88. Stamatovic, S.; Keep, R.; Andjelkovic, A. Brain endothelial cell-cell junctions: How to "open" the blood brain barrier. *Curr. Neuropharmacol.* **2008**, *6*, 179–192. [CrossRef]
89. Malle, E.; De Beer, F.C. Human serum amyloid A (SAA) protein: A prominent acute-phase reactant for clinical practice. *Eur. J. Clin. Investig.* **1996**, *26*, 427–435. [CrossRef]
90. Gabay, C.; Kushner, I. Acute-phase proteins and other systemic responses to inflammation. *N. Engl. J. Med.* **1999**, *340*, 448–454. [CrossRef]
91. Hsuchou, H.; Kastin, A.J.; Mishra, P.K.; Pan, W. C-reactive protein increases BBB permeability: Implications for obesity and neuroinflammation. *Cell Physiol. Biochem.* **2012**, *30*, 1109–1119. [CrossRef]
92. Sack, G.H. Serum amyloid A—A review. *Mol. Med.* **2018**, *24*, 46. [CrossRef] [PubMed]
93. Sorić Hosman, I.; Kos, I.; Lamot, L. Serum Amyloid A in Inflammatory Rheumatic Diseases: A Compendious Review of a Renowned Biomarker. *Front. Immunol.* **2021**, *11*, 631299. [CrossRef] [PubMed]
94. Matsumoto, J.; Dohgu, S.; Takata, F.; Iwao, T.; Kimura, I.; Tomohiro, M.; Aono, K.; Kataoka, Y.; Yamauchi, A. Serum amyloid A-induced blood-brain barrier dysfunction associated with decreased claudin-5 expression in rat brain endothelial cells and its inhibition by high-density lipoprotein in vitro. *Neurosci. Lett.* **2020**, *738*, 135352. [CrossRef]
95. MacKellar, M.; Vigerust, D.J. Role of Haptoglobin in Health and Disease: A Focus on Diabetes. *Clin. Diabetes* **2016**, *34*, 148–157. [CrossRef] [PubMed]
96. Zhao, X.; Song, S.; Sun, G.; Strong, R.; Zhang, J.; Grotta, J.C.; Aronowski, J. Neuroprotective Role of Haptoglobin after Intracerebral Hemorrhage. *J. Neurosci.* **2009**, *29*, 15819–15827. [CrossRef]
97. Vanuytsel, T.; Vermeire, S.; Cleynen, I. The role of Haptoglobin and its related protein, Zonulin, in inflammatory bowel disease. *Tissue Barriers* **2013**, *1*, e27321. [CrossRef]
98. Stuart, C.M.; Varatharaj, A.; Winberg, M.E.; Galea, P.; Larsson, H.B.W.; Cramer, S.P.; Fasano, A.; Maherally, Z.; Pilkington, G.J.; Keita, Å.V.; et al. Zonulin and blood-brain barrier permeability are dissociated in humans. *Clin. Transl. Med.* **2022**, *12*, e965. [CrossRef]
99. Rahman, M.T.; Ghosh, C.; Hossain, M.; Linfield, D.; Rezaee, F.; Janigro, D.; Marchi, N.; van Boxel-Dezaire, A.H.H. IFN-γ, IL-17A, or zonulin rapidly increase the permeability of the blood-brain and small intestinal epithelial barriers: Relevance for neuro-inflammatory diseases. *Biochem. Biophys. Res. Commun.* **2018**, *507*, 274–279. [CrossRef]
100. Garland, P.; Morton, M.J.; Haskins, W.; Zolnourian, A.; Durnford, A.; Gaastra, B.; Toombs, J.; Heslegrave, A.J.; More, J.; Okemefuna, A.I.; et al. Haemoglobin causes neuronal damage in vivo which is preventable by haptoglobin. *Brain Commun.* **2020**, *2*, fcz053. [CrossRef]
101. van Oijen, M.; Witteman, J.C.; Hofman, A.; Koudstaal, P.J.; Breteler, M.M.B. Fibrinogen is associated with an increased risk of Alzheimer disease and vascular dementia. *Stroke* **2005**, *36*, 2637–2641. [CrossRef]
102. Miranda Acuña, J.; Hidalgo de la Cruz, M.; Ros, A.L.; Tapia, S.P.; Martínez Ginés, M.L.; de Andrés Frutos, C.D. Elevated plasma fibrinogen levels in multiple sclerosis patients during relapse. *Mult. Scler. Relat. Disord.* **2017**, *18*, 157–160. [CrossRef]
103. Kossmann, T.; Hans, V.H.; Imhof, H.G.; Stocker, R.; Grob, P.; Trentz, O.; Morganti-Kossmann, C. Intrathecal and serum interleukin-6 and the acute-phase response in patients with severe traumatic brain injuries. *Shock* **1995**, *4*, 311–317. [CrossRef] [PubMed]
104. del Zoppo, G.J.; Levy, D.E.; Wasiewski, W.W.; Pancioli, A.M.; Demchuk, A.M.; Trammel, J.; Demaerschalk, B.M.; Kaste, M.; Albers, G.W.; Ringelstein, E.B. Hyperfibrinogenemia and functional outcome from acute ischemic stroke. *Stroke* **2009**, *40*, 1687–1691. [CrossRef] [PubMed]
105. Kerlin, B.; Cooley, B.C.; Isermann, B.H.; Hernandez, I.; Sood, R.; Zogg, M.; Hendrickson, S.B.; Mosesson, M.W.; Lord, S.; Weiler, H. Cause-effect relation between hyperfibrinogenemia and vascular disease. *Blood* **2004**, *103*, 1728–1734. [CrossRef] [PubMed]
106. Kannel, W.B.; D'Agostino, R.B.; Belanger, A.J.; Silbershatz, H.; Tofler, G.T. Long-term influence of fibrinogen on initial and recurrent cardiovascular events in men and women. *Am. J. Cardiol.* **1996**, *78*, 90–92. [CrossRef]

107. Pedrazzani, C.; Mantovani, G.; Salvagno, G.L.; Baldiotti, E.; Ruzzenente, A.; Iacono, C.; Lippi, G.; Guglielmi, A. Elevated fibrinogen plasma level is not an independent predictor of poor prognosis in a large cohort of Western patients undergoing surgery for colorectal cancer. *World J. Gastroenterol.* **2016**, *22*, 9994–10001. [CrossRef]
108. Adams, R.; Schachtrup, C.; Davalos, D.; Tsigelny, I.; Akassoglou, K. Fibrinogen signal transduction as a mediator and therapeutic target in inflammation: Lessons from multiple sclerosis. *Curr. Med. Chem.* **2007**, *14*, 2925–2936. [CrossRef]
109. Sen, U.; Tyagi, N.; Patibandla, P.K.; Dean, W.L.; Tyagi, S.C.; Roberts, A.M.; Lominadze, D. Fibrinogen-induced endothelin-1 production from endothelial cells. *Am. J. Physiol. Cell Physiol.* **2009**, *296*, C840–C847. [CrossRef]
110. Davalos, D.; Akassoglou, K. Fibrinogen as a key regulator of inflammation in disease. *Semin. Immunopathol.* **2012**, *34*, 43–62. [CrossRef]
111. Muradashvili, N.; Tyagi, N.; Tyagi, R.; Munjal, C.; Lominadze, D. Fibrinogen alters mouse brain endothelial cell layer integrity affecting vascular endothelial cadherin. *Biochem. Biophys. Res. Commun.* **2011**, *413*, 509–514. [CrossRef]
112. Vos, C.M.P.; Geurts, J.J.G.; Montagne, L.; van Haastert, E.S.; Bö, L.; van der Valk, P.; Barkhof, F.; de Vries, H.E. Blood–brain barrier alterations in both focal and diffuse abnormalities on postmortem MRI in multiple sclerosis. *Neurobiol. Dis.* **2005**, *20*, 953–960. [CrossRef]
113. Plumb, J.; McQuaid, S.; Mirakhur, M.; Kirk, J. Abnormal endothelial tight junctions in active lesions and normal-appearing white matter in multiple sclerosis. *Brain Pathol.* **2002**, *12*, 154–169. [CrossRef] [PubMed]
114. Danesh, J.; Lewington, S.; Thompson, S.G.; Lowe, G.D.; Collins, R.; Kostis, J.B.; Wilson, A.C.; Folsom, A.R.; Wu, K.; Benderly, M.; et al. Plasma fibrinogen level and the risk of major cardiovascular diseases and nonvascular mortality: An individual participant meta-analysis. *JAMA* **2005**, *294*, 1799–1809. [PubMed]
115. Lowe, G.D.O. Fibrinogen and cardiovascular disease: Historical introduction. *Eur. Heart J.* **1995**, *16*, 2–5. [CrossRef]
116. Marton, Z.; Kesmarky, G.; Vekasi, J.; Cser, A.; Russai, R.; Horvath, B.; Toth, K. Red blood cell aggregation measurements in whole blood and in fibrinogen solutions by different methods. *Clin. Hemorheol. Microcirc.* **2001**, *24*, 75–83.
117. Bishop, J.J.; Popel, A.S.; Intaglietta, M.; Johnson, P.C. Rheological effects of red blood cell aggregation in the venous network: A review of recent studies. *Biorheology* **2001**, *38*, 263–274.
118. Farrell, D.H.; Thiagarajan, P.; Chung, D.W.; Davie, E.W. Role of fibrinogen alpha and gamma chain sites in platelet aggregation. *Proc. Natl. Acad. Sci. USA* **1992**, *89*, 10729–10732. [CrossRef]
119. Lominadze, D.; Joshua, I.G.; Schuschke, D.A. In vivo platelet thrombus formation in microvessels of spontaneously hypertensive rats. *Am. J. Hypertens.* **1997**, *10*, 1140–1146. [CrossRef]
120. Lominadze, D.; Joshua, I.G.; Catalfamo, J.L.; Schuschke, D.A. Platelet thrombus formation in microvessels of young spontaneously hypertensive rats. *Clin. Exp. Hypertens.* **1998**, *20*, 917–937. [CrossRef] [PubMed]
121. Languino, L.R.; Duperray, A.; Joganic, K.J.; Fornaro, M.; Thornton, G.B.; Altieri, D.C. Regulation of leukocyte-endothelium interaction and leukocyte transendothelial migration by intercellular adhesion molecule 1-fibrinogen recognition. *Proc. Natl. Acad. Sci. USA* **1995**, *92*, 1505–1509. [CrossRef]
122. Maeda, N.; Imaizumi, K.; Sekiya, M.; Shiga, T. Rheological characteristics of desialylated erythrocytes in relation to fibrinogen-induced aggregation. *Biochim. Biophys. Acta* **1984**, *776*, 151–158. [CrossRef] [PubMed]
123. Lominadze, D.; Dean, W.L. Involvement of fibrinogen specific binding in erythrocyte aggregation. *FEBS Lett.* **2002**, *517*, 41–44. [CrossRef]
124. Chien, S.; Usami, S.; Dellenback, R.J.; Gregersen, M.I.; Nanninga, L.B.; Guest, M.M. Blood Viscosity: Influence of Erythrocyte Aggregation. *Science* **1967**, *157*, 829–831. [CrossRef] [PubMed]
125. Letcher, R.L.; Chien, S.; Pickering, T.G.; Sealey, J.E.; Laragh, J.H. Direct relationship between blood pressure and blood viscosity in normal and hypertensive subjects. Role of fibrinogen and concentration. *Am. J. Med.* **1981**, *70*, 1195–1202. [CrossRef] [PubMed]
126. Lominadze, D.; Tsakadze, N.; Sen, U.; Falcone, J.C.; D'Souza, S.E. Fibrinogen- and fragment D-induced vascular constriction. *Am. J. Physiol.* **2005**, *288*, H1257–H1264. [CrossRef] [PubMed]
127. Muradashvili, N.; Benton, R.; Tyagi, R.; Tyagi, S.; Lominadze, D. Elevated level of fibrinogen increases caveolae formation; Role of matrix metalloproteinase-9. *Cell Biochem. Biophys.* **2014**, *69*, 283–294. [CrossRef]
128. Muradashvili, N.; Qipshidze, N.; Munjal, C.; Givvimani, S.; Benton, R.L.; Roberts, A.M.; Tyagi, S.C.; Lominadze, D. Fibrinogen-induced increased pial venular permeability in mice. *J. Cereb. Blood Flow Metab.* **2012**, *32*, 150–163. [CrossRef]
129. Clark, V.D.; Layson, A.; Charkviani, M.; Muradashvili, N.; Lominadze, D. Hyperfibrinogenemia-mediated astrocyte activation. *Brain Res.* **2018**, *1699*, 158–165. [CrossRef]
130. Sulimai, N.; Brown, J.; Lominadze, D. Fibrinogen Interaction with Astrocyte ICAM-1 and PrPC Results in the Generation of ROS and Neuronal Death. *Int. J. Mol. Sci.* **2021**, *22*, 2391. [CrossRef]
131. Ryu, J.K.; Petersen, M.A.; Murray, S.G.; Baeten, K.M.; Meyer-Franke, A.; Chan, J.P.; Vagena, E.; Bedard, C.; Machado, M.R.; Rios Coronado, P.E.; et al. Blood coagulation protein fibrinogen promotes autoimmunity and demyelination via chemokine release and antigen presentation. *Nat. Commun.* **2015**, *6*, 8164. [CrossRef]
132. Merlini, M.; Rafalski, V.A.; Rios Coronado, P.E.; Gill, T.M.; Ellisman, M.; Muthukumar, G.; Subramanian, K.S.; Ryu, J.K.; Syme, C.A.; Davalos, D.; et al. Fibrinogen induces microglia-mediated spine elimination and cognitive impairment in an Alzheimer's disease model. *Neuron* **2019**, *101*, 1099–1108.e1096. [CrossRef]
133. Redman, C.M.; Xia, H.U.I. Fibrinogen biosynthesis. *Ann. N. Y. Acad. Sci.* **2001**, *936*, 480–495. [CrossRef]

134. Rendu, F.; Marche, P.; Viret, J.; Maclouf, J.; Lebret, M.; Tenza, D.; Caen, J.; Levy-Toledano, S. Signal transduction in normal and pathological thrombin-stimulated human platelets. *Biochimie* **1987**, *69*, 305–313. [CrossRef]
135. Rendu, F.; Brohard-Bohn, B. The platelet release reaction: Granules' constituents, secretion and functions. *Platelets* **2001**, *12* 261–273. [CrossRef] [PubMed]
136. Jonnalagadda, D.; Izu, L.T.; Whiteheart, S.W. Platelet secretion is kinetically heterogeneous in an agonist-responsive manner *Blood* **2012**, *120*, 5209–5216. [CrossRef] [PubMed]
137. Ruggeri, Z.M. Platelets in atherothrombosis. *Nat. Med.* **2002**, *8*, 1227–1234. [CrossRef] [PubMed]
138. Krüger-Genge, A.; Sternitzky, R.; Pindur, G.; Rampling, M.; Franke, R.P.; Jung, F. Erythrocyte aggregation in relation to plasma proteins and lipids. *J. Cell. Biotechnol.* **2019**, *5*, 65–70. [CrossRef]
139. Evin, G.; Zhu, A.; Holsinger, R.M.D.; Masters, C.L.; Li, Q.-X. Proteolytic processing of the Alzheimer's disease amyloid precursor protein in brain and platelets. *J. Neurosci. Res.* **2003**, *74*, 386–392. [CrossRef]
140. Graff-Radford, N.R.; Crook, J.E.; Lucas, J.; Boeve, B.F.; Knopman, D.S.; Ivnik, R.J.; Smith, G.E.; Younkin, L.H.; Petersen, R.C., Younkin, S.G. Association of low plasma aβ42/aβ40 ratios with increased imminent risk for mild cognitive impairment and alzheimer disease. *Arch. Neurol.* **2007**, *64*, 354–362. [CrossRef]
141. Schupf, N.; Tang, M.X.; Fukuyama, H.; Manly, J.; Andrews, H.; Mehta, P.; Ravetch, J.; Mayeux, R. Peripheral Aβ subspecies as risk biomarkers of Alzheimer's disease. *Proc. Natl. Acad. Sci. USA* **2008**, *105*, 14052–14057. [CrossRef]
142. van Oijen, M.; Hofman, A.; Soares, H.D.; Koudstaal, P.J.; Breteler, M.M.B. Plasma Aβ1–40 and Aβ1–42 and the risk of dementia: A prospective case-cohort study. *Lancet Neurol.* **2006**, *5*, 655–660. [CrossRef] [PubMed]
143. Bu, X.L.; Xiang, Y.; Jin, W.S.; Wang, J.; Shen, L.L.; Huang, Z.L.; Zhang, K.; Liu, Y.H.; Zeng, F.; Liu, J.H.; et al. Blood-derived amyloid-β protein induces Alzheimer's disease pathologies. *Mol. Psychiatry* **2018**, *23*, 1948–1956. [CrossRef] [PubMed]
144. Zlokovic, B.V. Neurovascular mechanisms of Alzheimer's neurodegeneration. *Trends Neurosci.* **2005**, *28*, 202–208. [CrossRef] [PubMed]
145. Johnson, V.E.; Stewart, W.; Smith, D.H. Traumatic brain injury and amyloid-[beta] pathology: A link to Alzheimer's disease? *Nat. Rev. Neurosci.* **2010**, *11*, 361–370. [CrossRef]
146. Uryu, K.; Laurer, H.; McIntosh, T.; Praticò, D.; Martinez, D.; Leight, S.; Lee, V.; Trojanowski, J. Repetitive mild brain trauma accelerates Abeta deposition, lipid peroxidation, and cognitive impairment in a transgenic mouse model of Alzheimer amyloidosis. *J. Neurosci.* **2002**, *22*, 446–454. [CrossRef] [PubMed]
147. Deane, R.; Du Yan, S.; Submamaryan, R.K.; LaRue, B.; Jovanovic, S.; Hogg, E.; Welch, D.; Manness, L.; Lin, C.; Yu, J.; et al. RAGE mediates amyloid-β peptide transport across the blood-brain barrier and accumulation in brain. *Nat. Med.* **2003**, *9*, 907–913. [CrossRef] [PubMed]
148. Cortes-Canteli, M.; Strickland, S. Fibrinogen, a possible key player in Alzheimer's disease. *J. Thromb. Haemost.* **2009**, *7*, 146–150. [CrossRef]
149. Hardy, J.; Selkoe, D.J. The amyloid hypothesis of Alzheimer's disease: Progress and problems on the road to therapeutics. *Science* **2002**, *297*, 353–356. [CrossRef]
150. Cleary, J.P.; Walsh, D.M.; Hofmeister, J.J.; Shankar, G.M.; Kuskowski, M.A.; Selkoe, D.J.; Ashe, K.H. Natural oligomers of the amyloid-[beta] protein specifically disrupt cognitive function. *Nat. Neurosci.* **2005**, *8*, 79–84. [CrossRef]
151. Gimbel, D.A.; Nygaard, H.B.; Coffey, E.E.; Gunther, E.C.; Laurén, J.; Gimbel, Z.A.; Strittmatter, S.M. Memory impairment in transgenic Alzheimer mice requires cellular prion protein. *J. Neurosci.* **2010**, *30*, 6367–6374. [CrossRef]
152. Chung, E.; Ji, Y.; Sun, Y.; Kascsak, R.; Kascsak, R.; Mehta, P.; Strittmatter, S.; Wisniewski, T. Anti-PrPC monoclonal antibody infusion as a novel treatment for cognitive deficits in an Alzheimer's disease model mouse. *BMC Neurosci.* **2010**, *11*, 130. [CrossRef] [PubMed]
153. Rubenstein, R.; Chang, B.; Grinkina, N.; Drummond, E.; Davies, P.; Ruditzky, M.; Sharma, D.; Wang, K.; Wisniewski, T. Tau phosphorylation induced by severe closed head traumatic brain injury is linked to the cellular prion protein. *Acta Neuropathol. Commun.* **2017**, *5*, 30. [CrossRef]
154. Westergard, L.; Christensen, H.M.; Harris, D.A. The Cellular Prion Protein (PrP(C)): Its Physiological Function and Role in Disease. *Biochim. Biophys. Acta* **2007**, *1772*, 629–644. [CrossRef] [PubMed]
155. Starke, R.; Drummond, O.; MacGregor, I.; Biggerstaff, J.; Gale, R.; Camilleri, R.; Mackie, I.; Machin, S.; Harrison, P. The expression of prion protein by endothelial cells: A source of the plasma form of prion protein? *Br. J. Haematol.* **2002**, *119*, 863–873. [CrossRef]
156. Pflanzner, T.; Petsch, B.; Andre-Dohmen, B.; Muller-Schiffmann, A.; Tschickardt, S.; Weggen, S.; Stitz, L.; Korth, C.; Pietrzik, C.U. Cellular prion protein participates in amyloid-[beta] transcytosis across the blood-brain barrier. *J. Cereb. Blood Flow Metab.* **2012**, *32*, 628–632. [CrossRef]
157. Nygaard, H.B.; van Dyck, C.H.; Strittmatter, S.M. Fyn kinase inhibition as a novel therapy for Alzheimer's disease. *Alzheimer's Res. Ther.* **2014**, *6*, 8. [CrossRef]
158. Kaufman, A.C.; Salazar, S.V.; Haas, L.T.; Yang, J.; Kostylev, M.A.; Jeng, A.T.; Robinson, S.A.; Gunther, E.C.; van Dyck, C.H.; Nygaard, H.B.; et al. Fyn Inhibition Rescues Established Memory and Synapse Loss in Alzheimer Mice. *Ann. Neurol.* **2015**, *77*, 953–971. [CrossRef]
159. Reitz, C.; Honig, L.; Vonsattel, J.P.; Tang, M.-X.; Mayeux, R. Memory performance is related to amyloid and tau pathology in the hippocampus. *J. Neurol. Neurosurg. Psychiatry* **2009**, *80*, 715–721. [CrossRef]

160. Tripathi, T.; Khan, H. Direct Interaction between the β-Amyloid Core and Tau Facilitates Cross-Seeding: A Novel Target for Therapeutic Intervention. *Biochemistry* **2020**, *59*, 341–342. [CrossRef] [PubMed]
161. Lin, Y.-T.; Cheng, J.-T.; Yao, Y.-C.; Lo, Y.-K.; Lin, C.-H.; Ger, L.-P.; Lu, P.-J. Increased Total TAU But Not Amyloid-β 42; in Cerebrospinal Fluid Correlates with Short-Term Memory Impairment in Alzheimer's Disease. *J. Alzheimer's Dis.* **2009**, *18*, 907–918. [CrossRef]
162. Ekinci, F.J.; Shea, T.B. Beta-Amyloid-Induced Tau Phosphorylation does not Correlate with Degeneration in Cultured Neurons. *J. Alzheimer's Dis.* **2000**, *2*, 7–15. [CrossRef]
163. Zou, J.; Tao, S.; Johnson, A.; Tomljanovic, Z.; Polly, K.; Klein, J.; Razlighi, Q.R.; Brickman, A.M.; Lee, S.; Stern, Y.; et al. Microglial activation, but not tau pathology, is independently associated with amyloid positivity and memory impairment. *Neurobiol. Aging* **2020**, *85*, 11–21. [CrossRef]
164. Puzzo, D.; Argyrousi, E.K.; Staniszewski, A.; Zhang, H.; Calcagno, E.; Zuccarello, E.; Acquarone, E.; Fa, M.; Li Puma, D.D.; Grassi, C.; et al. Tau is not necessary for amyloid-β-induced synaptic and memory impairments. *J. Clin. Investig.* **2020**, *130*, 4831–4844. [CrossRef] [PubMed]
165. Euler, U.S.V.; Gaddum, J.H. An unidentified depressor substance in certain tissue extracts. *J. Physiol.* **1931**, *72*, 74–87. [CrossRef] [PubMed]
166. Donkin, J.J.; Turner, R.J.; Hassan, I.; Vink, R. Substance P in traumatic brain injury. In *Progress in Brain Research*; Weber, J.T., Maas, A.I.R., Eds.; Elsevier: Amsterdam, The Netherlands, 2007; Volume 161, pp. 97–109.
167. Vink, R.; Gabrielian, L.; Thornton, E. The Role of Substance P in Secondary Pathophysiology after Traumatic Brain Injury. *Front. Neurol.* **2017**, *8*, 304.
168. Yu, Y.; Zeng, C.; Shu, S.; Liu, X.; Li, C. Similar effects of substance P on learning and memory function between hippocampus and striatal marginal division. *Neural Regen. Res.* **2014**, *9*, 857–863.
169. Abbott, N.J. Inflammatory mediators and modulation of blood-brain barrier permeability. *Cell. Mol. Neurobiol.* **2000**, *20*, 131–147. [CrossRef]
170. Lu, H.; Kashani, A.H.; Arfanakis, K.; Caprihan, A.; DeCarli, C.; Gold, B.T.; Li, Y.; Maillard, P.; Satizabal, C.L.; Stables, L.; et al. MarkVCID cerebral small vessel consortium: II. Neuroimaging protocols. *Alzheimer's Dement.* **2021**, *17*, 716–725. [CrossRef]

biomolecules

MDPI

Review

The Potential Role of Integrin Signaling in Memory and Cognitive Impairment

Ifechukwude Joachim Biose [1], Saifudeen Ismael [1], Blake Ouvrier [1,2], Amanda Louise White [1,2] and Gregory Jaye Bix [1,3,4,5,6,*]

1 Clinical Neuroscience Research Center, Department of Neurosurgery, Tulane University School of Medicine, New Orleans, LA 70112, USA
2 Tulane Brain Institute, Tulane University, New Orleans, LA 70112, USA
3 School of Medicine, Tulane University, New Orleans, LA 70112, USA
4 Department of Neurology, Tulane University School of Medicine, New Orleans, LA 70112, USA
5 Department of Microbiology and Immunology, Tulane University School of Medicine, New Orleans, LA 70112, USA
6 School of Public Health and Tropical Medicine, Tulane University, New Orleans, LA 70122, USA
* Correspondence: gbix@tulane.edu; Tel.: +1-504-988-3564

Abstract: Dementia currently has no cure and, due to the increased prevalence and associated economic and personal burden of this condition, current research efforts for the development of potential therapies have intensified. Recently, targeting integrins as a strategy to ameliorate dementia and other forms of cognitive impairment has begun to gain traction. Integrins are major bidirectional signaling receptors in mammalian cells, mediating various physiological processes such as cell–cell interaction and cell adhesion, and are also known to bind to the extracellular matrix. In particular, integrins play a critical role in the synaptic transmission of signals, hence their potential contribution to memory formation and significance in cognitive impairment. In this review, we describe the physiological roles that integrins play in the blood–brain barrier (BBB) and in the formation of memories. We also provide a clear overview of how integrins are implicated in BBB disruption following cerebral pathology. Given that vascular contributions to cognitive impairment and dementia and Alzheimer's' disease are prominent forms of dementia that involve BBB disruption, as well as chronic inflammation, we present current approaches shown to improve dementia-like conditions with integrins as a central focus. We conclude that integrins are vital in memory formation and that their disruption could lead to various forms of cognitive impairment. While further research to understand the relationships between integrins and memory is needed, we propose that the translational relevance of research efforts in this area could be improved through the use of appropriately aged, comorbid, male and female animals.

Keywords: integrins; extracellular matrix; dementia; and blood–brain barrier

1. Introduction and Integrin Signaling Mechanics

Integrins are major receptors in mammalian cells, mediating various physiological processes such as cell–cell interaction and cell adhesion, and are also known to bind to the extracellular matrix (ECM). Recently, the interaction between integrins and the ECM has been ascribed to information processing and memory functions via its mechanosensory functions and maintenance of dendritic integrity at the synapse [1]. Whilst the role of integrins in cognitive impairment has been understudied, the main role of integrin interactions with the ECM allows them to function as signal transducers in synergy with other classes of receptors in the activation of intracellular processes to promote cell proliferation, cell differentiation, cell survival, and cell growth [2]. Understanding the mechanics of integrin functions is vital to determining molecular targets for potential therapeutic applications. To facilitate this, here we briefly describe the role and interactions of the ECM with integrins.

Further, we describe integrins' role in blood–brain barrier (BBB) integrity and memory formation, as well as their potential therapeutic applications in ameliorating dementia.

Most cells are anchored in the ECM, a complex of various proteins such as collagen IV, laminin, and perlecan, which allows for sufficient tensile and adhesive strength to support and maintain the structure and orientation of cells. The ECM conveys many benefits to the surrounding cells, such as allowing for mechanical, biochemical, and electrical forces to impose changes on the cells in different and highly specific ways [3]. Cell adhesion to the ECM, vital for functional multicellularity, is accomplished by integrins, which are bidirectional transmembrane receptors that allow for both mechanical and biochemical signaling between cells and the ECM [3,4]. Connecting within cells to the intracellular actin cytoskeleton, integrins allow for the mechano-transduction of signals to induce functional and conformational changes in both the ECM proteins and cellular components. Some of these signaling events that modulate cells are proliferation, shape, polarity, motility, gene expression, and differentiation [5]. Additionally, the protective effects of integrins are directly related to essential physiological processes such as cell survival and proliferation: to block apoptosis, via phosphatidylinositol 3-kinase and protein kinase B (also known as Akt; i.e., PI3K-Akt) signaling, and to stimulate cell cycle progression, via extracellular-signal-regulated kinase (ERK) and cyclin D1 signaling [4,5].

All these cellular conformational and functional changes hinge upon integrin binding and signaling. The integrin family is composed of 18 alpha and 8 beta subunits, which form 24 distinct integrin heterodimers. Integrins mediate two different kinds of signaling pathways: "inside-out" and "outside-in". The "inside-out" pathway occurs when an intracellular signal promotes the binding of specific proteins, which may induce conformational changes on the integrin and increase the affinity for ECM ligands. However, the "outside-in" pathway recruits protein complexes that regulate cell function such as proliferation and differentiation [4]. These "inside-out" and "outside-in" signaling roles of integrins are accomplished via mechano-transduction between the ECM and intracellular actin cytoskeleton [4]. Mechanical loads on tissue are perceived by cells as stimuli via the surrounding ECM through integrin signaling. The perception of these stimuli by cells is critical for the cell–matrix interactions that regulate the mechanical homeostasis of tissues [3,6]. Integrins also have some association with memory formation due to their abundant expression at the synapses as well as their unique facilitation of mechanical homeostasis.

Integrins and the Blood–Brain Barrier

In the absence of a constant mechanical stimuli, integrins can induce structural apoptosis in the surrounding parenchyma, which may increase vascular permeability and thereby weaken the integrity of the BBB [6,7]. Hence, integrins reserve an integral physiological function in the maintenance of the BBB and cerebral homeostasis. Integrins can potentially be a therapeutic target to maintain BBB integrity in the event of cerebral pathology.

Integrin signaling is also involved in the promotion and maintenance of the selective permeability of the BBB [8,9]. Many integrin knock-out (KO) mouse models are lethal or develop brain defects. For example, β1 KO mice have decreased BBB integrity [8,10]. Similarly, α5 KO mice show increased BBB leakage [9]. Understanding the specific role of each of the outlined integrin, not only with regard to their role in inflammation but specifically for their unique contribution to the BBB, is germane to maintaining a healthy brain as well as developing potential therapies.

A common insult to BBB integrity is cerebral ischemia and/or vascular-dysfunction-induced oligemia. Cerebral ischemia/oligemia occurs when blood flow to downstream microvasculature or brain parenchyma is impaired, generating a core of dead brain cells with a salvageable peri-infarct region, called the penumbra. The integrity of the BBB in the penumbral region is subject to pro-apoptotic factors in a physiological and compensatory effort by the brain towards angiogenesis. Although various other neurovascular factors play a significant role in the maintenance of the BBB integrity, we [11] and others [12] demonstrated a reduced level of tight junction protein expressions for up to 14 days in the ischemic core. Similarly, in the peri-infarct region, it was reported that tight junction protein expression is also lowered immediately and in the first 4 days following cerebral ischemia induction; a slow restoration of normal tight junction protein expressions ensues thereafter [12]. These findings are in tandem with angiogenesis following ischemic stroke and were associated with increased endothelial α5β1 integrin expression [12]. The disruption of the BBB may have a direct effect on cognitive deficit and long-term functional recovery following brain ischemia.

Integrins, such as α5β1 and α2β1 receptors, contribute highly to angiogenesis and vascular remodeling and are under intense research focus as therapeutic targets. The α5β1 integrin is largely expressed in the endothelial cells of the cerebral vasculature. This is of importance since the c-terminal ligand of the perlecan domain V (DV, an ECM protein, which is cleaved by proteases and richly expressed in the brain following cerebral ischemia) binds with the α5β1 integrin receptor resulting in a pro-angiogenic effect through VEGF upregulation and the ERK signaling pathway [13]. DV is an 85kDa c-terminal domain of perlecan consisting of three laminin-like globular domains (LG1-LG3) and LG3 is the c-terminal domain mostly released by proteolytic cleavage [14]. It is well documented that the increased expression of α5β1 integrin is associated with a similar increased brain expression of angiopoietin-1 (Ang1) following ischemic stroke [12,15]. Ang1 is a vascular ligand for Tie2. Tie2 expression induces endothelial progenitor/cell migration and protects from apoptosis via the upregulation of focal adhesion kinase and Akt signaling, resulting in vascular protection, cell migration, and tube formation only when α5β1 integrin is upregulated [9,16]. However, Ang1 signaling and angiogenesis interferes with the β1 conformation of the ECM components, which impairs BBB integrity and results in increased BBB permeability [17]. Indeed, β1 integrins promote coagulation and phagocytosis, which are essential steps in angiogenesis, and this is likely responsible for the BBB permeability following brain ischemia [18].

While we have shown that selectively inhibiting the α5β1 integrin receptor with the small peptide ATN-161 in acute ischemic stroke ameliorates BBB disruption [11], we reason that the tandem increase in Ang1 and α5 integrin expression beyond the acute phase of cerebral ischemia may help stabilize the BBB and improve local blood flow through promoted angiogenesis as shown by other groups [12,15]. Consequently, inhibiting α5β1 integrin during the acute phase of ischemic stroke, but potentially not in the long term, could help stabilize the BBB and ameliorate neuroinflammatory processes, which holds the key to post-stroke outcomes.

Since the modulation of integrins results in changes in the brain, there is sufficient reason to believe that integrins play a role in memory pathology, such as dementia, where BBB disruption is a commonality. Below we describe a more direct involvement of integrins in memory formation.

2. Integrins and Memory

The role of integrins in memory function and the formation of new memories lies at the synaptic connection. Memory is stored and processed from the pattern of molecular changes involved at the presynaptic and post-synaptic terminals during the transmission of neuronal signals. The unique biochemical conformation and interaction of the ECM, cytoskeletal structures, and integrins allow neurotransmitter release and trafficking, which, along with other biochemical processes, enable signal transduction and processing. Ultimately, synaptic signal transmission is facilitated by the adhesive and signaling functions of integrins. Long-term potentiation (LTP), which describes the extended increased synaptic transmission of signals between two or more neurons based on prior persistent patterns of biochemical processes at the synapse, has been attenuated in Drosophila by integrin inhibitors or other pharmacological compounds [19–22]. Similarly, disruption of the integrin-associated protein (IAP, also known as CD47), causes memory impairment in mice due to its relationship with one of the genes related to memory formation. Moreover, the inhibition of IAP in the dentate gyrus of rat hippocampus impairs both synaptic plasticity and behavioral plasticity resulting in reduced memory retention and LTP [23–25]. Since learning and memory are a result of the constant alterations at the synaptic connections, the interplay between integrins and the ECM may be directly linked to memory formation. Hence, pathological or abnormal changes in the conformation of integrins may result in lapses in memory formation and retention (Figure 1). The first direct evidence for the role of integrins in memory function emanated from the study of Chan et al. [26]. They found that the concurrent attenuated expression of α3, α5, and α8 integrin subunits resulted in spatial memory deficit during Morris water maze (MWM) tests. This early finding showed that integrin receptors, which are known to play a role in cell adhesion, may also mediate behavioral plasticity. Further, the same study and other reports demonstrated that the specific deletion of α3 or β1 integrins in the forebrain and excitatory neurons impairs working memory in the hippocampal-dependent test of T-maze [26–28]. Additionally, α8 and β8 integrins are abundant at the dendritic spines of pyramidal neurons and associated with post-synaptic density [29,30]. α5 integrins are also richly localized in the apical dendrites of the pyramidal cells of the cortex and hippocampus [31]. This suggests that integrins are actively involved in the processing of memory function.

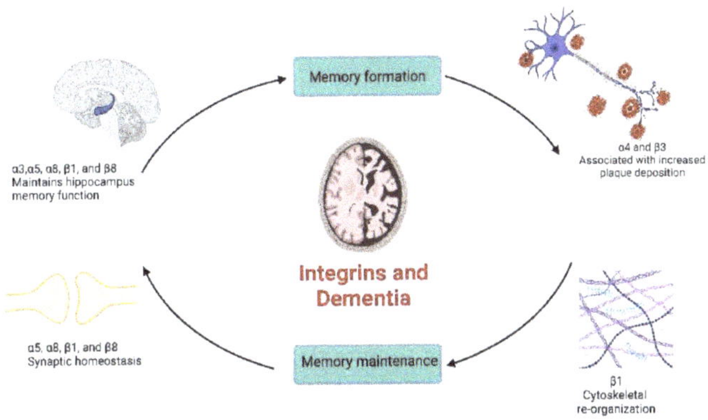

Figure 1. The potential role of integrins in memory formation and maintenance. Various integrins play vital roles in maintaining the number and size of the dendritic spine at the synapse as well as in functional hippocampal memory formation and retention. In addition, β1 integrins are responsible for the re-assembly of the cytoskeleton following synaptic transmission of signals. Other integrins are activated and associated with plaque deposition in human brain samples obtained from individuals with dementia-like symptoms. The inhibition/deletion of these integrins revealed functional loss of memory formation/retention; hence, targeting these integrins will be critical in illuminating their potential therapeutic values.

Using rodent hippocampal slices, several studies demonstrated the direct dependence of LTP on integrins. LTP was attenuated when α3 and α5 integrins were inhibited with antibodies from snake venoms [22,31]. Moreover, both LTP and the biochemical restoration of actin assembly were eliminated when β1-integrin inhibition was induced immediately after stimulation [32]. Despite the mounting body of evidence linking integrins to memory functions, it is not known whether integrins directly regulate aspects of memory formation and recall while performing their intracellular signaling or cell adhesion roles. This level of understanding will provide essential evidence to further integrin targeting to potentially improve memory functions.

The β-1 integrin subunit is the most commonly occurring subunit; this includes α4β1, α5β1, α6β1, αVβ1, αVβ3 α2β1, and α11β1. The downstream signaling functions of β1 integrins occurs via interactions with non-receptor tyrosine kinase Arg to modulate dendritic and synaptic plasticity in the hippocampal neurons [33]. Arg binds to and phosphorylates the intracellular tail of the β1 integrin at the dendritic spines where Arg is richly expressed [34–36]. Consequently, when the β1 integrin is conditionally knocked out in mice, hippocampal-dependent memory deficits were observed due to a significant reduction in the size and quantity of the dendritic spines and synapses [33]. A similar observation was noted in mice with homozygous deletion of Arg [36,37]. Moreover, the direct inhibition of the β1 integrin attenuated the quantity of synapses in the apical dendrites of CA1 pyramidal neurons [38]. These suggest that β1 is crucial not only for the early formation of synapses but also for the maintenance of hippocampal memory functions. Similarly, the deletion of the α5 integrin in the hippocampal neurons results in a reduction in synapses as well as dendritic spines [39].

Early studies reported the association of integrins with age-associated memory deficits. Aged human hippocampal and cortical neurons were immunoreactive for the α4 integrin subunit, which was not observed in samples from young adults [40]. Moreover, tau-positive plaques in samples from patients who had Alzheimer's disease (AD) reacted to antibodies for the α4 integrin subunit [40]. Relatedly, the senile plaques and neurofibrillary tangles in human brain samples from AD patients was highly reactive to antibodies for the β3 integrin [41]. This suggests that the increased deposition of plaques activated α4 and β3 integrins. The integrins' specific role(s) in AD or other diseases that impact memory functions is not well understood and calls for further investigations.

However, it has been shown that increased integrin expression at the site of tau-positive plaque formation could be the brain's attempt to rid itself of the plaques. Activated microglial cells in the region of amyloid plaques from the brain samples of dementia patients have a higher expression of α4β1 and αLβ2 integrins [42]. Experimental findings from rats demonstrate the colocalization of α1β1 and α5β1 integrins with β-amyloid precursor proteins in hippocampal neurons and cortical astrocytes [43,44]. Integrins are clearly implicated in the inflammatory response to abnormal brain processes that impair memory function. Hence, understanding the specific roles integrins play in cognitive dysfunction may be the first step towards the development of therapeutic strategies for dementia.

Given the importance and involvement of integrins in proper brain health and development, integrins could be a potential therapeutic target for dementia. In fact, coinciding with the growing aging population, incidences of dementia are also expected to increase and become an even greater healthcare burden [45,46]. In 2010, healthcare for dementia-related cases in the United States cost over 100 billion USD, and costs are predicted to double to over 250 billion USD by the year 2040 [47]. Population studies estimate that around 50 million people worldwide have been diagnosed with dementia and that dementia cases are only going to increase, with some estimates saying cases will triple, i.e., 150 million, by 2050 [48]. Therefore, increased efforts towards understanding the potential therapeutic roles of integrins may help reduce the economic impact of dementia. Below, we describe the two forms of dementia, vascular contributions to cognitive impairment and dementia (VCID) and AD, which may be well suited for increased integrin research endeavors.

2.1. *Vascular Contributions to Cognitive Impairment and Dementia (VCID)*

VCID is a term used to describe any degree of cognitive impairment caused by cerebrovascular dysfunction. VCID encompasses patients suffering from vascular cognitive impairment to the more severe diagnosis of vascular dementia. VCID represents a growing major health concern worldwide. VCID alone is the second leading cause of dementia and accounts for 20% of all dementia cases in Europe and North America and 30% in Asia [49]. While aging is one of the main risk factors, smoking, inflammation, hypertension, and stroke have all been shown to increase the risk of developing VCID [45,50].

VCID is a complex disease that can be caused by many different factors. In general, an event causing blood flow dysregulation or hypoperfusion in the brain leads to a decrease in glucose metabolism, vascular permeability, and ultimately to neuronal death. The processes that follow cerebral blood flow insufficiency are neuroinflammation, vascular remodeling, and BBB disruption. Chronic hypoperfusion has been shown to have a strong relationship in VCID pathogenesis likely leading to pathology such as infarcts, hemorrhages, and memory impairment in rodents [45]. In addition, cerebral small vessel disease, which results from chronic hypertension and cerebrovascular remodeling, leads to cognitive impairment [51]. Consequently, an excellent animal model is the spontaneously hypertensive rat (SHR), which develops cerebral small vessel disease early in life, developing white matter damage and a dysfunction of the BBB in later life [51].

Since BBB integrity is highly implicated in the pathogenesis of VCID, it is imperative to increase research efforts in understanding how integrins can be targeted to maintain BBB integrity and potentially ameliorate memory function. Only one study from our group has explored the expressions of the α5 integrin following 14 days of bilateral carotid artery stenosis in mice, a valid model of VCID [52]. We found an increase in the cortical and striatal expressions of the α5 integrin, a decrease in tight junction proteins, and a substantial BBB permeability 14 days after bilateral carotid artery stenosis in young adult males. This is suggestive of active angiogenesis, the brain's compensatory attempt to increase the number of blood vessels, which ultimately disrupts optimal BBB function and may potentially lead to cognitive impairments.

2.2. *Alzheimer's Disease (AD)*

In patients with AD, several integrins and integrin-binding factors are upregulated [18]. Chronic low-grade inflammation is known to play a critical role in the pathogenesis and progression of AD. Increased BBB permeability contributes to elevated leukocyte infiltration, particularly neutrophils, and thereby mediates vascular inflammation in the brain. Leukocytes attach to cerebral endothelial cells and migrate to the brain parenchyma, particularly in the hippocampus and other limbic structures [53,54]. Leukocyte infiltration is a multistep process that is mediated by adhesion molecules such as selectins, integrins, and the immunoglobulin superfamily [55]. The α4β1 integrin, also known as CD49d/CD29 or very late antigen-4 (VLA-4), is the most predominant β1 integrin expressed on leukocytes [56] and also plays an essential role in T cell trafficking during various inflammatory responses [57], as well as in CNS pathologies such as multiple sclerosis and experimental autoimmune encephalomyelitis [55,58]. Pietronigro et al. demonstrated that the α4β1 integrin is a pivotal mediator of leukocyte adhesion on activated endothelial cells and blocking the α4 chain with specific antibody inhibits rolling interactions in cortical venules in 3xtg-AD mice, indicating that VLA-4 promotes the leukocyte–vascular interactions in AD mice [56]. They found an age-dependent increase in the proportion of α4-integrin-expressing CD4+ cells in 3xTg-AD mice and an inhibition of α4β1 integrin improved cognitive function as evidenced by improved performance in a Y maze, contextual fear conditioning, and MWM tests [56]. Taken together, the study shows that the therapeutic potential of α4β1 integrin inhibition interferes with disease progression and cognitive impairment. Further, α4 integrins blockage attenuated leukocyte–endothelial interactions and thereby significantly inhibited neuropathological hallmarks such as Aβ deposition and tau hyperphosphorylation.

From the foregoing, integrins may be a worthwhile research focus as a therapeutic target in cognitive impairment and dementia.

3. Improving the Translational Perspective for Modulating Integrin Signaling in the Context of Cognitive Impairment

A number of potential therapeutic integrin targets are currently under investigation or have been shown previously to ameliorate the impact of cognitive impairment in the context of the two leading types of dementia. We will outline some of the findings (Table 1) and briefly indicate potential areas of growth in this area. C16 (KAFDITYVRLKF), a selective peptide inhibitor for αvβ3, was shown to interfere with the transmigration of leukocytes and inflammation [59]. Moreover, C16 has shown a beneficial effect on the ALS/Parkinsonism dementia complex (PDC), representing symptoms analogous to AD's such as dementia and Parkinsonism, when administrated along with angiopoietin 1, a nerve growth factor. A combination treatment of C16 with angiopoietin 1 improved oxidative stress, neuroinflammation, and cognitive function in a rat model of PDC induced by Beta-N-methylamino-L-alanine (L-BMAA) [60]. In addition to leukocytes, platelets also play a critical role in the development and progression of VCID and AD as they harbor amyloid precursor protein (APP) and secretases required to cleave the APP [61], and aberrant platelet activation has been reported in AD patients [62]. Lee et al. demonstrated that Aβ1–40 stimulated aberrant reactive oxygen species (ROS) production in human platelets and the activation of integrin αIIbβ3 through a PKC-δ-dependent mechanism [63]. Treatment with Rosmarinic acid, a phyto-polyphenolic compound, attenuated platelet adhesion through the modulation of ROS production and inhibition of αIIbβ3 signaling.

Recently, Ortiz-Sanz et al. demonstrated the therapeutic potential of the N-terminal signal peptide of β1 integrin localized at the first 20 amino acids, towards AD. The β1 integrin binds to Aβ oligomers and attenuated ROS generation in primary astrocyte cultures treated with Aβ oligomers [64]. Further, intrahippocampal administration of recombinant integrin β1 signal peptide prevented both astrogliosis and microgliosis and endoplasmic reticulum stress mediated by Aβ oligomers in vivo.

We have previously reported that the domain V (DV) 85-kDa protein fragment of the extracellular matrix proteogylcan perlecan is generated by proteolysis and could modulate α2β1 signaling induced by Aβ in vitro [65]. Perlecan DV is an α2 integrin ligand shown to inhibit Aβ-induced neurotoxicity in human cortical neurons in vitro through the α2β1 integrin receptor and a p-c-jun-dependent mechanism [66]. Aβ is a ligand for both α2β1 and αvβ1 confirming their role in AD pathology [67]. Further, DV (and its 25-kDa subfragment, LG3) administration has blocked Aβ toxicity in mouse fetal hippocampal neurons through the inhibition of c-Jun and caspase-3 [64] demonstrating the therapeutic potential of perlecan subunits.

Although the above stated studies have heralded the current thinking that integrins have a strategic role towards improving brain health and ameliorating dementia-like effects, the majority of preclinical studies continue to pursue translational relevance whilst not considering important factors associated with dementia [68–71]. For example, given that VCID and AD are diseases of advance age and affect both males and females, more work lies ahead in the use of appropriately aged models as well as the careful consideration of sex as a biological factor. Moreover, while we recognize that there exists no perfect model for translational studies, VCID more often results from hypertension and other comorbidities such as metabolic syndrome and diabetes mellitus [72–74]. Hence, the above translationally relevant factors ought to be considered for future studies seeking to develop therapeutic strategies for dementia whilst focusing on the roles of integrins in memory impairments.

Table 1. Summary of studies on modulation of integrin signaling in AD/dementia.

Disease Model	Inhibitor/Modulator	Inference	Reference
3xTg-AD mice	500 μg of the α4-integrin-specific antibody	Attenuated neuropathological hallmarks of AD, such as microgliosis, Aβ load, and tau hyperphosphorylation. The α4 integrin blocking attenuated leukocyte trafficking and improved cognitive impairment and AD neuropathology	Pietronigro et al., 2019 [56]
C57BL6/J mice with Intrahippocampal Aβ oligomers injection	recombinant integrin β1 N-terminal signal peptide	Inhibited Aβ-induced ROS generation in primary astrocytes Inhibited astrogliosis and ER stress in mouse of AD	Ortiz-Sanz et al., 2022 [64]
Rat model of ALS/PDC model (induced by L-BMAA)	C16 peptide KAFDITYVRLKF along with angiopetin 1 (Ang1)	Attenuated oxidative stress and inflammatory response Improved cognitive and motor function	Cai et al., 2018 [60]
Human and mouse cortical neurons treated with Aβ	Domain V and LG3 of perlecan	Inhibited Aβ-induced neurotoxicity in an α2 integrin and c-Jun dependent manner	Wright et al., 2010 [67]

Table 1. *Cont.*

Disease Model	Inhibitor/Modulator	Inference	Reference
Mouse hippocampal neurons treated with $A\beta_{42}$	DV and LG3	DV and LG3 inhibited the α2β1 integrin receptor and prevented Aβ from binding	Parham et al., 2016 [65]
Human platelets treated with Aβ1–40	Rosmarinic acid	Aβ1–40-induced platelet adhesion is ameliorated by RA through the inhibition of NADPH oxidase/ROS/PKC-δ/integrin αIIbβ3 signaling pathways	Lee et al., 2021 [63]

4. Conclusions

From the foregoing, the role of integrins in memory function and retention stems from their involvement in the formation and function of the dendritic spine. Several studies have shown that the targeted integrin inhibition of α3, α5, or β1 leads to a decrease in dendritic size and quantity, as well as LTP. Findings from multiple behavioral tests have also shown that the inhibition of these same integrins results in mice with impaired memory. Hence, β1, α5, and α3 integrins functioning at the synapse are crucial for proper memory function. However, it is still unclear if the observed deleterious effects on memory are a direct cause of integrin attenuation or due to the loss of communication between cells. Increasing research in this area could uncover if different integrins play different roles in memory formation and retention, as well as determine if there are distinct pathways for the creation of new memories and their recall.

Integrins and BBB disruption have both been linked with age-associated memory deficits. The disruption of α5 integrins mediating angiogenesis during the acute phase of vascular supply interruptions to the brain disrupts the integrity of the BBB and may be implicated in cognitive dysfunction. There is, currently, a lack of direct evidence on the specific roles of integrins in models of VCID and more studies are warranted to consider this direction given the global prevalence of VCID. Although reports have linked αVβ3, α4β1, and αLβ2 integrins to the pathology of AD, little is known of the role of these integrins in models with age-related comorbidities or in female models.

To better translate experimental findings, especially in the area of potential therapy that aims to target integrins, efforts must be intensified in modeling human disease conditions as well as in the inclusion of sex as a biological factor. Since experimental models are insufficient and to expand the true translational potential of bed-bench collaboration, more efforts should be directed towards integrin analyses of post-mortem brain tissues from individuals who were diagnosed with VCID and AD. One major step towards this endeavor will be to establish a subregional hippocampus tissue biobank and registry, which will preserve historical patient records as well as hippocampal brain specimens. Clearly, integrins are associated with memory function in VCID and AD pathobiology/models and an increased understanding of their specific roles and therapeutic potentials in ameliorating cognitive deficits is urgently warranted.

Author Contributions: I.J.B. conceptualized and planned the manuscript; I.J.B., S.I., B.O. and A.L.W. prepared sections of the manuscript; G.J.B. supervised manuscript drafts. All authors have read and agreed to the published version of the manuscript.

Funding: This research received no external funding.

Conflicts of Interest: The authors declare no conflict of interest.

References

1. Iskratsch, T.; Wolfenson, H.; Sheetz, M.P. Appreciating force and shape—The rise of mechanotransduction in cell biology. *Nat. Rev. Mol. Cell Biol.* **2014**, *15*, 825–833. [CrossRef] [PubMed]
2. Harburger, D.S.; Calderwood, D.A. Integrin signalling at a glance. *J. Cell Sci.* **2009**, *122*, 159–163. [CrossRef] [PubMed]
3. Sun, Z.; Guo, S.S.; Fässler, R. Integrin-mediated mechanotransduction. *J. Cell Biol.* **2016**, *215*, 445–456. [CrossRef] [PubMed]

4. Moreno-Layseca, P.; Icha, J.; Hamidi, H.; Ivaska, J. Integrin trafficking in cells and tissues. *Nat. Cell Biol.* **2019**, *21*, 122–132. [CrossRef]
5. Hynes, R.O. Integrins: Bidirectional, allosteric signaling machines. *Cell* **2002**, *110*, 673–687. [CrossRef]
6. Humphrey, J.D.; Dufresne, E.R.; Schwartz, M.A. Mechanotransduction and extracellular matrix homeostasis. *Nat. Rev. Mol. Cell Biol.* **2014**, *15*, 802–812. [CrossRef]
7. Ayloo, S.; Lazo, C.G.; Sun, S.; Zhang, W.; Cui, B.; Gu, C. Pericyte-to-endothelial cell signaling via vitronectin-integrin regulates blood-CNS barrier. *Neuron* **2022**, *110*, 1641–1655.e1646. [CrossRef]
8. Tang, J.; Kang, Y.; Huang, L.; Wu, L.; Peng, Y. TIMP1 preserves the blood-brain barrier through interacting with CD63/integrin β 1 complex and regulating downstream FAK/RhoA signaling. *Acta Pharm. Sin. B* **2020**, *10*, 987–1003. [CrossRef]
9. Milner, R.; Hung, S.; Wang, X.; Berg, G.I.; Spatz, M.; del Zoppo, G.J. Responses of endothelial cell and astrocyte matrix-integrin receptors to ischemia mimic those observed in the neurovascular unit. *Stroke* **2008**, *39*, 191–197. [CrossRef]
10. Giancotti, F.G.; Ruoslahti, E. Integrin signaling. *Science* **1999**, *285*, 1028–1032. [CrossRef]
11. Edwards, D.N.; Salmeron, K.; Lukins, D.E.; Trout, A.L.; Fraser, J.F.; Bix, G.J. Integrin α5β1 inhibition by ATN-161 reduces neuroinflammation and is neuroprotective in ischemic stroke. *J. Cereb. Blood Flow. Metab.* **2020**, *40*, 1695–1708. [CrossRef] [PubMed]
12. Sun, J.; Yu, L.; Huang, S.; Lai, X.; Milner, R.; Li, L. Vascular expression of angiopoietin1, α5β1 integrin and tight junction proteins is tightly regulated during vascular remodeling in the post-ischemic brain. *Neuroscience* **2017**, *362*, 248–256. [CrossRef] [PubMed]
13. Clarke, D.N.; Al Ahmad, A.; Lee, B.; Parham, C.; Auckland, L.; Fertala, A.; Kahle, M.; Shaw, C.S.; Roberts, J.; Bix, G.J. Perlecan Domain V induces VEGf secretion in brain endothelial cells through integrin α5β1 and ERK-dependent signaling pathways. *PLoS ONE* **2012**, *7*, e45257. [CrossRef] [PubMed]
14. Bix, G.; Iozzo, R.V. Matrix revolutions: "tails" of basement-membrane components with angiostatic functions. *Trends Cell Biol.* **2005**, *15*, 52–60. [CrossRef] [PubMed]
15. Wang, L.; Zhang, X.; Liu, X.; Feng, G.; Fu, Y.; Milner, R.; Li, L. Overexpression of α5β1 integrin and angiopoietin-1 co-operatively promote blood-brain barrier integrity and angiogenesis following ischemic stroke. *Exp. Neurol.* **2019**, *321*, 113042. [CrossRef]
16. Pang, D.; Wang, L.; Dong, J.; Lai, X.; Huang, Q.; Milner, R.; Li, L. Integrin α5β1-Ang1/Tie2 receptor cross-talk regulates brain endothelial cell responses following cerebral ischemia. *Exp. Mol. Med.* **2018**, *50*, 1–12. [CrossRef]
17. Izawa, Y.; Gu, Y.H.; Osada, T.; Kanazawa, M.; Hawkins, B.T.; Koziol, J.A.; Papayannopoulou, T.; Spatz, M.; Del Zoppo, G.J. β1-integrin-matrix interactions modulate cerebral microvessel endothelial cell tight junction expression and permeability. *J. Cereb. Blood Flow. Metab.* **2018**, *38*, 641–658. [CrossRef]
18. Grammas, P.; Sanchez, A.; Tripathy, D.; Luo, E.; Martinez, J. Vascular signaling abnormalities in Alzheimer disease. *Cleve Clin. J. Med.* **2011**, *78* (Suppl. 1), S50–S53. [CrossRef]
19. Bahr, B.A.; Staubli, U.; Xiao, P.; Chun, D.; Ji, Z.X.; Esteban, E.T.; Lynch, G. Arg-Gly-Asp-Ser-selective adhesion and the stabilization of long-term potentiation: Pharmacological studies and the characterization of a candidate matrix receptor. *J. Neurosci.* **1997**, *17*, 1320–1329. [CrossRef]
20. Kramár, E.A.; Bernard, J.A.; Gall, C.M.; Lynch, G. Alpha3 integrin receptors contribute to the consolidation of long-term potentiation. *Neuroscience* **2002**, *110*, 29–39. [CrossRef]
21. Stäubli, U.; Chun, D.; Lynch, G. Time-dependent reversal of long-term potentiation by an integrin antagonist. *J. Neurosci.* **1998**, *18*, 3460–3469. [CrossRef] [PubMed]
22. Chun, D.; Gall, C.M.; Bi, X.; Lynch, G. Evidence that integrins contribute to multiple stages in the consolidation of long term potentiation in rat hippocampus. *Neuroscience* **2001**, *105*, 815–829. [CrossRef] [PubMed]
23. Huang, A.M.; Wang, H.L.; Tang, Y.P.; Lee, E.H. Expression of integrin-associated protein gene associated with memory formation in rats. *J. Neurosci.* **1998**, *18*, 4305–4313. [CrossRef] [PubMed]
24. Chang, H.P.; Lindberg, F.P.; Wang, H.L.; Huang, A.M.; Lee, E.H. Impaired memory retention and decreased long-term potentiation in integrin-associated protein-deficient mice. *Learn. Mem.* **1999**, *6*, 448–457. [CrossRef] [PubMed]
25. Chang, H.P.; Ma, Y.L.; Wan, F.J.; Tsai, L.Y.; Lindberg, F.P.; Lee, E.H. Functional blocking of integrin-associated protein impairs memory retention and decreases glutamate release from the hippocampus. *Neuroscience* **2001**, *102*, 289–296. [CrossRef] [PubMed]
26. Chan, C.S.; Weeber, E.J.; Kurup, S.; Sweatt, J.D.; Davis, R.L. Integrin requirement for hippocampal synaptic plasticity and spatial memory. *J. Neurosci.* **2003**, *23*, 7107–7116. [CrossRef]
27. Huang, Z.; Shimazu, K.; Woo, N.H.; Zang, K.; Müller, U.; Lu, B.; Reichardt, L.F. Distinct roles of the beta 1-class integrins at the developing and the mature hippocampal excitatory synapse. *J. Neurosci.* **2006**, *26*, 11208–11219. [CrossRef] [PubMed]
28. Chan, C.S.; Levenson, J.M.; Mukhopadhyay, P.S.; Zong, L.; Bradley, A.; Sweatt, J.D.; Davis, R.L. Alpha3-integrins are required for hippocampal long-term potentiation and working memory. *Learn. Mem.* **2007**, *14*, 606–615. [CrossRef]
29. Einheber, S.; Schnapp, L.M.; Salzer, J.L.; Cappiello, Z.B.; Milner, T.A. Regional and ultrastructural distribution of the alpha 8 integrin subunit in developing and adult rat brain suggests a role in synaptic function. *J. Comp. Neurol.* **1996**, *370*, 105–134. [CrossRef]
30. Nishimura, S.L.; Boylen, K.P.; Einheber, S.; Milner, T.A.; Ramos, D.M.; Pytela, R. Synaptic and glial localization of the integrin alphavbeta8 in mouse and rat brain. *Brain Res.* **1998**, *791*, 271–282. [CrossRef]
31. Bi, X.; Lynch, G.; Zhou, J.; Gall, C.M. Polarized distribution of alpha5 integrin in dendrites of hippocampal and cortical neurons. *J. Comp. Neurol.* **2001**, *435*, 184–193. [CrossRef] [PubMed]

32. Kramár, E.A.; Lin, B.; Rex, C.S.; Gall, C.M.; Lynch, G. Integrin-driven actin polymerization consolidates long-term potentiation. *Proc. Natl. Acad. Sci. USA* **2006**, *103*, 5579–5584. [CrossRef] [PubMed]
33. Warren, M.S.; Bradley, W.D.; Gourley, S.L.; Lin, Y.C.; Simpson, M.A.; Reichardt, L.F.; Greer, C.A.; Taylor, J.R.; Koleske, A.J. Integrin β1 signals through Arg to regulate postnatal dendritic arborization, synapse density, and behavior. *J. Neurosci.* **2012**, *32*, 2824–2834. [CrossRef] [PubMed]
34. Koleske, A.J.; Gifford, A.M.; Scott, M.L.; Nee, M.; Bronson, R.T.; Miczek, K.A.; Baltimore, D. Essential roles for the Abl and Arg tyrosine kinases in neurulation. *Neuron* **1998**, *21*, 1259–1272. [CrossRef] [PubMed]
35. Simpson, M.A.; Bradley, W.D.; Harburger, D.; Parsons, M.; Calderwood, D.A.; Koleske, A.J. Direct interactions with the integrin β1 cytoplasmic tail activate the Abl2/Arg kinase. *J. Biol. Chem.* **2015**, *290*, 8360–8372. [CrossRef]
36. Moresco, E.M.; Donaldson, S.; Williamson, A.; Koleske, A.J. Integrin-mediated dendrite branch maintenance requires Abelson (Abl) family kinases. *J. Neurosci.* **2005**, *25*, 6105–6118. [CrossRef]
37. Sfakianos, M.K.; Eisman, A.; Gourley, S.L.; Bradley, W.D.; Scheetz, A.J.; Settleman, J.; Taylor, J.R.; Greer, C.A.; Williamson, A.; Koleske, A.J. Inhibition of Rho via Arg and p190RhoGAP in the postnatal mouse hippocampus regulates dendritic spine maturation, synapse and dendrite stability, and behavior. *J. Neurosci.* **2007**, *27*, 10982–10992. [CrossRef]
38. Nikonenko, I.; Toni, N.; Moosmayer, M.; Shigeri, Y.; Muller, D.; Sargent Jones, L. Integrins are involved in synaptogenesis, cell spreading, and adhesion in the postnatal brain. *Brain Res. Dev. Brain Res.* **2003**, *140*, 185–194. [CrossRef]
39. Webb, D.J.; Zhang, H.; Majumdar, D.; Horwitz, A.F. alpha5 integrin signaling regulates the formation of spines and synapses in hippocampal neurons. *J. Biol. Chem.* **2007**, *282*, 6929–6969.935. [CrossRef]
40. Van Gool, D.; Carmeliet, G.; Triau, E.; Cassiman, J.J.; Dom, R. Appearance of localized immunoreactivity for the alpha 4 integrin subunit and for fibronectin in brains from Alzheimer's, Lewy body dementia patients and aged controls. *Neurosci. Lett.* **1994**, *170*, 71–73. [CrossRef]
41. Akiyama, H.; Kawamata, T.; Dedhar, S.; McGeer, P.L. Immunohistochemical localization of vitronectin, its receptor and beta-3 integrin in Alzheimer brain tissue. *J. Neuroimmunol.* **1991**, *32*, 19–28. [CrossRef] [PubMed]
42. Preciado-Patt, L.; Hershkoviz, R.; Fridkin, M.; Lider, O. Serum amyloid A binds specific extracellular matrix glycoproteins and induces the adhesion of resting CD4+ T cells. *J. Immunol.* **1996**, *156*, 1189–1195. [CrossRef] [PubMed]
43. Yamazaki, T.; Koo, E.H.; Selkoe, D.J. Cell surface amyloid beta-protein precursor colocalizes with beta 1 integrins at substrate contact sites in neural cells. *J. Neurosci.* **1997**, *17*, 1004–1010. [CrossRef] [PubMed]
44. Matter, M.L.; Zhang, Z.; Nordstedt, C.; Ruoslahti, E. The alpha5beta1 integrin mediates elimination of amyloid-beta peptide and protects against apoptosis. *J. Cell Biol.* **1998**, *141*, 1019–1030. [CrossRef] [PubMed]
45. Iadecola, C.; Duering, M.; Hachinski, V.; Joutel, A.; Pendlebury, S.T.; Schneider, J.A.; Dichgans, M. Vascular Cognitive Impairment and Dementia: JACC Scientific Expert Panel. *J. Am. Coll Cardiol.* **2019**, *73*, 3326–3344. [CrossRef] [PubMed]
46. Raz, L.; Knoefel, J.; Bhaskar, K. The neuropathology and cerebrovascular mechanisms of dementia. *J. Cereb. Blood Flow. Metab.* **2016**, *36*, 172–186. [CrossRef] [PubMed]
47. Hurd, M.D.; Martorell, P.; Delavande, A.; Mullen, K.J.; Langa, K.M. Monetary costs of dementia in the United States. *N. Engl. J. Med.* **2013**, *368*, 1326–1334. [CrossRef] [PubMed]
48. Estimation of the global prevalence of dementia in 2019 and forecasted prevalence in 2050: An analysis for the Global Burden of Disease Study 2019. *Lancet Public Health* **2022**, *7*, e105–e125. [CrossRef]
49. Bir, S.C.; Khan, M.W.; Javalkar, V.; Toledo, E.G.; Kelley, R.E. Emerging Concepts in Vascular Dementia: A Review. *J. Stroke Cerebrovasc. Dis.* **2021**, *30*, 105864. [CrossRef]
50. O'Brien, J.T.; Erkinjuntti, T.; Reisberg, B.; Roman, G.; Sawada, T.; Pantoni, L.; Bowler, J.V.; Ballard, C.; DeCarli, C.; Gorelick, P.B.; et al. Vascular cognitive impairment. *Lancet Neurol.* **2003**, *2*, 89–98. [CrossRef]
51. Gooch, J.; Wilcock, D.M. Animal Models of Vascular Cognitive Impairment and Dementia (VCID). *Cell Mol. Neurobiol.* **2016**, *36*, 233–239. [CrossRef] [PubMed]
52. Roberts, J.M.; Maniskas, M.E.; Bix, G.J. Bilateral carotid artery stenosis causes unexpected early changes in brain extracellular matrix and blood-brain barrier integrity in mice. *PLoS ONE* **2018**, *13*, e0195765. [CrossRef] [PubMed]
53. Togo, T.; Akiyama, H.; Iseki, E.; Kondo, H.; Ikeda, K.; Kato, M.; Oda, T.; Tsuchiya, K.; Kosaka, K. Occurrence of T cells in the brain of Alzheimer's disease and other neurological diseases. *J. Neuroimmunol.* **2002**, *124*, 83–92. [CrossRef] [PubMed]
54. Itagaki, S.; McGeer, P.L.; Akiyama, H. Presence of T-cytotoxic suppressor and leucocyte common antigen positive cells in Alzheimer's disease brain tissue. *Neurosci. Lett.* **1988**, *91*, 259–264. [CrossRef] [PubMed]
55. Rossi, B.; Angiari, S.; Zenaro, E.; Budui, S.L.; Constantin, G. Vascular inflammation in central nervous system diseases: Adhesion receptors controlling leukocyte-endothelial interactions. *J. Leukoc. Biol.* **2011**, *89*, 539–556. [CrossRef] [PubMed]
56. Pietronigro, E.; Zenaro, E.; Bianca, V.D.; Dusi, S.; Terrabuio, E.; Iannoto, G.; Slanzi, A.; Ghasemi, S.; Nagarajan, R.; Piacentino, G.; et al. Blockade of α4 integrins reduces leukocyte-endothelial interactions in cerebral vessels and improves memory in a mouse model of Alzheimer's disease. *Sci. Rep.* **2019**, *9*, 12055. [CrossRef] [PubMed]
57. Kitayama, J.; Fuhlbrigge, R.C.; Puri, K.D.; Springer, T.A. P-selectin, L-selectin, and alpha 4 integrin have distinct roles in eosinophil tethering and arrest on vascular endothelial cells under physiological flow conditions. *J. Immunol.* **1997**, *159*, 3929–3939. [CrossRef]
58. Engelhardt, B.; Ransohoff, R.M. Capture, crawl, cross.s: The T cell code to breach the blood-brain barriers. *Trends Immunol.* **2012**, *33*, 579–589. [CrossRef]

59. Jiang, H.; Zhang, F.; Yang, J.; Han, S. Angiopoietin-1 ameliorates inflammation-induced vascular leakage and improves functional impairment in a rat model of acute experimental autoimmune encephalomyelitis. *Exp. Neurol.* **2014**, *261*, 245–257. [CrossRef]
60. Cai, H.Y.; Tian, K.W.; Zhang, Y.Y.; Jiang, H.; Han, S. Angiopoietin-1 and αvβ3 integrin peptide promote the therapeutic effects of L-serine in an amyotrophic lateral sclerosis/Parkinsonism dementia complex model. *Aging (Albany NY)* **2018**, *10*, 3507–3527. [CrossRef]
61. Li, Q.X.; Berndt, M.C.; Bush, A.I.; Rumble, B.; Mackenzie, I.; Friedhuber, A.; Beyreuther, K.; Masters, C.L. Membrane-associated forms of the beta A4 amyloid protein precursor of Alzheimer's disease in human platelet and brain: Surface expression on the activated human platelet. *Blood* **1994**, *84*, 133–142. [CrossRef] [PubMed]
62. Laske, C.; Sopova, K.; Stellos, K. Platelet activation in Alzheimer's disease: From pathophysiology to clinical value. *Curr. Vasc. Pharmacol.* **2012**, *10*, 626–630. [CrossRef] [PubMed]
63. Lee, B.K.; Jee, H.J.; Jung, Y.S. Aβ(1-40)-Induced Platelet Adhesion Is Ameliorated by Rosmarinic Acid through Inhibition of NADPH Oxidase/PKC-δ/Integrin α(IIb)β(3) Signaling. *Antioxidants* **2021**, *10*, 1671. [CrossRef] [PubMed]
64. Ortiz-Sanz, C.; Llavero, F.; Zuazo-Ibarra, J.; Balantzategi, U.; Quintela-López, T.; Wyssenbach, A.; Capetillo-Zarate, E.; Matute, C.; Alberdi, E.; Zugaza, J.L. Recombinant Integrin β1 Signal Peptide Blocks Gliosis Induced by Aβ Oligomers. *Int. J. Mol. Sci.* **2022**, *23*, 5747. [CrossRef] [PubMed]
65. Parham, C.L.; Shaw, C.; Auckland, L.D.; Dickeson, S.K.; Griswold-Prenner, I.; Bix, G. Perlecan Domain V Inhibits Amyloid-β Induced Activation of the α2β1 Integrin-Mediated Neurotoxic Signaling Cascade. *J. Alzheimers Dis.* **2016**, *54*, 1629–1647. [CrossRef]
66. Wright, S.; Parham, C.; Lee, B.; Clarke, D.; Auckland, L.; Johnston, J.; Lawrence, A.L.; Dickeson, S.K.; Santoro, S.A.; Griswold-Prenner, I.; et al. Perlecan domain V inhibits α2 integri.in-mediated amyloid-β neurotoxicity. *Neurobiol. Aging* **2012**, *33*, 1379–1388. [CrossRef]
67. Wright, S.; Malinin, N.L.; Powell, K.A.; Yednock, T.; Rydel, R.E.; Griswold-Prenner, I. Alpha2beta1 and alphaVbeta1 integrin signaling pathways mediate amyloid-beta-induced neurotoxicity. *Neurobiol. Aging* **2007**, *28*, 226–237. [CrossRef]
68. Qin, T.; Prins, S.; Groeneveld, G.J.; Van Westen, G.; de Vries, H.E.; Wong, Y.C.; Bischoff, L.J.M.; de Lange, E.C.M. Utility of Animal Models to Understand Human Alzheimer's Disease, Using the Mastermind Research Approach to Avoid Unnecessary Further Sacrifices of Animals. *Int. J. Mol. Sci.* **2020**, *21*, 3158. [CrossRef]
69. Vitek, M.P.; Araujo, J.A.; Fossel, M.; Greenberg, B.D.; Howell, G.R.; Rizzo, S.J.S.; Seyfried, N.T.; Tenner, A.J.; Territo, P.R.; Windisch, M.; et al. Translational animal models for Alzheimer's disease: An Alzheimer's Association Business Consortium Think Tank. *Alzheimers Dement. (N Y)* **2020**, *6*, e12114. [CrossRef]
70. Santiago, J.A.; Potashkin, J.A. The Impact of Disease Comorbidities in Alzheimer's Disease. *Front. Aging Neurosci.* **2021**, *13*, 631770. [CrossRef]
71. Veening-Griffioen, D.H.; Ferreira, G.S.; van Meer, P.J.K.; Boon, W.P.C.; Gispen-de Wied, C.C.; Moors, E.H.M.; Schellekens, H. Are some animal models more equal than others? A case study on the translational value of animal models of efficacy for Alzheimer's disease. *Eur. J. Pharmacol.* **2019**, *859*, 172524. [CrossRef] [PubMed]
72. Hainsworth, A.H.; Allan, S.M.; Boltze, J.; Cunningham, C.; Farris, C.; Head, E.; Ihara, M.; Isaacs, J.D.; Kalaria, R.N.; Lesnik Oberstein, S.A.; et al. Translational models for vascular cognitive impairment: A review including larger species. *BMC Med.* **2017**, *15*, 16. [CrossRef] [PubMed]
73. Muratoglu, S.C.; Charette, M.F.; Galis, Z.S.; Greenstein, A.S.; Daugherty, A.; Joutel, A.; Kozel, B.A.; Wilcock, D.M.; Collins, E.C.; Sorond, F.A.; et al. Perspectives on Cognitive Phenotypes and Models of Vascular Disease. *Arterioscler Thromb. Vasc. Biol.* **2022**, *42*, 831–838. [CrossRef] [PubMed]
74. Zheng, B.; Su, B.; Udeh-Momoh, C.; Price, G.; Tzoulaki, I.; Vamos, E.P.; Majeed, A.; Riboli, E.; Ahmadi-Abhari, S.; Middleton, L.T. Associations of Cardiovascular and Non-Cardiovascular Comorbidities with Dementia Risk in Patients with Diabetes: Results from a Large UK Cohort Study. *J. Prev. Alzheimers Dis.* **2022**, *9*, 86–91. [CrossRef] [PubMed]

biomolecules

MDPI

Review

Microvascular Contributions to Alzheimer Disease Pathogenesis: Is Alzheimer Disease Primarily an Endotheliopathy?

Rawan Tarawneh

Department of Neurology, Center for Memory and Aging, University of New Mexico, Albuquerque, NM 87106, USA; raltarawneh@salud.unm.edu; Tel.: +1-505-272-5429

Abstract: Alzheimer disease (AD) models are based on the notion that abnormal protein aggregation is the primary event in AD, which begins a decade or longer prior to symptom onset, and culminates in neurodegeneration; however, emerging evidence from animal and clinical studies suggests that reduced blood flow due to capillary loss and endothelial dysfunction are early and primary events in AD pathogenesis, which may precede amyloid and tau aggregation, and contribute to neuronal and synaptic injury via direct and indirect mechanisms. Recent data from clinical studies suggests that endothelial dysfunction is closely associated with cognitive outcomes in AD and that therapeutic strategies which promote endothelial repair in early AD may offer a potential opportunity to prevent or slow disease progression. This review examines evidence from clinical, imaging, neuropathological, and animal studies supporting vascular contributions to the onset and progression of AD pathology. Together, these observations support the notion that the onset of AD may be primarily influenced by vascular, rather than neurodegenerative, mechanisms and emphasize the importance of further investigations into the vascular hypothesis of AD.

Keywords: endothelium; pericyte; capillary; neurodegeneration; Alzheimer disease

1. Introduction

Alzheimer disease (AD) is the most common cause of dementia, accounting for over 70% of dementia cases in individuals above the age of 65 years [1]. The two main pathological hallmarks of AD are extracellular deposits of the amyloid-β (Aβ) protein in the form of amyloid plaques and intracellular aggregates of hyperphosphorylated tau protein in the form of neurofibrillary tangles [2]. Current disease models are based on the notion that abnormal protein aggregation is the primary event in AD, which begins a decade or longer prior to symptom onset, and ultimately leads to synaptic injury and neurodegeneration [3].

Vascular disease, including arteriolosclerosis, atherosclerosis, microinfarcts, and cerebral amyloid angiopathy (CAA), is a common co-pathology which is observed in 20–80% of AD brains at autopsy [4–7]. Furthermore, almost all AD brains display evidence of endothelial and capillary degeneration even in the absence of other forms of macrovascular pathology [8–12]. Significant and bidirectional interactions between AD and various forms of vascular pathology have been well documented [13]; amyloid and tau toxicity disrupts the blood–brain barrier (BBB) and alters vascular permeability, and structural or functional damage to cerebral vasculature impairs amyloid clearance and promotes tau aggregation. Previous neuropathological studies examining vascular pathology in AD have focused primarily on pathology within the small- and medium-sized arteries and arterioles; however, there is growing evidence to suggest that "micro"-vascular disease (i.e., at the capillary level) and alterations to specific vascular constituents, such as endothelium and pericytes, play an important role in AD pathogenesis [8,9,14–18]. From a functional standpoint, cross-sectional and longitudinal clinical studies and animal models that utilize advanced vascular imaging methods suggest that impaired cerebral blood flow (CBF) is a common and early predictor of AD pathology, which possibly precedes abnormal protein aggregation and directly contributes to neuronal and synaptic loss in even the earliest pre-symptomatic stages of the disease [8,19].

This review will examine evidence from clinical, imaging, neuropathological, and animal studies supporting vascular contributions to the onset and progression of AD pathology, including amyloid and tau aggregation, as well as direct vascular contributions to synaptic injury, impaired axonal repair, and neurodegeneration. Importantly, the present review discusses evidence supporting the role of specific vascular constituents (e.g., pericytes, endothelium, astrocytes, and basement membrane proteins) and vascular abnormalities at the capillary level (i.e., "microvascular" disease) in mediating AD onset and progression. Together, these observations support the notion that the onset of AD may be primarily influenced by vascular, rather than neurodegenerative, mechanisms and emphasize the importance of further investigations into the vascular hypothesis of AD.

2. Evidence to Support Vascular Contributions to AD Pathogenesis

Several lines of evidence support the presence of vascular contributions to AD pathogenesis including associations of vascular pathology (e.g., arteriolosclerosis, microinfarcts, and CAA) with mechanisms involved in AD pathogenesis, including amyloid and tau aggregation, oxidative stress, mitochondrial dysfunction, and vascular inflammation, which cumulatively contribute to synaptic loss and neurodegeneration [8,9,13–15,19,20]. Importantly, there is recent evidence to suggest associations of specific vascular constituents (e.g., endothelium or pericytes) with molecular mechanisms or pathways directly involved in synaptic plasticity, axonal growth, neuronal repair, and memory or learning functions (e.g., long-term potentiation) [21,22].

The vascular hypothesis of AD, which was first proposed over 30 years ago, is based on the notion that the first inciting event in AD pathogenesis is vascular dysregulation, which then initiates a cascade of molecular and neuropathological changes leading to neuronal dysfunction and the onset of AD pathologies, such as amyloid and tau aggregation [23–26]. A revised version of the vascular hypothesis suggests a "two-hit" process, in which amyloid-independent microvascular damage (hit one) occurs due to several mechanisms (described below), resulting in impaired amyloid clearance and then subsequent amyloid toxicity (hit two) exacerbates microvascular injury via a positive feedback loop [20]. In various versions of the vascular hypothesis, the inciting event in AD onset is vascular dysregulation and reduced CBF, which precedes, causes, and then further exacerbates protein aggregation and neuronal or synaptic injury, the latter driving the cognitive and behavioral decline observed in the symptomatic stages [8].

This review summarizes the literature supporting this hypothesis and presents new data suggesting additional direct links between specific vascular components and cognition. Importantly, this current review expands the vascular hypothesis beyond associations of macrovascular pathology with AD to highlight the importance of "micro"-vascular constituents at the cellular and molecular level in AD onset and progression.

3. Vascular Risk Factors Increase the Risk for AD

The association of vascular risk factors, including hypertension, hyperlipidemia, diabetes mellitus (DM), obesity, and metabolic syndrome, with AD has been well documented in several clinical and epidemiological studies [8,27–33]. Low insulin levels [34] and DM [35] have been linked to AD risk in several other large studies, with hazard ratios of 1.31 and 1.62, respectively. A systematic review of 15 epidemiological studies reported a pooled adjusted hazard ratio risk of 1.57 with DM [36,37]. Increased AD risk in individuals with DM was found to be almost comparable to, and synergistic with, that of the Apolipoprotein E4 (*APOE4*) genotype, the most significant genetic risk factor for AD [35,38]. Apolipoprotein E, a major cholesterol transporter in the brain, has three isoforms: *E2*, *E3*, and *E4*. The presence of 1 or 2 *APOE4* alleles increases AD risk by 2–3-fold and 10–15-fold, respectively, due to increased amyloid and tau aggregation, dysregulated cholesterol metabolism, and impaired synaptic plasticity [39–41]. Carriers of the *APOE4* allele have a higher risk for both late-onset AD and cardiovascular disease, further supporting an important link between vascular disease and AD pathogenesis [42].

Other studies have shown associations of AD with midlife hypertension, while associations of AD with late-life hypertension have been inconsistent. In the longitudinal Honolulu-Asia Aging Study, midlife hypertension was associated with a higher risk for AD in middle-aged Japanese American men followed over 25 years, who demonstrated neuropathological evidence of increased hippocampal tau pathology and brain atrophy over follow-up, compared to normotensive controls [43]. Interestingly, midlife hypertension increased the risk of mild cognitive impairment (MCI) to a similar degree as *APOE4* [44–46]. AD risk increases with higher levels of low-density-lipoprotein (LDL) cholesterol and lower levels of high-density-lipoprotein (HDL) cholesterol [47,48]. Further, early onset AD has been linked to coding variants in *APOB*, which codes for apolipoprotein B, an important component of LDL [49]. High LDL has been linked to AD pathogenesis by increasing the risk for cerebrovascular disease, vascular inflammation, increased Aβ synthesis, Aβ-mediated synaptic toxicity, and tau aggregation [50–52].

The Rotterdam study, which examined over 7000 older adults, reported a higher risk for AD in association with vascular risk factors, such as DM, thrombotic episodes, high fibrinogen or homocysteine levels, atrial fibrillation, smoking, alcoholism, and atherosclerosis [8,53]. Participants with severe atherosclerosis had a three-times higher risk of developing AD or vascular dementia [53]. In the large longitudinal Atherosclerosis Risk in Communities (ARIC) study, which followed a multiracial cohort of middle-aged adults for 25 years, vascular risk factors, such as midlife smoking, diabetes, prehypertension, and hypertension, were associated with a higher risk for dementia including AD [38]. Elevated homocysteine levels, which are independent risk factors for vascular disease, are also associated with a higher risk for AD, brain atrophy, and tau pathology [54,55]. High homocysteine has been linked to impaired post-synaptic GABA (γ-aminobutyric acid) signaling, matrix metalloproteinase-9 (MMP-9) activation, and increased oxidative injury, which disrupts the BBB and contributes to AD pathogenesis [56].

Furthermore, vascular risk factors appear to have additive or synergistic effects on AD risk, with a higher risk being reported in the presence of multiple vascular risk factors [57,58]. One study found that diabetes and current smoking were the strongest risk factors in isolation or in clusters, while hypertension and heart disease were related to a higher risk of AD when clustered together or with either diabetes or smoking [59].

4. Concomitant Macrovascular Disease Is Present in Most, and Microvascular Disease in Almost All, AD Brains

Vascular disease, including subcortical white matter lesions or lacunes, is a common co-pathology in AD, being reported in 18–84% of AD brains [5,7]; however, the degree to which these forms of vascular disease contribute to neurodegeneration is yet to be determined. Despite their co-occurrence, AD and vascular pathology often do not localize in the same brain regions [5]. Furthermore, it remains unclear whether the severity of vascular disease correlates with the severity of neuronal or synaptic loss in AD brains [5]. While several studies from various cohorts have shown that white matter hyperintensities (WMH), which are used as an imaging surrogate of small vessel disease, are closely associated with cognitive outcomes, other studies suggest that mild degrees of small vessel disease (e.g., subcortical lacunes and white matter lesions) are not strongly associated with cognition; however, their presence may be associated with a lower threshold for cognitive impairment due to AD pathology [5], and hence, a higher risk for AD dementia [5,13]. Further, studies suggest that WMH are common in AD, and when present in parietal brain regions, are frequently associated with AD pathology rather than arteriolosclerosis risk [60]. Consistent with these findings, a recent study from the Harvard Aging Brain cohort showed that vascular risk and amyloid pathology interacted to predict more severe longitudinal atrophy in certain brain regions in AD; however, this interaction was not related to WMH.

In individuals with a high amyloid burden, gray matter atrophy mediated the associations of vascular risk with cognitive outcomes [61].

Together, these findings suggest that vascular risk and Aβ likely have independent, yet overlapping [62], effects on brain atrophy and cognition [63] and that these effects are not strongly influenced by imaging estimates of small vessel disease [61]. Conversely, microvascular pathology in the form of capillary, endothelial, and pericyte abnormalities is ubiquitous in AD brains and may occur independently of vascular pathology in larger vessels [10,11]. There is growing evidence to support an important direct association of microvascular alterations with cognitive impairment and the onset or progression of AD pathology [8].

4.1. Brain Capillary Degeneration

Structural distortions to brain capillaries, including increased tortuosity, looping or kinking, basement membrane thickening, and luminal buckling or narrowing, are consistently observed in AD brains [10,16–18] and may have been reported in AD brains by Tuke decades prior to the first description of amyloid plaques and tau tangles [8,64]. It has been proposed that microvascular alterations in AD contribute to neuronal loss via reduced cerebral blood flow and impaired tissue oxygenation [8]. Microvascular alterations in AD display regional and laminar patterns that parallel those of neuronal loss [65], are preferentially localized in the vicinity of dystrophic neurites [66,67], and occur with a high prevalence in the hippocampus [65,68]. Animal studies have shown that significant alterations to brain microvasculature are evident in young AD transgenic mice and precede vascular or brain parenchymal amyloid pathology [15,69,70]. Furthermore, capillary abnormalities in AD do not correlate with the severity of AD pathological changes (i.e., Braak staging) [71,72], which suggests that microvascular alterations in AD are not solely a consequence of other AD pathologies.

4.2. Endothelium

Endothelial dysfunction is observed in almost all AD brains at autopsy [12], including the majority of AD brains that display no pathological evidence of macrovascular disease (i.e., arteriolosclerosis or atherosclerosis) [10,11]. Endothelial dysfunction refers to characteristic structural or functional alterations that occur at the intracellular or cell membrane level and interfere with endothelial cell integrity or function [9–11,16,65–67,73,74]. Structural changes consistent with endothelial dysfunction, such as increased pinocytic vesicles, laminin deposition in the basement membrane, collagen accumulation, and focal necrotic changes, are common in AD [9,11,16,73,74]. Endothelial cell loss in AD is demonstrated by reduced staining for endothelial markers, such as CD31 and CD34 [10], and higher plasma levels of endothelial markers, such as vascular cell adhesion molecule (VCAM-1), reflective of endothelial injury [75]. A recent study in a large well-characterized cohort of cognitively normal individuals with biomarker evidence of AD (i.e., "preclinical" or "pre-symptomatic" AD) has shown that cerebrospinal fluid (CSF) levels of the endothelial protein, vascular endothelial-cadherin (VE-Cadherin), are increased in even the earliest pre-symptomatic stages of AD compared to healthy controls; correlate with markers of amyloid, tau, and neurodegeneration; and are associated with cognitive outcomes independently of imaging measures of small vessel disease [21]. Furthermore, findings from this study suggest that endothelial injury partially mediates effects of amyloid and tau on cognitive outcomes [21].

Endothelial alterations have previously been considered secondary to amyloid or tau toxicity on cerebral vasculature; however, emerging evidence from animal studies suggests that endothelial dysfunction is an early and primary process in AD that may precede amyloid and tau pathologies and promotes abnormal protein aggregation and synaptic loss [12,15,69]. RNA-sequencing (RNA-seq) analyses suggest that endothelial pathways are among the most differentially expressed in human AD brains [76]. Consistent with these reports, 30 out of 45 AD risk genes are expressed in vascular cell types, and several of these have their highest expression levels in endothelial structures [77] (Table 1). Animal studies, including those that have examined the three-dimensional (3D) arrangement of brain vessels, suggest that significant morphological changes in cerebral vessels and endothelial dysfunction precede amyloid deposition and cognitive deficits in AD transgenic [69] or insulin-resistant mice [78]. While the mechanisms that underly these changes are yet to be elucidated, studies suggest that tumor necrosis factor (TNF)-α-mediated inflammation, oxidative stress, and leukocyte recruitment cause endothelial injury and loss of BBB integrity in insulin-resistant and diabetic mice [78,79]. It has been proposed that, in AD transgenic mice, impaired endothelial clearance of neuronal Aβ triggers the accumulation of Aβ and other proteins near endothelial membranes, which then form endothelial pores, further exacerbating endothelial dysfunction. Early endothelial damage correlates with cognitive deficits in these models and precedes the formation of larger vascular or parenchymal amyloid plaques [69]. Blocking endothelial interactions with leukocytes reduces Aβ deposition and tau hyperphosphorylation and improves memory in AD mouse models [80].

Table 1. AD Risk Genes Expressed in Brain Vascular Cell Types.

AD Risk Gene	Vascular Cell Types
APOE	Smooth muscle cells, Meningeal fibroblasts
PICALM *	Brain endothelium (arterial, capillary, and venous)
CLU	Meningeal fibroblasts, Ependymal cells
ABCA7	T cells
PTK2B	T cells
PLCG2 *	Brain endothelium (arterial)
HLA-DRB1 *	Brain endothelium (arterial)
CD2AP *	Brain endothelium (arterial, capillary, and venous)
SLC24A4	Ependymal cells
RIN3	T cells
ADAMTS1 *	Smooth muscle cells, Pericytes, Brain endothelium (arterial)
ADAMTS4	Smooth muscle cells, Pericytes
FERMT2	Smooth muscle cells, Pericytes
SCIMP	Ependymal cells
CLNK	T cell
ECHDC3	Perivascular fibroblasts
TNIP1 *	Brain endothelium (capillary and venous), T cells
ABCA1 *	Brain endothelium (venous), Perivascular fibroblasts
USP6NL *	Brain endothelium (capillary and venous)
INPP5D *	Brain endothelium (capillary), T cells
ACE *	Brain endothelium (arterial and capillary)
IQCK	Ependymal cells
ABI3	T cells
HESX1	Meningeal fibroblasts
FHL2	Perivascular fibroblasts
CHRNE	Perivascular fibroblasts, T cells
AGRN	Pericytes

Table 1. *Cont.*

AD Risk Gene	Vascular Cell Types
IL34	Smooth muscle cells, Pericytes, Meningeal fibroblasts
NYAP1	Ependymal cells
CASS4 *	Brain endothelium (capillary and venous)

Of the 45 top AD risk genes, 30 genes (shown above) are expressed in vascular cell types (excluding perivascular macrophages), including 11 that are expressed in the brain endothelium (marked with an asterisk). The remaining 15 genes include 6 genes that are solely expressed in perivascular macrophages (*CR1, HAVCR2, CD33, TREM2, MS4A6A,* and *SPI1*). The other 9 top AD risk genes were *BIN1, SORL1, ADAM10, CNTNAP2, WWOX, APH1B, HS3ST1, KAT8,* and *CCDC6*. Adapted from Yang et al. [77].

4.3. Pericytes

Pericytes are specialized cells located in the capillary basement membrane, which have contractile properties, and are uniquely positioned to interact closely with other cellular components of the neurovascular unit, such as endothelium, neurons, and astrocytes [81,82]. Pericytes play an important role in maintaining microvascular stability of the BBB and regulating neurovascular coupling and CBF [81]. Pericytes also demonstrate important phagocytic functions; under normal physiological conditions, these cells are involved in Aβ clearance and degradation through their surface expression of the low-density-lipoprotein receptor-related protein-1 (LRP-1) in an *APOE*-isoform-specific manner [83]; however, the phagocytic capacity of pericytes is limited in the presence of toxic Aβ levels, eventually leading to pericyte loss and further exacerbation of amyloid aggregation [82]. Pericyte loss is observed in AD and CAA mouse models and correlates with Aβ burden, white matter degeneration, and BBB disruption in the cortex and hippocampus [84–87].

Pericyte dysfunction has been associated with *APOE4*-mediated disruption of the neurovascular unit in clinical and animal studies. *APOE4* expression results in upregulation of the LRP-1-dependent proinflammatory cyclophilin A (CypA)–nuclear factor κB (NF-κB)–MMP-9 pathway in pericytes and brain endothelium. Activated MMP-9 from pericytes leads to endothelial toxicity, BBB breakdown, and neuronal injury [88]. An age-dependent disruption of the BBB is observed in *APOE4*-carriers years prior to the onset of cognitive decline and is associated with increased CSF levels of MMP-9 derived from pericytes [89].

Pericytes also appear to contribute to tau aggregation in AD; pericyte loss is associated with neuronal accumulation of hyperphosphorylated tau, caspase-cleaved tau, and tau aggregation in animal studies [90,91]. Interestingly, these changes are only observed in animal models that also had evidence of excess Aβ [91]. In a study by Sagare et al. [91], pericyte-deficient (*Pdgfrβ*$^{+/-}$) mice were crossed with mice overexpressing the Swedish mutation of the human amyloid precursor protein (APP; *APP*$^{sw/0}$ mice) [91]. While neither *Pdgfrβ*$^{+/-}$ nor *APP*$^{sw/0}$ mice demonstrated tau pathology, *APP*$^{sw/0}$/Pdgfrβ$^{+/-}$ mice had evidence of tau pathology as early as 9 months. Neuronal loss and impaired hippocampal functions were observed in pericyte-deficient mice, although these changes appeared to be more severe when excess Aβ was also present [90,91]. Other animal studies have shown tau accumulation in endothelium and vascular smooth muscle cells in AD [92]. While impaired caveolin-mediated endocytosis in degenerating pericytes and vascular smooth muscle has been implicated in vascular tau accumulation in animal models of traumatic brain injury (TBI), it remains unclear whether similar mechanisms also occur in AD [93].

4.4. Astrocytic End-Foot Processes

Astrocytes are specialized glia that interact with other components of the neurovascular unit through long cellular processes (i.e., astrocytic end-feet) [94]. Several lines of evidence from in vitro and in vivo animal studies implicate astrocytes early in AD pathogenesis via several mechanisms [94], including increased Aβ synthesis and release, neuronal and astrocytic tau aggregation, neuroinflammation, increased oxidative stress, glutamate excitotoxicity due to disrupted astrocyte–neuron signaling, and impaired energy metabolism, which ultimately lead to neuronal and synaptic dysfunction.

Reactive astrocytes in AD may contribute to amyloid pathology through both *APOE4*-dependent and *APOE4*-independent pathways [95]. Under physiological conditions, astrocytes can internalize and degrade Aβ through enzymatic cleavage by neprilysin, insulin-degrading enzyme, and MMPs [95,96]. Astrocytes also contribute to Aβ clearance through the glymphatic system [97]. Astrocytic end-foot processes are rich in aquaporin-4 (AQP4), a transmembrane protein involved in regulating water transport and interstitial fluid/CSF exchange, including Aβ clearance. Loss of astrocytic AQP4 in AD is associated with impaired Aβ clearance [98].

Conversely, inflammatory conditions with high TNF-α and interferon-γ (INF-γ) levels are associated with increased Aβ synthesis and release by astrocytes [99,100]. Excess Aβ is toxic to astrocytes and activates several proinflammatory pathways, such as the receptor for advanced glycation end-products (RAGE) and NF-κB, which result in the release of inflammatory mediators into the neuronal milieu and promote Aβ synthesis [94]. *APOE4* expression by astrocytes also stimulates their transformation into the pro-inflammatory phenotype, which further contributes to Aβ aggregation [99].

Astrocytes have been linked to increased tau phosphorylation and tau-mediated neurodegeneration in AD, including tau accumulation within astrocytes and the release of tau oligomers through exosome-dependent pathways [101]. Astrocytic release of glypican-4 is implicated in *APOE4*-dependent tau hyperphosphorylation [102]. Several AD risk genes are expressed in astrocytes (*APOE4*, *WWOX*, *CLU*, *CDK2AP1*, *MEF2C*, and *IQCK*) and are linked to both amyloid and tau metabolism [95,103].

4.5. Other Vascular Constituents

Other vascular constituents, such as the extracellular matrix (ECM) and basement membrane proteins, may have a role in AD pathogenesis [104]. It has recently been shown that the ECM protein decorin is significantly increased in *APP* knock-in mice at an early age and that CSF decorin levels correlate with Aβ plaque load and decorin levels in the choroid plexus [104]. Consistent with these reports, CSF decorin levels are increased in preclinical AD and correlate with CSF Aβ42 levels [104]. Elevated CSF decorin levels appear to differentiate a subtype of AD characterized by activation of the innate immune system and possibly choroid plexus dysfunction [104].

5. Impaired Cerebral Perfusion Is a Common and Early Event in AD Which Strongly Predicts Future Cognitive Impairment in Clinical and Animal Studies

Impaired CBF, measured by arterial spin labeling magnetic resonance imaging (MRI), is well documented early in AD, including preclinical stages, as well as in animal studies [105–109]. Furthermore, CBF abnormalities predict future cognitive impairment and disease progression in symptomatic individuals [110]. CBF reductions of 10% to 25% are observed early in AD [111]. Studies utilizing single photon emission computed tomography (SPECT)-image reconstruction found that older adults with symptomatic AD have significantly reduced hippocampal perfusion compared to age-matched controls [112]. In one study, individuals with subjective memory loss who had significant hypoperfusion in the hippocampus and amygdala on SPECT were more likely to be diagnosed with AD over follow-up [113]. In another study of non-demented individuals, higher CBF was associated with larger hippocampal and amygdala volumes and lower risk for progression to dementia [114]. Consistent with these findings, lower volumetric flow rates in the internal and middle cerebral arteries, measured using four-dimensional (4D) MRI, were associated with more severe brain atrophy, and low flow rates in the internal carotid artery with amyloid pathology [115]. Temporoparietal, hippocampal–parahippocampal, posterior cingulate, and inferior parietal lobe hypoperfusion is more common in MCI or early symptomatic AD compared to controls, and several of these regional hypoperfusion abnormalities are associated with a higher risk for progression from MCI to AD dementia [112,116,117]. Importantly, recent data from large longitudinal AD cohorts enrolled in the Alzheimer's

Disease Neuroimaging Initiative (ADNI) suggests that vascular flow abnormalities occur very early in AD and may precede biomarker indicators of amyloid and tau pathology [19].

Consistent with these findings, animal models of chronic brain hypoperfusion (e.g., bilateral carotid artery occlusion) demonstrate AD pathology, including amyloid and tau aggregation and capillary changes in the CA1 region of the hippocampus that are almost identical to those seen in human AD brains [72]. In these models, CBF reductions to 25–33% of baseline were associated with several metabolic changes leading to cognitive and behavioral impairment and neuronal loss in the absence of strokes, hemorrhages, or hypertension [118,119]. These changes included loss of microtubule-associated protein-2 in CA1 [120], reduced hippocampal cytochrome oxidase [121], impaired monoamine neurotransmitter turnover [108], altered glucose utilization [122], reduced post-synaptic cholinergic activity [123], increased heme oxygenase [124] and matrix metalloproteinase (MMP)-2 expression [125], and reactive gliosis, leading to a state of impaired energy metabolism, disturbed protein synthesis, and increased oxidative stress, which were detectable months prior to the onset of hippocampal neuronal loss or cognitive impairment. Other animal studies of chronic cerebral hypoperfusion show increased hippocampal and frontal Aβ pathology due to activation of β- and γ-secretases [126,127] (two enzymes in the amyloidogenic pathway of *APP* processing that sequentially cleave *APP* to produce the Aβ42 peptide), and impaired Aβ clearance due to reduced LRP-1 and increased RAGE expression in the brain endothelium [128].

While neuronal dysfunction and subsequent hypometabolism may contribute to reduced CBF in AD, several observations suggest that hypoperfusion in AD cannot be fully attributed to hypometabolism [129]; CBF alterations are detected before neurodegeneration [19] and the degree of hypoperfusion in AD exceeds what is expected based on the degree of atrophy alone [130]. Importantly, structural and functional changes to brain capillaries which are observed in AD, including capillary constrictions near amyloid deposits, and contraction of pericytes and endothelium, have been shown to reduce blood vessel diameter by up to 50% and thereby provide a mechanistic basis for reduced CBF in AD beyond what can be explained by neuronal or blood vessel loss [131,132]. Other possible mechanisms for capillary constriction in AD include regional vasoconstriction and increased vascular resistance in response to Aβ [133], capillary blockage by neutrophils [134], fibrin deposits [135], and focal constriction sites caused by pericyte soma [133]. Administration of an antineutrophil antibody (anti-Ly6G), or treatment with the anticoagulant dabigatran, are associated with increased CBF and improved memory functions in transgenic AD models [134,135]. High levels of reactive oxygen species (ROS) produced by microglia and pericytes, and endothelin-1 produced by the endothelium, are associated with capillary constriction in vitro [133]. Endothelial dysfunction may contribute to reduced CBF due to imbalance between vasodilators (e.g., nitric oxide (NO), bradykinin, and prostacyclin) and vasoconstrictors (e.g., endothelin-1 and thromboxane A2) [14]. Oligomeric Aβ further exacerbates capillary constriction and reduced CBF via the activation of microglial and astrocytic inflammasomes and the release of interleukin (IL)-1β. Furthermore, low levels of neuroglobin, a protein with ROS-scavenging and anti-apoptotic activities, may contribute to increased sensitivity of the hippocampus to hypoxia and, subsequently, an increased regional vulnerability to AD pathology due to impaired oxidative metabolism [136]. Together, these various mechanisms lead to hypoxia and reduce CBF, which contributes to neuronal dysfunction, synaptic loss, and hypometabolism [129,130].

Additional support for the importance of vascular contributions to AD pathogenesis comes from studies of individuals with trisomy 21 (Down syndrome) who have an additional copy of the *APP* gene. Capillary degeneration [65] and reduced brain perfusion are observed in the brains of individuals with Down syndrome years prior to any evidence of amyloid plaques or neurofibrillary tangles [137,138]. Oxidative stress appears to precede Aβ deposition by many years in individuals with Down syndrome who die in their teens and twenties [139], a finding that indicates that AD-like pathology is not likely the trigger of neuronal metabolic disruption in these individuals.

6. A Forward-Feedback Loop between Hypoxia and Endothelial Dysfunction Contributes to AD Pathogenesis

The brain endothelium is a highly active interface that regulates the neuronal milieu via diffusible mediators and, as a part of the neurovascular unit, controls blood flow and neuronal metabolism [14,140]. Brain endothelial cells produce several extracellular matrix proteins (e.g., proteoglycans, laminin, and fibronectin), growth factors (e.g., endothelium-derived growth factor (EDGF), platelet-derived growth factor (PDGF), and fibroblast growth factors (FGF)), proteases (e.g., thrombin and MMPs), vasoconstrictors (e.g., endothelin-1, leukotrienes, and thromboxane A2) and vasodilators (e.g., NO, prostacyclin/prostaglandin E2, and endothelium-derived hyperpolarizing factor (EDHF)), antithrombotic (e.g., antithrombin III) and prothrombotic factors (e.g., thromboplastin and platelet-activating factor), and various inflammatory mediators (e.g., IL-1, IL-6, IL-8, monocyte chemoattractant proteins (MCP) 1-2, leukotrienes, and cell adhesion molecules) [14,140].

Several studies suggest that hypoxia causes structural and functional alterations in the brain endothelium via different mechanisms, including rearrangement of the endothelial cytoskeleton, altering intercellular adhesions of endothelial cells, and promoting endothelial constriction, apoptosis, and degeneration [141]. Studies of intermittent hypoxia suggest that endothelial cell resistance is reduced due to ROS-dependent activation of the extracellular signal-regulated kinase (ERK1/2) pathway and c-Jun N-terminal kinase (JNK)-mediated re-arrangement of the endothelial cytoskeleton, including the redistribution of actin stress fibers and VE-cadherin, leading to abnormal endothelial barrier function [142,143]. Hypoxia stimulates the release of inflammatory mediators (e.g., thrombin, histamine, and bradykinin), which contribute to increased endothelial cell contractibility and permeability [144]. In vitro studies suggest that hypoxia increases endothelial apoptosis and endothelial cell degeneration by stimulating their transition into fibroblast-like cells [145]. Hypoxia is also associated with activation of the pro-inflammatory Toll-like receptor-4 (TLR4)–NF-κB pathway [146] and interferes with the physiologic anti-inflammatory and anti-oxidant properties of the endothelial membrane constituent, high-molecular-weight hyaluronic acid [147]. Conversely, endothelial dysfunction further exacerbates CBF reductions and cellular hypoxia through several mechanisms such as impaired angiogenesis, dysregulated thrombosis, local immune dysregulation, vascular inflammation, and neurovascular uncoupling [14]. Together, these findings suggest that hypoxia-ischemia and endothelial dysfunction act in a forward-feedback loop to trigger and exacerbate AD pathology, including abnormal protein aggregation, vascular inflammation, and neuronal/synaptic injury, via several converging mechanisms that are described below (Figure 1).

6.1. Increased Amyloidogenic APP Processing

Data from animal and human studies suggests that capillary constriction in AD is sufficient to reduce CBF by up to 50%, leading to upregulation of the β-secretase1 (*BACE1*) enzyme involved in Aβ synthesis [107]. *BACE1* levels are increased in human AD brains, including those in the preclinical or early symptomatic stages (i.e., MCI), and may correlate with Aβ plaque burden and cognitive decline [148]. Mechanisms by which reduced CBF increases *BACE1* expression in AD include a hypoxia-inducible factor (HIF)-1α-mediated increase in *BACE1* mRNA expression and enzymatic activity [149]. The *BACE1* gene promoter contains a functional hypoxia-response element (HRE), and HIF-1α can effectively bind to the *BACE1* promoter [149]. Significant reductions in cortical and hippocampal *BACE1* levels are observed in HIF-1α knockout mice. *BACE1* upregulation due to hypoxia is also mediated by caspase-3, which cleaves GGA3 (Golgi-associated, gamma adaptin ear containing, ARF binding protein-3) involved in *BACE1* trafficking, thereby reducing *BACE1* degradation [150–152]. As a γ-secretase-activating protein, HIF-1α also interacts directly with γ-secretase and increases its activity [153]. The APP intracellular domain (AICD) peptide, a product of *APP* cleavage by γ-secretase, stimulates the transcription of the *HIF-1α* gene, potentially creating a positive forward loop between HIF-1α and γ-secretase [154]. Hypoxia alters several proteins that interact with γ-secretase, including voltage-dependent anion channel 1 (VDAC1), a γ-secretase-activating protein whose expression is increased in human AD brains and AD mouse models [155]. Reduced VDAC1 expression is associated with lower brain expression of *APP*, *Presenilin-1* (*PS1*), *Presenilin-2* (*PS2*), and *BACE1* [156]. Other studies have shown that hypoxia inhibits α-secretase, an enzyme in the non-amyloidogenic pathway of *APP* processing, which reduces the production of the extracellular s*APP*α fragment, and results in subsequent loss of its neuroprotective and neurotrophic effects [149]. Studies in neuronal cultures suggest that hypoxia increases the production of an alternative splice variant of *PSEN2* (aka *PS2*) that lacks exon 5, referred to as *PSEN2V* [157], via activation of the high mobility group A protein 1a (HMGA1a). The *PSEN2V* variant is abundantly expressed in the hippocampus of AD brains, promotes Aβ synthesis, and increases the vulnerability of the endoplasmic reticulum (ER) to stress [157].

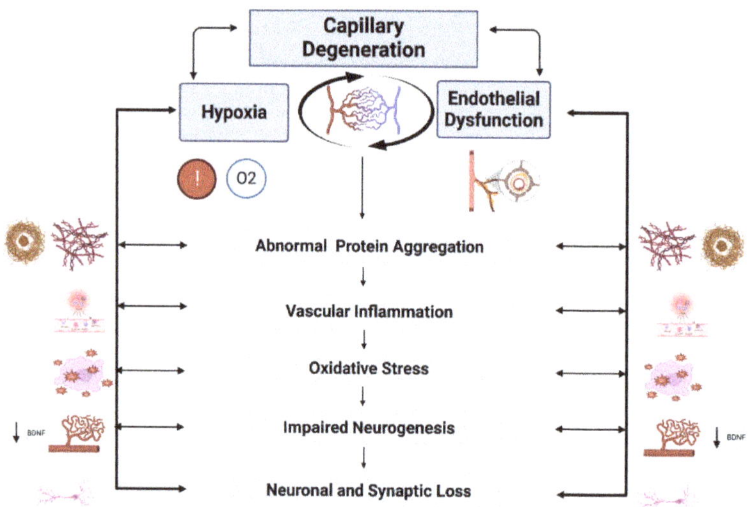

Figure 1. Bidirectional interactions between hypoxia and endothelial dysfunction in Alzheimer Disease. Capillary degeneration leads to reduced cerebral blood flow and hypoxia, which causes endothelial injury, further exacerbating tissue hypoxia. A forward-feedback loop between hypoxia and endothelial dysfunction initiates a cascade of events, including amyloid and tau aggregation, vascular inflammation, oxidative stress, impaired neurogenesis, and neuronal and/or synaptic injury. Created with Biorender.com.

6.2. Activation of Tau Kinases

Tau hyperphosphorylation is increased under hypoxic-ischemic conditions in an Aβ-independent manner [158,159]. Tau hyperphosphorylation is observed in hypertensive rats that do not display Aβ pathology [158] and can be triggered by unilateral carotid artery occlusion even when Aβ levels are not elevated [159]. Proposed mechanisms include activation of Cdk5 (cyclin-dependent kinase-5), calpain-mediated cleavage of the Cdk5 regulatory subunit p35 [160,161], and activation of glycogen synthase kinase (GSK)-3β via reduced activity of the phosphatidylinositol 3-kinase/Akt pathway [162].

6.3. Endoplasmic Reticulum Stress

ER stress and impaired autophagy also occur due to hypoxia and are associated with increased *PSEN1* (aka *PS1*) expression and γ-secretase activity, contributing to increased amyloid pathology [149], and m-calpain activation which promotes tau phosphorylation [163]. Two major pathways which are involved in the ER-stress-induced unfolded protein response (UPR), PERK–eIF2α–ATF4 and IRE1–X-box binding protein 1 (XBP1) [164], are activated in the setting of hypoxia-ischemia. ATF4 binds to the *PSEN1* gene promoter and induces its transcription [165], and PERK-eIF2α signaling increases *BACE1* expression [166]. ER stress can also suppress autophagy through different mechanisms (e.g., increased degradation of the autophagy inducer FOXO1 by XBP1), resulting in increased Aβ synthesis [149,167].

Endothelial dysfunction also contributes to increased amyloid and tau aggregation via different mechanisms, including ineffective Aβ and tau clearance, dysregulated lipid raft endocytosis, and reduced endothelial nitric oxide synthase (eNOS), which are described below.

6.4. Impaired Endothelial Aβ Clearance

Two neuronal cell surface receptors involved in Aβ clearance are impaired in AD: the downregulation of LRP-1, which transports free parenchymal Aβ into the systemic circulation, and the upregulation of RAGE, which transports Aβ across the BBB into the brain [168]. Several reports from animal and neuropathological studies suggest that vascular LRP-1, expressed by endothelium, pericytes, and vascular smooth muscle, is also downregulated in AD and contributes to impaired Aβ clearance [169]. Further, Aβ toxicity is associated with increased proteasomal degradation of LRP-1 in endothelium [170]. Inhibition or knock-down of vascular LRP-1 in animal models is associated with increased parenchymal and vascular Aβ deposits [171]. Aβ accumulation in AD is further exacerbated by reduced levels of the endothelial molecule, PICALM (phosphatidylinositol-binding clathrin assembly protein), which is important for LRP-1-mediated Aβ-transport [172]. Conversely, vascular RAGE expression is increased in AD, especially in areas with high Aβ pathology, and contributes to increased brain amyloid load. RAGE inhibition reduces Aβ pathology and improves cognition in AD mouse models [173]. Brain endothelial cells also express Aβ-degrading enzymes, such as neprilysin and insulin-degrading enzyme whose expression or activity is altered in AD, further promoting Aβ aggregation [174]. Proteoglycans on the endothelial cell surface interact directly with Aβ peptides via electrostatic interactions promoting Aβ fibrillization and plaque formation and promote the uptake of extracellular *APOE4*-containing lipoproteins [175].

6.5. Dysregulated Endothelial Nitric Oxide Synthase (eNOS)

Endothelial NOS (eNOS) plays an essential role in tau homeostasis and regulating tau phosphorylation by balancing the activity of tau kinases and phosphatases [176]. Crossing *eNOS* knockout ($eNOS^{-/-}$) mice with a mouse model of amyloid pathology (*APP/PS1* mice) was associated with increased phosphorylated tau (p-tau), which was attributed to an increase in Cdk5 activation by p25, while other kinases, such as GSK-3β, were unchanged [176,177]. Interestingly, neuronal p25 levels and Cdk5 activity were increased in the $eNOS^{-/-}$ and the $APP/PS1/eNOS^{-/-}$ mice, supporting the notion that the brain endothelium can influence neuronal mechanisms via soluble mediators [176,177]. Another important finding reported by these studies was that *eNOS* knockout alone was not sufficient to induce tau pathology, as increased p-tau was only observed in *eNOS* knockout mice that also had amyloid pathology (i.e., $APP/PS1/eNOS^{-/-}$ mice) [176,177]. This is consistent with the current AD model, which suggests that Aβ pathology precedes tau pathology [3] and is a prerequisite for AD pathogenesis [178]; however, tau-dependent processes are needed for Aβ-mediated synaptic toxicity [179]. These findings propose additional mechanisms by which endothelial dysfunction may contribute to AD pathology via increased tau phosphorylation and enhancing Aβ-mediated toxicity.

6.6. Altered Lipid Raft Endocytosis

Altered lipid raft endocytosis has been implicated in abnormal protein aggregation in AD and other neurodegenerative disorders and has been proposed as an additional mechanism by which hypoxia or endothelial dysfunction contribute to AD pathogenesis [180,181]. Lipid rafts are specialized microdomains of cell membranes rich in cholesterol, sphingolipids, and saturated fatty acids and play an important role in regulating membrane trafficking, ligand binding, axonal growth, and synaptic maintenance [182]. Profound changes in neuronal lipid raft composition and dynamics are detectable in even the earliest preclinical stages of AD [180]. Lipid raft endocytosis facilitates Aβ synthesis by β- and γ-secretases [183,184] and promotes amyloid aggregation [185]. Several lines of evidence implicate altered lipid raft endocytosis in abnormal Aβ accumulation in AD, including increased β- and γ-secretase *APP* processing within the lipid rafts [183,184], and increased translocation of active *PS1* and *NOTCH3* into lipid drafts due to oxidative stress [186]. Aβ in lipid rafts is associated with increased recruitment of *APOE4* and p-tau to the membrane in an age-dependent manner [187]. Interestingly, while tau is considered a neuronal protein, most of tau is secreted in a vesicle-free form which can interact with the cell membrane via LRP-1, heparan sulfate proteoglycans, and direct binding to membrane lipids [188], and p-tau filaments have also been shown to accumulate within lipid rafts [188].

Lipid raft clusters are also present in endothelial cells and contribute to vascular inflammation [189]. Further, vascular endothelial growth factor receptor-2 (VEGFR2), an angiogenic receptor that is expressed on endothelial cells, colocalizes with endothelial lipid rafts to regulate its activation [181]. Lipid rafts are important in modulating the stability of non-activated VEGFR2, thereby stabilizing vascular endothelial growth factor (VEGF)-mediated activation of the ERK pathway, which is important for angiogenesis [181]. Oxidative stress activates apoptotic pathways and increases lipid raft clustering in endothelial cells which stimulates NAD(P)H-oxidase-generated ROS and leads to endothelial dysfunction [190].

6.7. Impaired Angiogenesis

Angiogenesis and vascular remodeling are impaired in AD as evidenced by altered expression profiles of several angiogenic factors, such as VEGF, vascular-restricted mesenchyme homeobox-2 (MEOX2), MMP-9, angiopoietin-2, integrins (αvβ3 and αvβ5), and the transferrin receptor, leading to vascular regression and abnormal vessel sprouting [191,192]. Several inflammatory mediators (e.g., IL-1β) stimulate angiogenesis through pathways that overlap with hypoxia signaling, suggesting important links between hypoxia, inflammation, and angiogenesis in AD [14].

Although several angiogenic factors are upregulated in AD, Aβ toxicity and low MEOX-2 levels in the brain endothelium, which impair angiogenesis, are also present. Hypoxia suppresses MEOX-2 expression in brain endothelium, and MEOX-2-deficient mice demonstrate vascular regression [193]. Therefore, new vessel formation in AD is ineffective and dysregulated, causing aberrant vessel formation, premature vessel pruning, and reduced microvascular density [194,195]. Hypoxia may promote aberrant angiogenesis by upregulation of HIF-1α expression, and impaired angiogenesis further exacerbates reduced CBF and its subsequent effects on amyloid and tau aggregation. Endothelial dysfunction is implicated in dysregulated angiogenesis due to loss of regulatory feedback signals, leading to a state of chronic endothelial activation, which further contributes to hypoxia and neuronal damage [14].

VEGF is an important pro-angiogenic factor that is present in the vessel wall, reactive astrocytes, and perivascular amyloid deposits [196]. While a few animal studies suggest that VEGF is downregulated in AD transgenic mice [197], most animal and clinical studies demonstrate that VEGF brain expression levels are increased in AD and are associated with cognitive decline [198]. Elevated intrathecal VEGF levels correlate with amyloid pathology and worse cognitive outcomes [199]. Several polymorphisms in the VEGF gene promoter have been associated with a higher AD risk [200]. Despite increased VEGF expression, studies suggest that it is ineffective in promoting angiogenesis in AD [201], as Aβ toxicity interferes with VEGF binding to its receptor, VEGFR-2, on endothelial cells, thereby interfering with its pro-angiogenic role. Paradoxically, increased VEGF expression in AD appears to exacerbate CBF reductions by causing capillary stalls and reductions in the tight junction protein occludin, leading to increased vascular permeability, the subsequent activation of endothelial inflammatory pathways, and the recruitment of leukocytes towards injured endothelium, which exacerbates capillary blockage and the stalling of blood flow [202].

6.8. Dysregulated Thrombosis

Thrombin is an endothelium-derived factor with important roles in angiogenesis and inflammation and contributes to tau pathology and neurodegeneration in AD. High levels of thrombin and the thrombin receptor, protease-activated receptor-1 (PAR-1) [203,204], and low levels of the thrombin inhibitor protease nexin-1 [205], are observed in AD brains. High thrombin levels in AD are associated with the upregulation of several endothelial proteins including VEGF receptors, angiopoietin-2, αvβ3, MMPs, IL-1β, and IL-8 [206–208]. Thrombin is directly toxic to neurons in vivo and in vitro [209,210]. Intracerebral injection of thrombin or overexpression of the thrombin receptor PAR-1 are associated with neurotoxicity and cognitive deficits in animal studies [211] via several mechanisms, including increased tau aggregation, apoptosis, oxidative stress, and microglial or astrocytic activation [212–215]. Thrombin activates pro-inflammatory pathways in microglia and astrocytes; these include the JAK2–STAT3 pathways, which increase the production of TNF-α and inducible NOS in microglia, and regulation of ERK1/2 leading to increased expression of MMP-9 in astrocytes [215–217].

6.9. Vascular Inflammation and Immune Dysregulation

Recent studies have shown that endothelial cells and pericytes demonstrate immunoregulatory functions and influence the neuronal milieu through the release of inflammatory and immune mediators. The brain endothelium in AD expresses high levels of inflammatory mediators such as MCP-1, intercellular adhesion molecule-1 (ICAM-1; CD54), and cationic antimicrobial protein 37 kDa (CAP37) [14,218,219]. Microvessels from AD brains secrete significantly higher levels of IL-1β, IL-6, IL-8, MMPs, thrombin, TNF-α, and transforming growth factor (TGF)-β compared to microvessels from healthy brains [14,218,219]. Aβ upregulates C-C chemokine receptor type 5 (CCR5) expression by the brain endothelium via its interaction with RAGE receptors and the activation of JNK-, ERK-, and PI3K-signaling pathways [220]. E-selectin (CD62) is an inducible endothelium-specific protein that is exclusively expressed in response to endothelial cell activation by cytokines [221,222]. Stimulation of endothelial Toll-like receptor 2 (TLR2) is associated with the upregulation of E-selectin and the recruitment of immune cells [223]. A few studies have shown that CSF E-selectin levels are elevated in dementia and correlate with AD biomarkers [224]. Furthermore, inhibition of vascular inflammation by sunitinib has been shown to reduce vascular Aβ pathology and improve cognitive deficits in AD animal models [225]. Endothelial cells may also act as antigen-presenting cells and express major histocompatibility complex (MHC) proteins I and II [226], and a subset of endothelial cells has been shown to perform immune functions, such as phagocytosis and scavenging, when stimulated by Aβ, including surface expression of the scavenger receptor CD36 [227,228].

6.10. Neuronal and Synaptic Dysfunction

Endothelial dysfunction indirectly contributes to neuronal and synaptic loss via increased protein aggregation, vascular inflammation, impaired angiogenesis, and dysregulated thrombosis; however, it has become increasingly recognized that endothelial injury can also cause neuronal and synaptic loss via other mechanisms, including loss of endothelial GLUT-1 (glucose transporter protein type 1) which leads to impaired neuronal glucose uptake and metabolism, impaired neurovascular coupling, and potential direct endothelial interactions with pathways involved in synaptic plasticity and axonal growth and repair [21,229].

There is growing evidence from animal and translational AD studies that endothelial dysfunction exerts direct effects on synaptic and neuronal functions that are independent of amyloid or tau [14]. For example, endothelial dysfunction is associated with reduced expression of several presynaptic (e.g., synaptosomal-associated protein-25 (SNAP-25) and growth associated protein-43 (GAP-43)) and postsynaptic (e.g., PSD95) proteins in human AD brains [230–232]. Proposed mechanisms by which endothelial dysfunction predisposes to neuronal/synaptic loss include impaired release of neurotrophic growth factors (e.g., VEGF and FGF), and reduced hippocampal expression of NO, an important mediator of synaptic transmission and long-term potentiation [12,14].

Recent studies highlight several pathways by which endothelial dysfunction influences axonal repair, neuronal integrity, and synaptic plasticity independently of amyloid and tau, including interactions of endothelial proteins with semaphorin-3A involved in memory or learning functions [21,22]. Endothelial proteins also interact with the neuropilin/plexin-A1 (NRP-1/PLXNA) complex and Nr-CAM (Figure 2), which are expressed on neuronal surfaces and regulate axonal growth during development or in response to injury [233]. Interestingly, recent studies in animal stroke models suggest that endothelial cells regulate astrocytic differentiation into neural progenitor cells and improve behavioral recovery through the upregulation of the pro-neural factor Ascl1 by the endothelium [234].

While the mechanisms that underly neurovascular uncoupling in AD are complex, there is ample evidence to suggest that brain endothelial dysfunction plays an important role in this process [235,236]. NADPH oxidase and the generation of ROS impair endothelial NO production and cause an imbalance between endothelial vasodilator and vasoconstrictor responses [237]. Further, insulin-like growth factor (IGF-1) deficiency has been shown to promote endothelial and astrocytic dysfunction in AD, including dysregulation of metabotropic-glutamate-receptor expression in astrocytes and impaired CBF responses mediated by eicosanoids [238]. Neurovascular uncoupling due to IGF-1 deficiency in animal studies is associated with deficits in hippocampal-dependent spatial memory, suggesting an important association between neurovascular coupling and cognition [238].

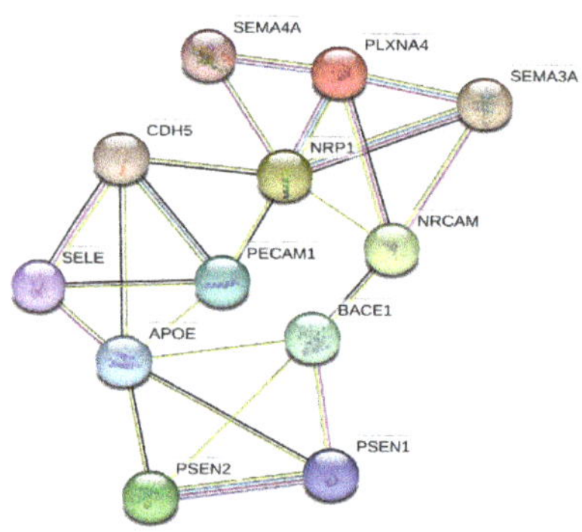

Figure 2. Endothelial proteins interact with molecular pathways involved in synaptic plasticity and axonal repair. Functional pathway analyses of endothelial and neuronal proteins in AD using STRING. These functional pathway analyses suggest significant interactions between endothelial and neuronal proteins. Network nodes represent proteins; splice isoforms or post-translational modifications are collapsed so that each node represents all the proteins produced by a single protein-coding gene locus. Edges represent protein–protein associations that are meant to be specific and meaningful. Color coding for the interactions are as follows: known interactions from curated databases (teal); known experimentally determined interactions (purple); predicted interactions from gene neighborhood (green), gene fusions (red), or gene co-occurrence (blue); predicted interactions from text-mining (yellow), co-expression (black), and protein homology (lavender). *APOE*, apolipoprotein E; *BACE*, β-secretase; *CDH5*, cadherin-5, also known as vascular endothelial-cadherin (VE-cadherin); *NRCAM*, neuronal cell adhesion molecule; *NRP-1*, neuropilin-1; *PECAM1*, platelet endothelial cell adhesion molecule-1, also known as CD31; *PSEN1*, presenilin-1 (aka PS1); *PSEN2*, presenilin-2 (aka PS2); *PLXNA4*, Plexin A4; *SELE*, E-Selectin; *SEMA3A*, semaphorin-3A; *SEMA4A*, semaphorin-4A. Figure created with STRINGv11.5 for functional protein association networks (https://string-db.org) [239–241].

6.11. Impaired Neurogenesis

In addition to neurons, brain endothelial cells also produce brain-derived neurotrophic factor (BDNF), which plays an important role in neurogenesis and neuroplasticity [242]. One mechanism by which endothelial dysfunction can contribute to neurodegeneration is through impaired neurogenesis and loss of the neurotrophic and neuroprotective effects of endothelial BDNF. Animal models of DM and hypertension demonstrate reduced neuronal and endothelial BDNF expression, which is considered a marker of endothelial dysfunction [242–244]. Loss of endothelial-derived BDNF decreases neuronal resistance to conditions associated with oxidative stress such as hypoxia, glucose deprivation, and Aβ toxicity [245,246].

Proposed mechanisms by which endothelial BDNF promotes neuroplasticity include direct binding to neuronal tyrosine receptor kinase B (TrKB) receptors or sustained activation of endothelial TrKB receptors, which results in increased NO production and induces long-term potentiation, an essential process for memory and learning [247,248]. The latter suggests a direct link between endothelial dysfunction and cognitive impairment. Further, some of the effects of endothelial BDNF on neurons may be mediated by astrocytes at tripartite synapses. Astrocytes express the p75 neurotrophin receptor (p75NTR), which binds with high affinity to the BDNF precursor, proBDNF, leading to the internalization of this complex into astrocytes and the maturation of proBDNF into mature BDNF, which is then released into the neuronal milieu [249,250]. Therefore, endothelial BDNF can modulate synaptic plasticity through astrocyte-dependent or astrocyte-independent pathways (Figure 3).

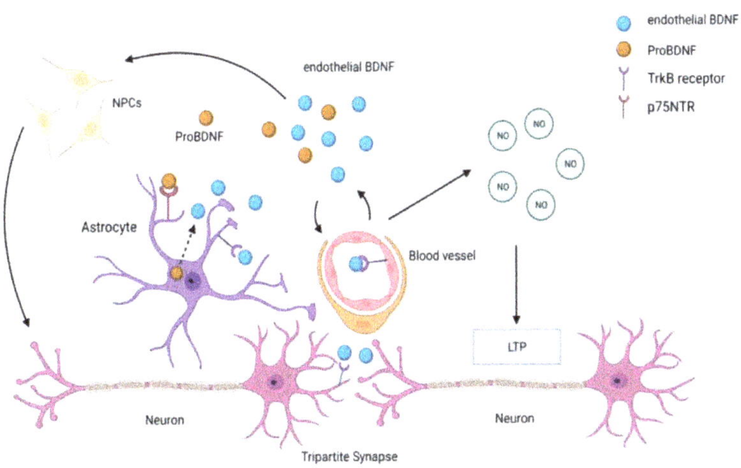

Figure 3. Schematic diagram of the putative role of endothelial-derived BDNF in neurogenesis. Endothelial cells produce BDNF, which binds to TrkB receptors on endothelium, neurons, and astrocytes. Endothelial-derived BDNF can influence neuronal functions and differentiation through astrocyte-dependent and -independent pathways. Endothelial cell activation by BDNF increases nitric oxide (NO) production, which is important for long-term potentiation (LTP) and memory functions. Endothelial-derived BDNF stimulates the differentiation of neural progenitor cells (NPCs) into neurons. Astrocytes at tripartite synapses can also internalize the BDNF precursor pro-BDNF, through the p75 neurotrophin receptor (p75 NTR), and transform it into mature BDNF before releasing it into the extracellular space. Created with Biorender.com.

Together, these findings support the role of dysfunctional endothelium in mediating neurodegeneration in AD. Further support for this comes from studies that show that the direct co-culture of neurons with microvessels from AD brains, or the exposure of cultured neurons to conditioned mediums from AD microvessels but not microvessels from normal aging brains, is associated with neurotoxicity [251], and that brain endothelial mechanisms (e.g., endothelial BDNF) are directly associated with neuroplasticity and cognition. Importantly, data from animal and clinical studies suggests that endothelial and vascular alterations, leading to reduced CBF and capillary loss, may be the inciting event in AD pathogenesis, which then triggers the AD cascade (Figure 1). This posits the notion that AD may primarily be an endotheliopathy; however, additional mechanistic studies in animal models and cell cultures will be needed to support this hypothesis. Further, longitudinal clinical studies that measure fluid and imaging biomarkers of various AD pathologies (including endothelial injury) in well-characterized AD cohorts will provide important insight into the significance of endothelial dysfunction and the temporal ordering of endothelial dysfunction in relation to other AD pathologies and the onset of cognitive impairment.

Finally, improving brain endothelial health has the potential to support neuronal regeneration and repair. Endothelial progenitor cells (EPCs), which express the surface markers CD34, VEGFR2, and CD133, are bone-marrow derived cells that can differentiate into mature endothelial cells and produce angiogenic growth factors to support endothelial repair [252]. Reduced EPC counts and functionality (e.g., reduced mobility or adhesion capacity) have both been reported in AD and correlate with the severity of cognitive impairment [253,254]. EPCs are targeted by Aβ toxicity, further exacerbating endothelial dysfunction by impairing endothelial repair [254]. Conversely, transplantation of EPCs is associated with upregulation of tight-junction proteins, increased angiogenesis, higher hippocampal neuronal survival due to anti-apoptotic effects, reduced hippocampal and cortical Aβ, and improved memory and learning functions [255]. Animal studies provide exciting preliminary evidence to support the utility of EPC administration in restoring the BBB and promoting angiogenesis in animal models of stroke and TBI.

7. Conclusions and Future Directions

Although the current hypothesis of neurodegeneration in AD is centered around abnormal amyloid and tau aggregation, emerging evidence suggests that endothelial dysfunction is an early and primary event in AD pathogenesis that may precede abnormal protein aggregation and directly contribute to neurodegeneration and synaptic injury. Together, studies examining endothelial contributions to neuronal and synaptic dysfunction in AD suggest it may be a potential therapeutic target in AD. Brain endothelial health has both direct and indirect effects on neuronal and axonal repair mechanisms and synaptic plasticity via endothelial–neuronal interactions as well as diffusible endothelial mediators that regulate neuronal functions. Therapeutic interventions, which improve endothelial health and promote endothelial repair, have the potential to improve neurovascular coupling, reduce vascular inflammation and thrombosis, and decrease amyloid and tau aggregation, ultimately leading to restoring CBF and improving cognition. Specifically, the implementation of EPCs may offer an exciting opportunity for endothelial repair and mitigating the deleterious effects of endothelial dysfunction early in the disease process. Further investigations into the mechanisms by which endothelial repair may improve neuronal and synaptic plasticity, and the development of novel AD therapies targeting endothelial injury in AD, are needed.

Funding: This work is supported by the National Institute of Health Grants R21-AG067755 (RT), P20AG068077 (The University of New Mexico exploratory Alzheimer Disease Research Center) and the University of New Mexico Grand Challenges Initiative.

Conflicts of Interest: The author declares no conflict of interest.

References

1. Tarawneh, R.; Holtzman, D.M. The clinical problem of symptomatic Alzheimer disease and mild cognitive impairment. *Cold Spring Harb. Perspect. Med.* **2012**, *2*, a006148. [CrossRef] [PubMed]
2. Price, J.L.; Davis, P.B.; Morris, J.C.; White, D.L. The distribution of tangles, plaques and related immunohistochemical markers in healthy aging and Alzheimer's disease. *Neurobiol. Aging* **1991**, *12*, 295–312. [CrossRef]

3. Jack, C.R., Jr.; Knopman, D.S.; Jagust, W.J.; Petersen, R.C.; Weiner, M.W.; Aisen, P.S.; Shaw, L.M.; Vemuri, P.; Wiste, H.J.; Weigand, S.D.; et al. Tracking pathophysiological processes in Alzheimer's disease: An updated hypothetical model of dynamic biomarkers. *Lancet Neurol.* **2013**, *12*, 207–216. [CrossRef] [PubMed]
4. Esiri, M.M.; Joachim, C.; Sloan, C.; Christie, S.; Agacinski, G.; Bridges, L.R.; Wilcock, G.K.; Smith, A.D. Cerebral Subcortical Small Vessel Disease in Subjects with Pathologically Confirmed Alzheimer Disease: A Clinicopathologic Study in the Oxford Project to Investigate Memory and Ageing (OPTIMA). *Alzheimer Dis. Assoc. Disord.* **2014**, *28*, 30–35. [CrossRef]
5. Attems, J.; Jellinger, K.A. The overlap between vascular disease and Alzheimer's disease-lessons from pathology. *BMC Med.* **2014**, *12*, 206. [CrossRef] [PubMed]
6. Launer, L.J.; Petrovitch, H.; Ross, G.W.; Markesbery, W.; White, L.R. AD brain pathology: Vascular origins? Results from the HAAS autopsy study. *Neurobiol. Aging* **2008**, *29*, 1587–1590. [CrossRef]
7. Barker, W.W.; Luis, C.A.; Kashuba, A.; Luis, M.; Harwood, D.G.; Loewenstein, D.; Waters, C.; Jimison, P.; Shepherd, E.; Sevush, S.; et al. Relative frequencies of Alzheimer disease, Lewy body, vascular and frontotemporal dementia, and hippocampal sclerosis in the State of Florida Brain Bank. *Alzheimer Dis. Assoc. Disord.* **2002**, *16*, 203–212. [CrossRef]
8. de la Torre, J.C. Alzheimer disease as a vascular disorder: Nosological evidence. *Stroke* **2002**, *33*, 1152–1162. [CrossRef]
9. de la Torre, J.C. Cerebromicrovascular pathology in Alzheimer's disease compared to normal aging. *Gerontology* **1997**, *43*, 26–43. [CrossRef]
10. Kalaria, R.N.; Hedera, P. Differential degeneration of the cerebral microvasculature in Alzheimer's disease. *Neuroreport* **1995**, *6*, 477–480. [CrossRef]
11. Kalaria, R.N. Cerebral vessels in ageing and Alzheimer's disease. *Pharmacol. Ther.* **1996**, *72*, 193–214. [CrossRef] [PubMed]
12. Kelleher, R.J.; Soiza, R.L. Evidence of endothelial dysfunction in the development of Alzheimer's disease: Is Alzheimer's a vascular disorder? *Am. J. Cardiovasc. Dis.* **2013**, *3*, 197–226. [PubMed]
13. O'Brien, J.T.; Markus, H.S. Vascular risk factors and Alzheimer's disease. *BMC Med.* **2014**, *12*, 218. [CrossRef] [PubMed]
14. Grammas, P. Neurovascular dysfunction, inflammation and endothelial activation: Implications for the pathogenesis of Alzheimer's disease. *J. Neuroinflamm.* **2011**, *8*, 26. [CrossRef] [PubMed]
15. de la Torre, J.C. Is Alzheimer's disease a neurodegenerative or a vascular disorder? Data, dogma, and dialectics. *Lancet. Neurol.* **2004**, *3*, 184–190. [CrossRef] [PubMed]
16. Claudio, L. Ultrastructural features of the blood-brain barrier in biopsy tissue from Alzheimer's disease patients. *Acta Neuropathol.* **1996**, *91*, 6–14. [CrossRef]
17. Miyakawa, T.; Kuramoto, R. Ultrastructural study of senile plaques and microvessels in the brain with Alzheimer's disease and Down's syndrome. *Ann. Med.* **1989**, *21*, 99–102. [CrossRef] [PubMed]
18. Yamashita, K.; Miyakawa, T.; Katsuragi, S. Vascular changes in the brains with Alzheimer's disease. *Jpn. J. Psychiatry Neurol.* **1991**, *45*, 79–84. [CrossRef]
19. Iturria-Medina, Y.; Sotero, R.C.; Toussaint, P.J.; Mateos-Pérez, J.M.; Evans, A.C.; Weiner, M.W.; Aisen, P.; Petersen, R.; Jack, C.R.; Jagust, W.; et al. Early role of vascular dysregulation on late-onset Alzheimer's disease based on multifactorial data-driven analysis. *Nat. Commun.* **2016**, *7*, 11934. [CrossRef]
20. Eisenmenger, L.B.; Peret, A.; Famakin, B.M.; Spahic, A.; Roberts, G.S.; Bockholt, J.H.; Johnson, K.M.; Paulsen, J.S. Vascular contributions to Alzheimer's disease. *Transl. Res.* **2022**, *254*, 41–53. [CrossRef]
21. Tarawneh, R.; Kasper, R.S.; Sanford, J.; Phuah, C.L.; Hassenstab, J.; Cruchaga, C. Vascular endothelial-cadherin as a marker of endothelial injury in preclinical Alzheimer disease. *Ann. Clin. Transl. Neurol.* **2022**, *9*, 1926–1940. [CrossRef]
22. Jitsuki-Takahashi, A.; Jitsuki, S.; Yamashita, N.; Kawamura, M.; Abe, M.; Sakimura, K.; Sano, A.; Nakamura, F.; Goshima, Y.; Takahashi, T. Activity-induced secretion of semaphorin 3A mediates learning. *Eur. J. Neurosci.* **2021**, *53*, 3279–3293. [CrossRef]
23. de la Torre, J.C.; Mussivand, T. Can disturbed brain microcirculation cause Alzheimer's disease? *Neurol. Res.* **1993**, *15*, 146–153. [CrossRef] [PubMed]
24. de la Torre, J.C. Impaired brain microcirculation may trigger Alzheimer's disease. *Neurosci. Biobehav. Rev.* **1994**, *18*, 397–401. [CrossRef] [PubMed]
25. de la Torre, J.C. Critical threshold cerebral hypoperfusion causes Alzheimer's disease? *Acta Neuropathol.* **1999**, *98*, 1–8. [CrossRef] [PubMed]
26. de la Torre, J.C. Hemodynamic consequences of deformed microvessels in the brain in Alzheimer's disease. *Ann. N. Y. Acad. Sci.* **1997**, *826*, 75–91. [CrossRef]
27. Stewart, R.; Prince, M.; Mann, A. Vascular risk factors and Alzheimer's disease. *Aust. N. Z. J. Psychiatry* **1999**, *33*, 809–813. [CrossRef]
28. Breteler, M.M. Vascular risk factors for Alzheimer's disease: An epidemiologic perspective. *Neurobiol. Aging* **2000**, *21*, 153–160. [CrossRef]
29. Cechetto, D.F.; Hachinski, V.; Whitehead, S.N. Vascular risk factors and Alzheimer's disease. *Expert Rev. Neurother.* **2008**, *8*, 743–750. [CrossRef]
30. Shi, J.; Perry, G.; Smith, M.A.; Friedland, R.P. Vascular abnormalities: The insidious pathogenesis of Alzheimer's disease. *Neurobiol. Aging* **2000**, *21*, 357–361. [CrossRef]
31. Sparks, D.L.; Martin, T.A.; Gross, D.R.; Hunsaker, J.C., 3rd. Link between heart disease, cholesterol, and Alzheimer's disease: A review. *Microsc. Res. Tech.* **2000**, *50*, 287–290. [CrossRef] [PubMed]

32. Rhodin, J.A.; Thomas, T. A vascular connection to Alzheimer's disease. *Microcirculation* **2001**, *8*, 207–220. [CrossRef] [PubMed]
33. Marx, J. Alzheimer's disease. Bad for the heart, bad for the mind? *Science* **2001**, *294*, 508–509. [CrossRef] [PubMed]
34. Ronnemaa, E.; Zethelius, B.; Sundelof, J.; Sundstrom, J.; Degerman-Gunnarsson, M.; Berne, C.; Lannfelt, L.; Kilander, L. Impaired insulin secretion increases the risk of Alzheimer disease. *Neurology* **2008**, *71*, 1065–1071. [CrossRef] [PubMed]
35. Irie, F.; Fitzpatrick, A.L.; Lopez, O.L.; Kuller, L.H.; Peila, R.; Newman, A.B.; Launer, L.J. Enhanced risk for Alzheimer disease in persons with type 2 diabetes and APOE epsilon4: The Cardiovascular Health Study Cognition Study. *Arch. Neurol.* **2008**, *65*, 89–93. [CrossRef]
36. Vagelatos, N.T.; Eslick, G.D. Type 2 diabetes as a risk factor for Alzheimer's disease: The confounders, interactions, and neuropathology associated with this relationship. *Epidemiol. Rev.* **2013**, *35*, 152–160. [CrossRef] [PubMed]
37. Lee, H.J.; Seo, H.I.; Cha, H.Y.; Yang, Y.J.; Kwon, S.H.; Yang, S.J. Diabetes and Alzheimer's Disease: Mechanisms and Nutritional Aspects. *Clin. Nutr. Res.* **2018**, *7*, 229–240. [CrossRef]
38. Gottesman, R.F.; Albert, M.S.; Alonso, A.; Coker, L.H.; Coresh, J.; Davis, S.M.; Deal, J.A.; McKhann, G.M.; Mosley, T.H.; Sharrett, A.R.; et al. Associations Between Midlife Vascular Risk Factors and 25-Year Incident Dementia in the Atherosclerosis Risk in Communities (ARIC) Cohort. *JAMA Neurol.* **2017**, *74*, 1246–1254. [CrossRef]
39. Liu, C.C.; Liu, C.C.; Kanekiyo, T.; Xu, H.; Bu, G. Apolipoprotein E and Alzheimer disease: Risk, mechanisms and therapy. *Nat. Rev. Neurol.* **2013**, *9*, 106–118. [CrossRef]
40. Corder, E.H.; Saunders, A.M.; Strittmatter, W.J.; Schmechel, D.E.; Gaskell, P.C.; Small, G.W.; Roses, A.D.; Haines, J.L.; Pericak-Vance, M.A. Gene dose of apolipoprotein E type 4 allele and the risk of Alzheimer's disease in late onset families. *Science* **1993**, *261*, 921–923. [CrossRef]
41. Troutwine, B.R.; Hamid, L.; Lysaker, C.R.; Strope, T.A.; Wilkins, H.M. Apolipoprotein E and Alzheimer's disease. *Acta Pharm. Sinica. B* **2022**, *12*, 496–510. [CrossRef] [PubMed]
42. Mahley, R.W. Apolipoprotein E: From cardiovascular disease to neurodegenerative disorders. *J. Mol. Med.* **2016**, *94*, 739–746. [CrossRef] [PubMed]
43. Launer, L.J.; Ross, G.W.; Petrovitch, H.; Masaki, K.; Foley, D.; White, L.R.; Havlik, R.J. Midlife blood pressure and dementia: The Honolulu-Asia aging study. *Neurobiol. Aging* **2000**, *21*, 49–55. [CrossRef] [PubMed]
44. DeCarli, C.; Miller, B.L.; Swan, G.E.; Reed, T.; Wolf, P.A.; Carmelli, D. Cerebrovascular and brain morphologic correlates of mild cognitive impairment in the National Heart, Lung, and Blood Institute Twin Study. *Arch. Neurol.* **2001**, *58*, 643–647. [CrossRef]
45. Saunders, A.M.; Roses, A.D. Apolipoprotein E4 allele frequency, ischemic cerebrovascular disease, and Alzheimer's disease. *Stroke* **1993**, *24*, 1416–1417. [CrossRef]
46. Roses, A.D.; Saunders, A.M. ApoE, Alzheimer's disease, and recovery from brain stress. *Ann. N. Y. Acad. Sci.* **1997**, *826*, 200–212. [CrossRef]
47. Zhang, X.; Tong, T.; Chang, A.; Ang, T.F.A.; Tao, Q.; Auerbach, S.; Devine, S.; Qiu, W.Q.; Mez, J.; Massaro, J.; et al. Midlife lipid and glucose levels are associated with Alzheimer's disease. *Alzheimer's Dement.* **2023**, *19*, 181–193. [CrossRef]
48. Iwagami, M.; Qizilbash, N.; Gregson, J.; Douglas, I.; Johnson, M.; Pearce, N.; Evans, S.; Pocock, S. Blood cholesterol and risk of dementia in more than 1.8 million people over two decades: A retrospective cohort study. *Lancet Healthy Longev.* **2021**, *2*, e498–e506. [CrossRef]
49. Wingo, T.S.; Cutler, D.J.; Wingo, A.P.; Le, N.A.; Rabinovici, G.D.; Miller, B.L.; Lah, J.J.; Levey, A.I. Association of Early-Onset Alzheimer Disease With Elevated Low-Density Lipoprotein Cholesterol Levels and Rare Genetic Coding Variants of APOB. *JAMA Neurol.* **2019**, *76*, 809–817. [CrossRef]
50. Chen, X.; Hui, L.; Geiger, J.D. Role of LDL cholesterol and endolysosomes in amyloidogenesis and Alzheimer's disease. *J. Neurol. Neurophysiol.* **2014**, *5*, 236. [CrossRef]
51. Feringa, F.M.; van der Kant, R. Cholesterol and Alzheimer's Disease; From Risk Genes to Pathological Effects. *Front. Aging Neurosci.* **2021**, *13*, 690372. [CrossRef] [PubMed]
52. Di Paolo, G.; Kim, T.-W. Linking lipids to Alzheimer's disease: Cholesterol and beyond. *Nat. Rev. Neurosci.* **2011**, *12*, 284–296. [CrossRef] [PubMed]
53. Hofman, A.; Ott, A.; Breteler, M.M.; Bots, M.L.; Slooter, A.J.; van Harskamp, F.; van Duijn, C.N.; Van Broeckhoven, C.; Grobbee, D.E. Atherosclerosis, apolipoprotein E, and prevalence of dementia and Alzheimer's disease in the Rotterdam Study. *Lancet* **1997**, *349*, 151–154. [CrossRef]
54. den Heijer, T.; Vermeer, S.E.; Clarke, R.; Oudkerk, M.; Koudstaal, P.J.; Hofman, A.; Breteler, M.M.B. Homocysteine and brain atrophy on MRI of non-demented elderly. *Brain* **2002**, *126*, 170–175. [CrossRef] [PubMed]
55. Shirafuji, N.; Hamano, T.; Yen, S.H.; Kanaan, N.M.; Yoshida, H.; Hayashi, K.; Ikawa, M.; Yamamura, O.; Kuriyama, M.; Nakamoto, Y. Homocysteine Increases Tau Phosphorylation, Truncation and Oligomerization. *Int. J. Mol. Sci.* **2018**, *19*, 891. [CrossRef] [PubMed]
56. Lominadze, D.; Tyagi, N.; Sen, U.; Ovechkin, A.; Tyagi, S.C. Homocysteine alters cerebral microvascular integrity and causes remodeling by antagonizing GABA-A receptor. *Mol. Cell. Biochem.* **2012**, *371*, 89–96. [CrossRef]
57. Peila, R.; White, L.R.; Petrovich, H.; Masaki, K.; Ross, G.W.; Havlik, R.J.; Launer, L.J. Joint effect of the APOE gene and midlife systolic blood pressure on late-life cognitive impairment: The Honolulu-Asia aging study. *Stroke* **2001**, *32*, 2882–2889. [CrossRef] [PubMed]

58. Kivipelto, M.; Helkala, E.L.; Hanninen, T.; Laakso, M.P.; Hallikainen, M.; Alhainen, K.; Soininen, H.; Tuomilehto, J.; Nissinen, A. Midlife vascular risk factors and late-life mild cognitive impairment: A population-based study. *Neurology* **2001**, *56*, 1683–1689. [CrossRef]
59. Luchsinger, J.A.; Reitz, C.; Honig, L.S.; Tang, M.X.; Shea, S.; Mayeux, R. Aggregation of vascular risk factors and risk of incident Alzheimer disease. *Neurology* **2005**, *65*, 545–551. [CrossRef]
60. Phuah, C.L.; Chen, Y.; Strain, J.F.; Yechoor, N.; Laurido-Soto, O.J.; Ances, B.M.; Lee, J.M.; for the Alzheimer's Disease Neuroimaging, I. Association of Data-Driven White Matter Hyperintensity Spatial Signatures with Distinct Cerebral Small Vessel Disease Etiologies. *Neurology* **2022**, *99*, e2535–e2547. [CrossRef]
61. Rabin, J.S.; Pruzin, J.; Scott, M.; Yang, H.S.; Hampton, O.; Hsieh, S.; Schultz, A.P.; Buckley, R.F.; Hedden, T.; Rentz, D.; et al. Association of beta-Amyloid and Vascular Risk on Longitudinal Patterns of Brain Atrophy. *Neurology* **2022**, *99*, e270–e280. [CrossRef] [PubMed]
62. Rizvi, B.; Lao, P.J.; Chesebro, A.G.; Dworkin, J.D.; Amarante, E.; Beato, J.M.; Gutierrez, J.; Zahodne, L.B.; Schupf, N.; Manly, J.J.; et al. Association of Regional White Matter Hyperintensities with Longitudinal Alzheimer-like Pattern of Neurodegeneration in Older Adults. *JAMA Netw. Open* **2021**, *4*, e2125166. [CrossRef] [PubMed]
63. Brickman, A.M. Contemplating Alzheimer's disease and the contribution of white matter hyperintensities. *Curr. Neurol. Neurosci. Rep.* **2013**, *13*, 415. [CrossRef] [PubMed]
64. Tuke, J.B. On the Morbid Histology of the Brain and Spinal Cord as Observed in the Insane. *Br. Foreign Med. Chir. Rev.* **1873**, *51*, 450–460. [PubMed]
65. Buée, L.; Hof, P.R.; Bouras, C.; Delacourte, A.; Perl, D.P.; Morrison, J.H.; Fillit, H.M. Pathological alterations of the cerebral microvasculature in Alzheimer's disease and related dementing disorders. *Acta Neuropathol.* **1994**, *87*, 469–480. [CrossRef]
66. Miyakawa, T.; Shimoji, A.; Kuramoto, R.; Higuchi, Y. The relationship between senile plaques and cerebral blood vessels in Alzheimer's disease and senile dementia. Morphological mechanism of senile plaque production. *Virchows Archiv. B Cell. Pathol. Incl. Mol. Pathol.* **1982**, *40*, 121–129. [CrossRef]
67. Araki, K.; Miyakawa, T.; Katsuragi, S. Ultrastructure of senile plaque using thick sections in the brain with Alzheimer's disease. *Jpn. J. Psychiatry Neurol.* **1991**, *45*, 85–89. [CrossRef]
68. Mancardi, G.L.; Perdelli, F.; Rivano, C.; Leonardi, A.; Bugiani, O. Thickening of the basement membrane of cortical capillaries in Alzheimer's disease. *Acta Neuropathol.* **1980**, *49*, 79–83. [CrossRef]
69. Meyer, E.P.; Ulmann-Schuler, A.; Staufenbiel, M.; Krucker, T. Altered morphology and 3D architecture of brain vasculature in a mouse model for Alzheimer's disease. *Proc. Natl. Acad. Sci. USA* **2008**, *105*, 3587–3592. [CrossRef]
70. Bell, R.D.; Zlokovic, B.V. Neurovascular mechanisms and blood-brain barrier disorder in Alzheimer's disease. *Acta Neuropathol.* **2009**, *118*, 103–113. [CrossRef]
71. De Jong, G.I.; Farkas, E.; Stienstra, C.M.; Plass, J.R.; Keijser, J.N.; de la Torre, J.C.; Luiten, P.G. Cerebral hypoperfusion yields capillary damage in the hippocampal CA1 area that correlates with spatial memory impairment. *Neuroscience* **1999**, *91*, 203–210. [CrossRef] [PubMed]
72. De Jong, G.I.; De Vos, R.A.; Steur, E.N.; Luiten, P.G. Cerebrovascular hypoperfusion: A risk factor for Alzheimer's disease? Animal model and postmortem human studies. *Ann. N. Y. Acad. Sci.* **1997**, *826*, 56–74. [CrossRef] [PubMed]
73. Christov, A.; Ottman, J.; Hamdheydari, L.; Grammas, P. Structural changes in Alzheimer's disease brain microvessels. *Curr. Alzheimer Res.* **2008**, *5*, 392–395. [CrossRef] [PubMed]
74. Farkas, E.; Luiten, P.G. Cerebral microvascular pathology in aging and Alzheimer's disease. *Prog. Neurobiol.* **2001**, *64*, 575–611. [CrossRef]
75. Zuliani, G.; Cavalieri, M.; Galvani, M.; Passaro, A.; Munari, M.R.; Bosi, C.; Zurlo, A.; Fellin, R. Markers of endothelial dysfunction in older subjects with late onset Alzheimer's disease or vascular dementia. *J. Neurol. Sci.* **2008**, *272*, 164–170. [CrossRef]
76. Lau, S.-F.; Cao, H.; Fu, A.K.Y.; Ip, N.Y. Single-nucleus transcriptome analysis reveals dysregulation of angiogenic endothelial cells and neuroprotective glia in Alzheimer's disease. *Proc. Natl. Acad. Sci. USA* **2020**, *117*, 25800. [CrossRef] [PubMed]
77. Yang, A.C.; Vest, R.T.; Kern, F.; Lee, D.P.; Agam, M.; Maat, C.A.; Losada, P.M.; Chen, M.B.; Schaum, N.; Khoury, N.; et al. A human brain vascular atlas reveals diverse mediators of Alzheimer's risk. *Nature* **2022**, *603*, 885–892. [CrossRef]
78. Takechi, R.; Lam, V.; Brook, E.; Giles, C.; Fimognari, N.; Mooranian, A.; Al-Salami, H.; Coulson, S.H.; Nesbit, M.; Mamo, J.C.L. Blood-Brain Barrier Dysfunction Precedes Cognitive Decline and Neurodegeneration in Diabetic Insulin Resistant Mouse Model: An Implication for Causal Link. *Front. Aging Neurosci.* **2017**, *9*, 399. [CrossRef] [PubMed]
79. Shah, G.N.; Morofuji, Y.; Banks, W.A.; Price, T.O. High glucose-induced mitochondrial respiration and reactive oxygen species in mouse cerebral pericytes is reversed by pharmacological inhibition of mitochondrial carbonic anhydrases: Implications for cerebral microvascular disease in diabetes. *Biochem. Biophys. Res. Commun.* **2013**, *440*, 354–358. [CrossRef]
80. Pietronigro, E.; Zenaro, E.; Bianca, V.D.; Dusi, S.; Terrabuio, E.; Iannoto, G.; Slanzi, A.; Ghasemi, S.; Nagarajan, R.; Piacentino, G.; et al. Blockade of α4 integrins reduces leukocyte-endothelial interactions in cerebral vessels and improves memory in a mouse model of Alzheimer's disease. *Sci. Rep.* **2019**, *9*, 12055. [CrossRef]
81. ElAli, A.; Theriault, P.; Rivest, S. The role of pericytes in neurovascular unit remodeling in brain disorders. *Int. J. Mol. Sci.* **2014**, *15*, 6453–6474. [CrossRef]
82. Alcendor, D.J. Interactions between Amyloid-Beta Proteins and Human Brain Pericytes: Implications for the Pathobiology of Alzheimer's Disease. *J. Clin. Med.* **2020**, *9*, 1490. [CrossRef]

83. Ma, Q.; Zhao, Z.; Sagare, A.P.; Wu, Y.; Wang, M.; Owens, N.C.; Verghese, P.B.; Herz, J.; Holtzman, D.M.; Zlokovic, B.V. Blood-brain barrier-associated pericytes internalize and clear aggregated amyloid-β42 by LRP1-dependent apolipoprotein E isoform-specific mechanism. *Mol. Neurodegener.* **2018**, *13*, 57. [CrossRef] [PubMed]
84. Verbeek, M.M.; Van Nostrand, W.E.; Otte-Holler, I.; Wesseling, P.; De Waal, R.M. Amyloid-beta-induced degeneration of human brain pericytes is dependent on the apolipoprotein E genotype. *Ann. N. Y. Acad. Sci.* **2000**, *903*, 187–199. [CrossRef]
85. Verbeek, M.M.; de Waal, R.M.; Schipper, J.J.; Van Nostrand, W.E. Rapid degeneration of cultured human brain pericytes by amyloid beta protein. *J. Neurochem.* **1997**, *68*, 1135–1141. [CrossRef]
86. Sengillo, J.D.; Winkler, E.A.; Walker, C.T.; Sullivan, J.S.; Johnson, M.; Zlokovic, B.V. Deficiency in mural vascular cells coincides with blood-brain barrier disruption in Alzheimer's disease. *Brain Pathol.* **2013**, *23*, 303–310. [CrossRef]
87. Montagne, A.; Nikolakopoulou, A.M.; Zhao, Z.; Sagare, A.P.; Si, G.; Lazic, D.; Barnes, S.R.; Daianu, M.; Ramanathan, A.; Go, A., et al. Pericyte degeneration causes white matter dysfunction in the mouse central nervous system. *Nat. Med.* **2018**, *24*, 326–337 [CrossRef] [PubMed]
88. Halliday, M.R.; Rege, S.V.; Ma, Q.; Zhao, Z.; Miller, C.A.; Winkler, E.A.; Zlokovic, B.V. Accelerated pericyte degeneration and blood-brain barrier breakdown in apolipoprotein E4 carriers with Alzheimer's disease. *J. Cereb. Blood Flow Metab.* **2016**, *36*, 216–227. [CrossRef] [PubMed]
89. Montagne, A.; Nation, D.A.; Sagare, A.P.; Barisano, G.; Sweeney, M.D.; Chakhoyan, A.; Pachicano, M.; Joe, E.; Nelson, A.R.; D'Orazio, L.M.; et al. APOE4 leads to blood-brain barrier dysfunction predicting cognitive decline. *Nature* **2020**, *581*, 71–76. [CrossRef]
90. Winkler, E.A.; Sagare, A.P.; Zlokovic, B.V. The pericyte: A forgotten cell type with important implications for Alzheimer's disease? *Brain Pathol.* **2014**, *24*, 371–386. [CrossRef] [PubMed]
91. Sagare, A.P.; Bell, R.D.; Zhao, Z.; Ma, Q.; Winkler, E.A.; Ramanathan, A.; Zlokovic, B.V. Pericyte loss influences Alzheimer-like neurodegeneration in mice. *Nat. Commun.* **2013**, *4*, 2932. [CrossRef] [PubMed]
92. Castillo-Carranza, D.L.; Nilson, A.N.; Van Skike, C.E.; Jahrling, J.B.; Patel, K.; Garach, P.; Gerson, J.E.; Sengupta, U.; Abisambra, J.; Nelson, P.; et al. Cerebral Microvascular Accumulation of Tau Oligomers in Alzheimer's Disease and Related Tauopathies. *Aging Dis.* **2017**, *8*, 257–266. [CrossRef] [PubMed]
93. Ojo, J.; Eisenbaum, M.; Shackleton, B.; Lynch, C.; Joshi, U.; Saltiel, N.; Pearson, A.; Ringland, C.; Paris, D.; Mouzon, B.; et al. Mural cell dysfunction leads to altered cerebrovascular tau uptake following repetitive head trauma. *Neurobiol. Dis.* **2021**, *150*, 105237. [CrossRef]
94. González-Reyes, R.E.; Nava-Mesa, M.O.; Vargas-Sánchez, K.; Ariza-Salamanca, D.; Mora-Muñoz, L. Involvement of Astrocytes in Alzheimer's Disease from a Neuroinflammatory and Oxidative Stress Perspective. *Front. Mol. Neurosci.* **2017**, *10*, 427. [CrossRef] [PubMed]
95. Rodriguez-Giraldo, M.; Gonzalez-Reyes, R.E.; Ramirez-Guerrero, S.; Bonilla-Trilleras, C.E.; Guardo-Maya, S.; Nava-Mesa, M.O. Astrocytes as a Therapeutic Target in Alzheimer's Disease-Comprehensive Review and Recent Developments. *Int. J. Mol. Sci.* **2022**, *23*, 13630. [CrossRef] [PubMed]
96. Davis, N.; Mota, B.C.; Stead, L.; Palmer, E.O.C.; Lombardero, L.; Rodriguez-Puertas, R.; de Paola, V.; Barnes, S.J.; Sastre, M. Pharmacological ablation of astrocytes reduces Abeta degradation and synaptic connectivity in an ex vivo model of Alzheimer's disease. *J. Neuroinflamm.* **2021**, *18*, 73. [CrossRef] [PubMed]
97. Silva, I.; Silva, J.; Ferreira, R.; Trigo, D. Glymphatic system, AQP4, and their implications in Alzheimer's disease. *Neurol. Res. Pract.* **2021**, *3*, 5. [CrossRef]
98. Iliff, J.J.; Wang, M.; Liao, Y.; Plogg, B.A.; Peng, W.; Gundersen, G.A.; Benveniste, H.; Vates, G.E.; Deane, R.; Goldman, S.A.; et al. A paravascular pathway facilitates CSF flow through the brain parenchyma and the clearance of interstitial solutes, including amyloid beta. *Sci. Transl. Med.* **2012**, *4*, 147ra111. [CrossRef]
99. Arnaud, L.; Benech, P.; Greetham, L.; Stephan, D.; Jimenez, A.; Jullien, N.; García-González, L.; Tsvetkov, P.O.; Devred, F.; Sancho-Martinez, I.; et al. APOE4 drives inflammation in human astrocytes via TAGLN3 repression and NF-κB activation. *Cell Rep.* **2022**, *40*, 111200. [CrossRef]
100. Avila-Muñoz, E.; Arias, C. When astrocytes become harmful: Functional and inflammatory responses that contribute to Alzheimer's disease. *Ageing Res. Rev.* **2014**, *18*, 29–40. [CrossRef]
101. Amro, Z.; Yool, A.J.; Collins-Praino, L.E. The potential role of glial cells in driving the prion-like transcellular propagation of tau in tauopathies. *Brain Behav. Immun.-Health* **2021**, *14*, 100242. [CrossRef] [PubMed]
102. Saroja, S.R.; Gorbachev, K.; Julia, T.; Goate, A.M.; Pereira, A.C. Astrocyte-secreted glypican-4 drives APOE4-dependent tau hyperphosphorylation. *Proc. Natl. Acad. Sci. USA* **2022**, *119*, e2108870119. [CrossRef]
103. Smith, A.M.; Davey, K.; Tsartsalis, S.; Khozoie, C.; Fancy, N.; Tang, S.S.; Liaptsi, E.; Weinert, M.; McGarry, A.; Muirhead, R.C.J.; et al. Diverse human astrocyte and microglial transcriptional responses to Alzheimer's pathology. *Acta Neuropathol.* **2022**, *143*, 75–91. [CrossRef] [PubMed]
104. Jiang, R.; Smailovic, U.; Haytural, H.; Tijms, B.M.; Li, H.; Haret, R.M.; Shevchenko, G.; Chen, G.; Abelein, A.; Gobom, J.; et al. Increased CSF-decorin predicts brain pathological changes driven by Alzheimer's Aβ amyloidosis. *Acta Neuropathol. Commun.* **2022**, *10*, 96. [CrossRef]
105. Prohovnik, I.; Mayeux, R.; Sackeim, H.A.; Smith, G.; Stern, Y.; Alderson, P.O. Cerebral perfusion as a diagnostic marker of early Alzheimer's disease. *Neurology* **1988**, *38*, 931–937. [CrossRef] [PubMed]

106. Mattsson, N.; Tosun, D.; Insel, P.S.; Simonson, A.; Jack, C.R., Jr.; Beckett, L.A.; Donohue, M.; Jagust, W.; Schuff, N.; Weiner, M.W.; et al. Association of brain amyloid-beta with cerebral perfusion and structure in Alzheimer's disease and mild cognitive impairment. *Brain* **2014**, *137 Pt 5*, 1550–1561. [CrossRef]
107. Korte, N.; Nortley, R.; Attwell, D. Cerebral blood flow decrease as an early pathological mechanism in Alzheimer's disease. *Acta Neuropathol.* **2020**, *140*, 793–810. [CrossRef]
108. Tanaka, K.; Wada, N.; Ogawa, N. Chronic cerebral hypoperfusion induces transient reversible monoaminergic changes in the rat brain. *Neurochem. Res.* **2000**, *25*, 313–320. [CrossRef]
109. Johnson, K.A.; Albert, M.S. Perfusion abnormalities in prodromal AD. *Neurobiol. Aging* **2000**, *21*, 289–292. [CrossRef]
110. Duan, W.; Zhou, G.D.; Balachandrasekaran, A.; Bhumkar, A.B.; Boraste, P.B.; Becker, J.T.; Kuller, L.H.; Lopez, O.L.; Gach, H.M.; Dai, W. Cerebral Blood Flow Predicts Conversion of Mild Cognitive Impairment into Alzheimer's Disease and Cognitive Decline: An Arterial Spin Labeling Follow-up Study. *J. Alzheimer's Dis.* **2021**, *82*, 293–305. [CrossRef]
111. Bracko, O.; Cruz Hernandez, J.C.; Park, L.; Nishimura, N.; Schaffer, C.B. Causes and consequences of baseline cerebral blood flow reductions in Alzheimer's disease. *J. Cereb. Blood Flow Metab.* **2021**, *41*, 1501–1516. [CrossRef] [PubMed]
112. Rodriguez, G.; Vitali, P.; Calvini, P.; Bordoni, C.; Girtler, N.; Taddei, G.; Mariani, G.; Nobili, F. Hippocampal perfusion in mild Alzheimer's disease. *Psychiatry Res.* **2000**, *100*, 65–74. [CrossRef] [PubMed]
113. Johnson, K.A.; Jones, K.; Holman, B.L.; Becker, J.A.; Spiers, P.A.; Satlin, A.; Albert, M.S. Preclinical prediction of Alzheimer's disease using SPECT. *Neurology* **1998**, *50*, 1563–1571. [CrossRef] [PubMed]
114. Ruitenberg, A.; den Heijer, T.; Bakker, S.L.; van Swieten, J.C.; Koudstaal, P.J.; Hofman, A.; Breteler, M.M. Cerebral hypoperfusion and clinical onset of dementia: The Rotterdam Study. *Ann. Neurol.* **2005**, *57*, 789–794. [CrossRef] [PubMed]
115. Berman, S.E.; Rivera-Rivera, L.A.; Clark, L.R.; Racine, A.M.; Keevil, J.G.; Bratzke, L.C.; Carlsson, C.M.; Bendlin, B.B.; Rowley, H.A.; Blennow, K.; et al. Intracranial Arterial 4D-Flow is Associated with Metrics of Brain Health and Alzheimer's Disease. *Alzheimer's Dement.* **2015**, *1*, 420–428. [CrossRef]
116. Kogure, D.; Matsuda, H.; Ohnishi, T.; Asada, T.; Uno, M.; Kunihiro, T.; Nakano, S.; Takasaki, M. Longitudinal evaluation of early Alzheimer's disease using brain perfusion SPECT. *J. Nucl. Med.* **2000**, *41*, 1155–1162.
117. Okamura, N.; Shinkawa, M.; Arai, H.; Matsui, T.; Nakajo, K.; Maruyama, M.; Hu, X.S.; Sasaki, H. Prediction of progression in patients with mild cognitive impairment using IMP-SPECT. *Nihon Ronen Igakkai Zasshi* **2000**, *37*, 974–978. [CrossRef]
118. Tanaka, K.; Wada, N.; Hori, K.; Asanuma, M.; Nomura, M.; Ogawa, N. Chronic cerebral hypoperfusion disrupts discriminative behavior in acquired-learning rats. *J. Neurosci. Methods* **1998**, *84*, 63–68. [CrossRef]
119. Ni, J.; Ohta, H.; Matsumoto, K.; Watanabe, H. Progressive cognitive impairment following chronic cerebral hypoperfusion induced by permanent occlusion of bilateral carotid arteries in rats. *Brain Res.* **1994**, *653*, 231–236. [CrossRef]
120. Abdollahian, N.P.; Cada, A.; Gonzalez-Lima, F.; de la Torre, J.C. Cytochrome oxidase: A predictive marker of neurodegeneration. In *Cytochrome Oxidase in Neuronal Metabolism and Alzheimer's Disease*; Gonzalez-Lima, F., Ed.; Plenum Press: New York, NY, USA, 1998; pp. 233–261.
121. de la Torre, J.C.; Cada, A.; Nelson, N.; Davis, G.; Sutherland, R.J.; Gonzalez-Lima, F. Reduced cytochrome oxidase and memory dysfunction after chronic brain ischemia in aged rats. *Neurosci. Lett.* **1997**, *223*, 165–168. [CrossRef]
122. Otori, T.; Katsumata, T.; Katayama, Y.; Terashi, A. Measurement of regional cerebral blood flow and glucose utilization in rat brain under chronic hypoperfusion conditions following bilateral carotid artery occlusion. Analyzed by autoradiographical methods. *Nihon Ika Daigaku Zasshi* **1997**, *64*, 428–439. [CrossRef] [PubMed]
123. Ouchi, Y.; Tsukada, H.; Kakiuchi, T.; Nishiyama, S.; Futatsubashi, M. Changes in cerebral blood flow and postsynaptic muscarinic cholinergic activity in rats with bilateral carotid artery ligation. *J. Nucl. Med.* **1998**, *39*, 198–202. [PubMed]
124. Pappas, B.A.; Davidson, C.M.; Bennett, S.A.; de la Torre, J.C.; Fortin, T.; Tenniswood, M.P. Chronic ischemia: Memory impairment and neural pathology in the rat. *Ann. N. Y. Acad. Sci.* **1997**, *826*, 498–501. [CrossRef] [PubMed]
125. Ihara, M.; Tomimoto, H.; Kinoshita, M.; Oh, J.; Noda, M.; Wakita, H.; Akiguchi, I.; Shibasaki, H. Chronic cerebral hypoperfusion induces MMP-2 but not MMP-9 expression in the microglia and vascular endothelium of white matter. *J. Cereb. Blood Flow Metab.* **2001**, *21*, 828–834. [CrossRef]
126. Cai, Z.; Liu, Z.; Xiao, M.; Wang, C.; Tian, F. Chronic Cerebral Hypoperfusion Promotes Amyloid-Beta Pathogenesis via Activating beta/gamma-Secretases. *Neurochem. Res.* **2017**, *42*, 3446–3455. [CrossRef]
127. Park, J.H.; Hong, J.H.; Lee, S.W.; Ji, H.D.; Jung, J.A.; Yoon, K.W.; Lee, J.I.; Won, K.S.; Song, B.I.; Kim, H.W. The effect of chronic cerebral hypoperfusion on the pathology of Alzheimer's disease: A positron emission tomography study in rats. *Sci. Rep.* **2019**, *9*, 14102. [CrossRef]
128. Shang, J.; Yamashita, T.; Tian, F.; Li, X.; Liu, X.; Shi, X.; Nakano, Y.; Tsunoda, K.; Nomura, E.; Sasaki, R.; et al. Chronic cerebral hypoperfusion alters amyloid-beta transport related proteins in the cortical blood vessels of Alzheimer's disease model mouse. *Brain Res.* **2019**, *1723*, 146379. [CrossRef]
129. Zhu, W.M.; Neuhaus, A.; Beard, D.J.; Sutherland, B.A.; DeLuca, G.C. Neurovascular coupling mechanisms in health and neurovascular uncoupling in Alzheimer's disease. *Brain* **2022**, *145*, 2276–2292. [CrossRef]
130. Smith, G.S.; de Leon, M.J.; George, A.E.; Kluger, A.; Volkow, N.D.; McRae, T.; Golomb, J.; Ferris, S.H.; Reisberg, B.; Ciaravino, J.; et al. Topography of cross-sectional and longitudinal glucose metabolic deficits in Alzheimer's disease. Pathophysiologic implications. *Arch. Neurol.* **1992**, *49*, 1142–1150. [CrossRef]

131. Hansra, G.K.; Popov, G.; Banaczek, P.O.; Vogiatzis, M.; Jegathees, T.; Goldsbury, C.S.; Cullen, K.M. The neuritic plaque in Alzheimer's disease: Perivascular degeneration of neuronal processes. *Neurobiol. Aging* **2019**, *82*, 88–101. [CrossRef]
132. Kimura, T.; Hashimura, T.; Miyakawa, T. Observations of microvessels in the brain with Alzheimer's disease by the scanning electron microscopy. *Jpn. J. Psychiatry Neurol.* **1991**, *45*, 671–676. [CrossRef]
133. Nortley, R.; Korte, N.; Izquierdo, P.; Hirunpattarasilp, C.; Mishra, A.; Jaunmuktane, Z.; Kyrargyri, V.; Pfeiffer, T.; Khennouf, L.; Madry, C.; et al. Amyloid beta oligomers constrict human capillaries in Alzheimer's disease via signaling to pericytes. *Science* **2019**, *365*, eaav9518. [CrossRef]
134. Cruz Hernandez, J.C.; Bracko, O.; Kersbergen, C.J.; Muse, V.; Haft-Javaherian, M.; Berg, M.; Park, L.; Vinarcsik, L.K.; Ivasyk, I.; Rivera, D.A.; et al. Neutrophil adhesion in brain capillaries reduces cortical blood flow and impairs memory function in Alzheimer's disease mouse models. *Nat. Neurosci.* **2019**, *22*, 413–420. [CrossRef] [PubMed]
135. Cortes-Canteli, M.; Kruyer, A.; Fernandez-Nueda, I.; Marcos-Diaz, A.; Ceron, C.; Richards, A.T.; Jno-Charles, O.C.; Rodriguez, I.; Callejas, S.; Norris, E.H.; et al. Long-Term Dabigatran Treatment Delays Alzheimer's Disease Pathogenesis in the TgCRND8 Mouse Model. *J. Am. Coll. Cardiol.* **2019**, *74*, 1910–1923. [CrossRef]
136. Burmester, T.; Weich, B.; Reinhardt, S.; Hankeln, T. A vertebrate globin expressed in the brain. *Nature* **2000**, *407*, 520–523. [CrossRef] [PubMed]
137. Jones, A.M.; Kennedy, N.; Hanson, J.; Fenton, G.W. A study of dementia in adults with Down's syndrome using 99Tc(m)-HMPAO SPET. *Nucl. Med. Commun.* **1997**, *18*, 662–667. [CrossRef] [PubMed]
138. Kao, C.H.; Wang, P.Y.; Wang, S.J.; Chou, K.T.; Hsu, C.Y.; Lin, W.Y.; Liao, S.Q.; Yeh, S.H. Regional cerebral blood flow of Alzheimer's disease-like pattern in young patients with Down's syndrome detected by 99Tcm-HMPAO brain SPECT. *Nucl. Med. Commun.* **1993**, *14*, 47–51. [CrossRef]
139. Nunomura, A.; Perry, G.; Pappolla, M.A.; Friedland, R.P.; Hirai, K.; Chiba, S.; Smith, M.A. Neuronal oxidative stress precedes amyloid-beta deposition in Down syndrome. *J. Neuropathol. Exp. Neurol.* **2000**, *59*, 1011–1017. [CrossRef]
140. Sumpio, B.E.; Riley, J.T.; Dardik, A. Cells in focus: Endothelial cell. *Int. J. Biochem. Cell Biol.* **2002**, *34*, 1508–1512. [CrossRef]
141. Li, M.M.; Zheng, Y.L.; Wang, W.D.; Lin, S.; Lin, H.L. Neuropeptide Y: An Update on the Mechanism Underlying Chronic Intermittent Hypoxia-Induced Endothelial Dysfunction. *Front. Physiol.* **2021**, *12*, 712281. [CrossRef]
142. Makarenko, V.V.; Usatyuk, P.V.; Yuan, G.; Lee, M.M.; Nanduri, J.; Natarajan, V.; Kumar, G.K.; Prabhakar, N.R. Intermittent hypoxia-induced endothelial barrier dysfunction requires ROS-dependent MAP kinase activation. *Am. J. Physiol. Cell Physiol.* **2014**, *306*, C745–C752. [CrossRef] [PubMed]
143. Prabhakar, N.R.; Kumar, G.K.; Nanduri, J.; Semenza, G.L. ROS signaling in systemic and cellular responses to chronic intermittent hypoxia. *Antioxid. Redox Signal.* **2007**, *9*, 1397–1403. [CrossRef] [PubMed]
144. Claesson-Welsh, L.; Dejana, E.; McDonald, D.M. Permeability of the Endothelial Barrier: Identifying and Reconciling Controversies. *Trends Mol. Med.* **2021**, *27*, 314–331. [CrossRef]
145. Kovacic, J.C.; Dimmeler, S.; Harvey, R.P.; Finkel, T.; Aikawa, E.; Krenning, G.; Baker, A.H. Endothelial to Mesenchymal Transition in Cardiovascular Disease: JACC State-of-the-Art Review. *J. Am. Coll. Cardiol.* **2019**, *73*, 190–209. [CrossRef]
146. Zeng, X.; Guo, R.; Dong, M.; Zheng, J.; Lin, H.; Lu, H. Contribution of TLR4 signaling in intermittent hypoxia-mediated atherosclerosis progression. *J. Transl. Med.* **2018**, *16*, 106. [CrossRef] [PubMed]
147. Meszaros, M.; Kis, A.; Kunos, L.; Tarnoki, A.D.; Tarnoki, D.L.; Lazar, Z.; Bikov, A. The role of hyaluronic acid and hyaluronidase-1 in obstructive sleep apnoea. *Sci. Rep.* **2020**, *10*, 19484. [CrossRef]
148. Hampel, H.; Vassar, R.; De Strooper, B.; Hardy, J.; Willem, M.; Singh, N.; Zhou, J.; Yan, R.; Vanmechelen, E.; De Vos, A.; et al. The beta-Secretase BACE1 in Alzheimer's Disease. *Biol. Psychiatry* **2021**, *89*, 745–756. [CrossRef]
149. Salminen, A.; Kauppinen, A.; Kaarniranta, K. Hypoxia/ischemia activate processing of Amyloid Precursor Protein: Impact of vascular dysfunction in the pathogenesis of Alzheimer's disease. *J. Neurochem.* **2017**, *140*, 536–549. [CrossRef]
150. Tesco, G.; Koh, Y.H.; Kang, E.L.; Cameron, A.N.; Das, S.; Sena-Esteves, M.; Hiltunen, M.; Yang, S.H.; Zhong, Z.; Shen, Y.; et al. Depletion of GGA3 stabilizes BACE and enhances beta-secretase activity. *Neuron* **2007**, *54*, 721–737. [CrossRef]
151. Xiong, M.; Zhang, T.; Zhang, L.M.; Lu, S.D.; Huang, Y.L.; Sun, F.Y. Caspase inhibition attenuates accumulation of beta-amyloid by reducing beta-secretase production and activity in rat brains after stroke. *Neurobiol. Dis.* **2008**, *32*, 433–441. [CrossRef]
152. Zhang, Y.; Xiong, M.; Yan, R.Q.; Sun, F.Y. Mutant ubiquitin-mediated beta-secretase stability via activation of caspase-3 is related to beta-amyloid accumulation in ischemic striatum in rats. *J. Cereb. Blood Flow Metab.* **2010**, *30*, 566–575. [CrossRef]
153. Villa, J.C.; Chiu, D.; Brandes, A.H.; Escorcia, F.E.; Villa, C.H.; Maguire, W.F.; Hu, C.J.; de Stanchina, E.; Simon, M.C.; Sisodia, S.S.; et al. Nontranscriptional role of Hif-1α in activation of γ-secretase and notch signaling in breast cancer. *Cell Rep.* **2014**, *8*, 1077–1092. [CrossRef] [PubMed]
154. Kaufmann, M.R.; Barth, S.; Konietzko, U.; Wu, B.; Egger, S.; Kunze, R.; Marti, H.H.; Hick, M.; Muller, U.; Camenisch, G.; et al. Dysregulation of hypoxia-inducible factor by presenilin/gamma-secretase loss-of-function mutations. *J. Neurosci.* **2013**, *33*, 1915–1926. [CrossRef]
155. Cuadrado-Tejedor, M.; Vilariño, M.; Cabodevilla, F.; Del Río, J.; Frechilla, D.; Pérez-Mediavilla, A. Enhanced expression of the voltage-dependent anion channel 1 (VDAC1) in Alzheimer's disease transgenic mice: An insight into the pathogenic effects of amyloid-β. *J. Alzheimer's Dis.* **2011**, *23*, 195–206. [CrossRef]
156. Manczak, M.; Sheiko, T.; Craigen, W.J.; Reddy, P.H. Reduced VDAC1 protects against Alzheimer's disease, mitochondria, and synaptic deficiencies. *J. Alzheimer's Dis.* **2013**, *37*, 679–690. [CrossRef]

157. Sato, N.; Imaizumi, K.; Manabe, T.; Taniguchi, M.; Hitomi, J.; Katayama, T.; Yoneda, T.; Morihara, T.; Yasuda, Y.; Takagi, T.; et al. Increased production of beta-amyloid and vulnerability to endoplasmic reticulum stress by an aberrant spliced form of presenilin 2. *J. Biol. Chem.* **2001**, *276*, 2108–2114. [CrossRef] [PubMed]
158. Raz, L.; Bhaskar, K.; Weaver, J.; Marini, S.; Zhang, Q.; Thompson, J.F.; Espinoza, C.; Iqbal, S.; Maphis, N.M.; Weston, L.; et al. Hypoxia promotes tau hyperphosphorylation with associated neuropathology in vascular dysfunction. *Neurobiol. Dis.* **2019**, *126*, 124–136. [CrossRef] [PubMed]
159. Qiu, L.; Ng, G.; Tan, E.K.; Liao, P.; Kandiah, N.; Zeng, L. Chronic cerebral hypoperfusion enhances Tau hyperphosphorylation and reduces autophagy in Alzheimer's disease mice. *Sci. Rep.* **2016**, *6*, 23964. [CrossRef] [PubMed]
160. Liu, S.L.; Wang, C.; Jiang, T.; Tan, L.; Xing, A.; Yu, J.T. The Role of Cdk5 in Alzheimer's Disease. *Mol. Neurobiol.* **2016**, *53*, 4328–4342. [CrossRef]
161. Seo, J.; Kritskiy, O.; Watson, L.A.; Barker, S.J.; Dey, D.; Raja, W.K.; Lin, Y.T.; Ko, T.; Cho, S.; Penney, J.; et al. Inhibition of p25/Cdk5 Attenuates Tauopathy in Mouse and iPSC Models of Frontotemporal Dementia. *J. Neurosci.* **2017**, *37*, 9917–9924. [CrossRef]
162. Mottet, D.; Dumont, V.; Deccache, Y.; Demazy, C.; Ninane, N.; Raes, M.; Michiels, C. Regulation of hypoxia-inducible factor-1alpha protein level during hypoxic conditions by the phosphatidylinositol 3-kinase/Akt/glycogen synthase kinase 3beta pathway in HepG2 cells. *J. Biol. Chem.* **2003**, *278*, 31277–31285. [CrossRef] [PubMed]
163. Wang, C.Y.; Xie, J.W.; Wang, T.; Xu, Y.; Cai, J.H.; Wang, X.; Zhao, B.L.; An, L.; Wang, Z.Y. Hypoxia-triggered m-calpain activation evokes endoplasmic reticulum stress and neuropathogenesis in a transgenic mouse model of Alzheimer's disease. *CNS Neurosci. Ther.* **2013**, *19*, 820–833. [CrossRef] [PubMed]
164. Badiola, N.; Penas, C.; Minano-Molina, A.; Barneda-Zahonero, B.; Fado, R.; Sanchez-Opazo, G.; Comella, J.X.; Sabria, J.; Zhu, C.; Blomgren, K.; et al. Induction of ER stress in response to oxygen-glucose deprivation of cortical cultures involves the activation of the PERK and IRE-1 pathways and of caspase-12. *Cell Death Dis.* **2011**, *2*, e149. [CrossRef] [PubMed]
165. Mitsuda, T.; Hayakawa, Y.; Itoh, M.; Ohta, K.; Nakagawa, T. ATF4 regulates gamma-secretase activity during amino acid imbalance. *Biochem. Biophys. Res. Commun.* **2007**, *352*, 722–727. [CrossRef] [PubMed]
166. O'Connor, T.; Sadleir, K.R.; Maus, E.; Velliquette, R.A.; Zhao, J.; Cole, S.L.; Eimer, W.A.; Hitt, B.; Bembinster, L.A.; Lammich, S.; et al. Phosphorylation of the translation initiation factor eIF2alpha increases BACE1 levels and promotes amyloidogenesis. *Neuron* **2008**, *60*, 988–1009. [CrossRef]
167. Rashid, H.O.; Yadav, R.K.; Kim, H.R.; Chae, H.J. ER stress: Autophagy induction, inhibition and selection. *Autophagy* **2015**, *11*, 1956–1977. [CrossRef]
168. Deane, R.; Wu, Z.; Zlokovic, B.V. RAGE (yin) versus LRP (yang) balance regulates alzheimer amyloid beta-peptide clearance through transport across the blood-brain barrier. *Stroke* **2004**, *35* (Suppl. 1), 2628–2631. [CrossRef]
169. Govindpani, K.; Vinnakota, C.; Waldvogel, H.J.; Faull, R.L.; Kwakowsky, A. Vascular dysfunction in Alzheimer's disease: A biomarker of disease progression and a potential therapeutic target. *Neural Regen. Res.* **2020**, *15*, 1030.
170. Deane, R.; Wu, Z.; Sagare, A.; Davis, J.; Du Yan, S.; Hamm, K.; Xu, F.; Parisi, M.; LaRue, B.; Hu, H.W.; et al. LRP/Amyloid β-Peptide Interaction Mediates Differential Brain Efflux of Aβ Isoforms. *Neuron* **2004**, *43*, 333–344. [CrossRef]
171. Kanekiyo, T.; Liu, C.C.; Shinohara, M.; Li, J.; Bu, G. LRP1 in brain vascular smooth muscle cells mediates local clearance of Alzheimer's amyloid-β. *J. Neurosci.* **2012**, *32*, 16458–16465. [CrossRef]
172. Zhao, Z.; Sagare, A.P.; Ma, Q.; Halliday, M.R.; Kong, P.; Kisler, K.; Winkler, E.A.; Ramanathan, A.; Kanekiyo, T.; Bu, G.; et al. Central role for PICALM in amyloid-beta blood-brain barrier transcytosis and clearance. *Nat. Neurosci.* **2015**, *18*, 978–987. [CrossRef] [PubMed]
173. Deane, R.; Du Yan, S.; Submamaryan, R.K.; LaRue, B.; Jovanovic, S.; Hogg, E.; Welch, D.; Manness, L.; Lin, C.; Yu, J.; et al. RAGE mediates amyloid-beta peptide transport across the blood-brain barrier and accumulation in brain. *Nat. Med.* **2003**, *9*, 907–913. [CrossRef] [PubMed]
174. Wang, S.; Wang, R.; Chen, L.; Bennett, D.A.; Dickson, D.W.; Wang, D.S. Expression and functional profiling of neprilysin, insulin-degrading enzyme, and endothelin-converting enzyme in prospectively studied elderly and Alzheimer's brain. *J. Neurochem.* **2010**, *115*, 47–57. [CrossRef] [PubMed]
175. Zhang, G.L.; Zhang, X.; Wang, X.M.; Li, J.P. Towards understanding the roles of heparan sulfate proteoglycans in Alzheimer's disease. *BioMed Res. Int.* **2014**, *2014*, 516028. [CrossRef]
176. Austin, S.A.; Katusic, Z.S. Loss of Endothelial Nitric Oxide Synthase Promotes p25 Generation and Tau Phosphorylation in a Murine Model of Alzheimer's Disease. *Circ. Res.* **2016**, *119*, 1128–1134. [CrossRef]
177. Iadecola, C. Untangling Neurons With Endothelial Nitric Oxide. *Circ. Res.* **2016**, *119*, 1052–1054. [CrossRef]
178. Jack, C.R., Jr.; Bennett, D.A.; Blennow, K.; Carrillo, M.C.; Dunn, B.; Haeberlein, S.B.; Holtzman, D.M.; Jagust, W.; Jessen, F.; Karlawish, J.; et al. NIA-AA Research Framework: Toward a biological definition of Alzheimer's disease. *Alzheimer's Dement.* **2018**, *14*, 535–562. [CrossRef]
179. Ittner, L.M.; Ke, Y.D.; Delerue, F.; Bi, M.; Gladbach, A.; van Eersel, J.; Wolfing, H.; Chieng, B.C.; Christie, M.J.; Napier, I.A.; et al. Dendritic function of tau mediates amyloid-beta toxicity in Alzheimer's disease mouse models. *Cell* **2010**, *142*, 387–397. [CrossRef]
180. Marin, R.; Fabelo, N.; Fernandez-Echevarria, C.; Canerina-Amaro, A.; Rodriguez-Barreto, D.; Quinto-Alemany, D.; Mesa-Herrera, F.; Diaz, M. Lipid Raft Alterations in Aged-Associated Neuropathologies. *Curr. Alzheimer Res.* **2016**, *13*, 973–984. [CrossRef]

181. Zabroski, I.O.; Nugent, M.A. Lipid Raft Association Stabilizes VEGF Receptor 2 in Endothelial Cells. *Int. J. Mol. Sci.* **2021**, *22*, 798 [CrossRef]
182. Quinn, P.J. A lipid matrix model of membrane raft structure. *Prog. Lipid Res.* **2010**, *49*, 390–406. [CrossRef] [PubMed]
183. Bhattacharyya, R.; Barren, C.; Kovacs, D.M. Palmitoylation of amyloid precursor protein regulates amyloidogenic processing in lipid rafts. *J. Neurosci.* **2013**, *33*, 11169–11183. [CrossRef] [PubMed]
184. Osenkowski, P.; Ye, W.; Wang, R.; Wolfe, M.S.; Selkoe, D.J. Direct and potent regulation of gamma-secretase by its lipid microenvironment. *J. Biol. Chem.* **2008**, *283*, 22529–22540. [CrossRef] [PubMed]
185. Rushworth, J.V.; Hooper, N.M. Lipid Rafts: Linking Alzheimer's Amyloid-β Production, Aggregation, and Toxicity at Neuronal Membranes. *Int. J. Alzheimer's Dis.* **2010**, *2011*, 603052. [CrossRef] [PubMed]
186. Danza, G.; Di Serio, C.; Ambrosio, M.R.; Sturli, N.; Lonetto, G.; Rosati, F.; Rocca, B.J.; Ventimiglia, G.; del Vecchio, M.T.; Prudovsky I.; et al. Notch3 is activated by chronic hypoxia and contributes to the progression of human prostate cancer. *Int. J. Cancer* **2013** *133*, 2577–2586. [CrossRef]
187. Kawarabayashi, T.; Shoji, M.; Younkin, L.H.; Wen-Lang, L.; Dickson, D.W.; Murakami, T.; Matsubara, E.; Abe, K.; Ashe, K.H. Younkin, S.G. Dimeric amyloid beta protein rapidly accumulates in lipid rafts followed by apolipoprotein E and phosphorylated tau accumulation in the Tg2576 mouse model of Alzheimer's disease. *J. Neurosci.* **2004**, *24*, 3801–3809. [CrossRef]
188. Bok, E.; Leem, E.; Lee, B.R.; Lee, J.M.; Yoo, C.J.; Lee, E.M.; Kim, J. Role of the Lipid Membrane and Membrane Proteins in Tau Pathology. *Front. Cell. Dev. Biol.* **2021**, *9*, 653815. [CrossRef]
189. Pilarczyk, M.; Mateuszuk, L.; Rygula, A.; Kepczynski, M.; Chlopicki, S.; Baranska, M.; Kaczor, A. Endothelium in spots–high-content imaging of lipid rafts clusters in db/db mice. *PLoS ONE* **2014**, *9*, e106065. [CrossRef]
190. Touyz, R.M. Lipid rafts take center stage in endothelial cell redox signaling by death receptors. *Hypertension* **2006**, *47*, 16–18 [CrossRef]
191. Thirumangalakudi, L.; Samany, P.G.; Owoso, A.; Wiskar, B.; Grammas, P. Angiogenic proteins are expressed by brain blood vessels in Alzheimer's disease. *J. Alzheimer's Dis.* **2006**, *10*, 111–118. [CrossRef]
192. Grammas, P.; Samany, P.G.; Thirumangalakudi, L. Thrombin and inflammatory proteins are elevated in Alzheimer's disease microvessels: Implications for disease pathogenesis. *J. Alzheimer's Dis.* **2006**, *9*, 51–58. [CrossRef]
193. Wu, Z.; Guo, H.; Chow, N.; Sallstrom, J.; Bell, R.D.; Deane, R.; Brooks, A.I.; Kanagala, S.; Rubio, A.; Sagare, A.; et al. Role of the MEOX2 homeobox gene in neurovascular dysfunction in Alzheimer disease. *Nat. Med.* **2005**, *11*, 959–965. [CrossRef] [PubMed]
194. Jellinger, K.A. Alzheimer disease and cerebrovascular pathology: An update. *J. Neural Transm.* **2002**, *109*, 813–836. [CrossRef] [PubMed]
195. Buee, L.; Hof, P.R.; Delacourte, A. Brain microvascular changes in Alzheimer's disease and other dementias. *Ann. N. Y. Acad. Sci.* **1997**, *826*, 7–24. [CrossRef] [PubMed]
196. Kalaria, R.N.; Cohen, D.L.; Premkumar, D.R.; Nag, S.; LaManna, J.C.; Lust, W.D. Vascular endothelial growth factor in Alzheimer's disease and experimental cerebral ischemia. *Brain Res. Mol. Brain Res.* **1998**, *62*, 101–105. [CrossRef]
197. Guo, H.; Xia, D.; Liao, S.; Niu, B.; Tang, J.; Hu, H.; Qian, H.; Cao, B. Vascular endothelial growth factor improves the cognitive decline of Alzheimer's disease via concurrently inducing the expression of ADAM10 and reducing the expression of beta-site APP cleaving enzyme 1 in Tg2576 mice. *Neurosci. Res.* **2019**, *142*, 49–57. [CrossRef]
198. Tang, H.; Mao, X.; Xie, L.; Greenberg, D.A.; Jin, K. Expression level of vascular endothelial growth factor in hippocampus is associated with cognitive impairment in patients with Alzheimer's disease. *Neurobiol. Aging* **2013**, *34*, 1412–1415. [CrossRef]
199. Tarkowski, E.; Issa, R.; Sjögren, M.; Wallin, A.; Blennow, K.; Tarkowski, A.; Kumar, P. Increased intrathecal levels of the angiogenic factors VEGF and TGF-β in Alzheimer's disease and vascular dementia. *Neurobiol. Aging* **2002**, *23*, 237–243. [CrossRef]
200. Del Bo, R.; Ghezzi, S.; Scarpini, E.; Bresolin, N.; Comi, G.P. VEGF genetic variability is associated with increased risk of developing Alzheimer's disease. *J. Neurol. Sci.* **2009**, *283*, 66–68. [CrossRef]
201. Paris, D.; Townsend, K.; Quadros, A.; Humphrey, J.; Sun, J.; Brem, S.; Wotoczek-Obadia, M.; DelleDonne, A.; Patel, N.; Obregon, D.F.; et al. Inhibition of angiogenesis by Abeta peptides. *Angiogenesis* **2004**, *7*, 75–85. [CrossRef]
202. Ali, M.; Bracko, O. VEGF Paradoxically Reduces Cerebral Blood Flow in Alzheimer's Disease Mice. *Neurosci. Insights* **2022**, *17*, 1–4. [CrossRef] [PubMed]
203. Nishino, A.; Suzuki, M.; Ohtani, H.; Motohashi, O.; Umezawa, K.; Nagura, H.; Yoshimoto, T. Thrombin may contribute to the pathophysiology of central nervous system injury. *J. Neurotrauma* **1993**, *10*, 167–179. [CrossRef] [PubMed]
204. Sokolova, E.; Reiser, G. Prothrombin/thrombin and the thrombin receptors PAR-1 and PAR-4 in the brain: Localization, expression and participation in neurodegenerative diseases. *Thromb. Haemost.* **2008**, *100*, 576–581. [CrossRef] [PubMed]
205. Vaughan, P.J.; Su, J.; Cotman, C.W.; Cunningham, D.D. Protease nexin-1, a potent thrombin inhibitor, is reduced around cerebral blood vessels in Alzheimer's disease. *Brain Res.* **1994**, *668*, 160–170. [CrossRef]
206. Naldini, A.; Carney, D.H.; Pucci, A.; Pasquali, A.; Carraro, F. Thrombin regulates the expression of proangiogenic cytokines via proteolytic activation of protease-activated receptor-1. *Gen. Pharm.* **2000**, *35*, 255–259. [CrossRef]
207. Dupuy, E.; Habib, A.; Lebret, M.; Yang, R.; Levy-Toledano, S.; Tobelem, G. Thrombin induces angiogenesis and vascular endothelial growth factor expression in human endothelial cells: Possible relevance to HIF-1alpha. *J. Thromb. Haemost.* **2003**, *1*, 1096–1102. [CrossRef]
208. Tsopanoglou, N.E.; Andriopoulou, P.; Maragoudakis, M.E. On the mechanism of thrombin-induced angiogenesis: Involvement of alphavbeta3-integrin. *Am. J. Physiol. Cell Physiol.* **2002**, *283*, C1501–C1510. [CrossRef]

209. Smirnova, I.V.; Zhang, S.X.; Citron, B.A.; Arnold, P.M.; Festoff, B.W. Thrombin is an extracellular signal that activates intracellular death protease pathways inducing apoptosis in model motor neurons. *J. Neurobiol.* **1998**, *36*, 64–80. [CrossRef]
210. Mhatre, M.; Nguyen, A.; Kashani, S.; Pham, T.; Adesina, A.; Grammas, P. Thrombin, a mediator of neurotoxicity and memory impairment. *Neurobiol. Aging* **2004**, *25*, 783–793. [CrossRef]
211. Turgeon, V.L.; Milligan, C.E.; Houenou, L.J. Activation of the protease-activated thrombin receptor (PAR)-1 induces motoneuron degeneration in the developing avian embryo. *J. Neuropathol. Exp. Neurol.* **1999**, *58*, 499–504. [CrossRef]
212. Suo, Z.; Wu, M.; Citron, B.A.; Palazzo, R.E.; Festoff, B.W. Rapid tau aggregation and delayed hippocampal neuronal death induced by persistent thrombin signaling. *J. Biol. Chem.* **2003**, *278*, 37681–37689. [CrossRef] [PubMed]
213. Park, K.W.; Jin, B.K. Thrombin-induced oxidative stress contributes to the death of hippocampal neurons: Role of neuronal NADPH oxidase. *J. Neurosci. Res.* **2008**, *86*, 1053–1063. [CrossRef] [PubMed]
214. Lee, D.Y.; Park, K.W.; Jin, B.K. Thrombin induces neurodegeneration and microglial activation in the cortex in vivo and in vitro: Proteolytic and non-proteolytic actions. *Biochem. Biophys. Res. Commun.* **2006**, *346*, 727–738. [CrossRef]
215. Huang, C.F.; Li, G.; Ma, R.; Sun, S.G.; Chen, J.G. Thrombin-induced microglial activation contributes to the degeneration of nigral dopaminergic neurons in vivo. *Neurosci. Bull.* **2008**, *24*, 66–72. [CrossRef] [PubMed]
216. Huang, C.; Ma, R.; Sun, S.; Wei, G.; Fang, Y.; Liu, R.; Li, G. JAK2-STAT3 signaling pathway mediates thrombin-induced proinflammatory actions of microglia in vitro. *J. Neuroimmunol.* **2008**, *204*, 118–125. [CrossRef] [PubMed]
217. Choi, M.S.; Kim, Y.E.; Lee, W.J.; Choi, J.W.; Park, G.H.; Kim, S.D.; Jeon, S.J.; Go, H.S.; Shin, S.M.; Kim, W.K.; et al. Activation of protease-activated receptor1 mediates induction of matrix metalloproteinase-9 by thrombin in rat primary astrocytes. *Brain Res. Bull.* **2008**, *76*, 368–375. [CrossRef]
218. Grammas, P.; Ovase, R. Inflammatory factors are elevated in brain microvessels in Alzheimer's disease. *Neurobiol. Aging* **2001**, *22*, 837–842. [CrossRef]
219. Grammas, P.; Ovase, R. Cerebrovascular transforming growth factor-beta contributes to inflammation in the Alzheimer's disease brain. *Am. J. Pathol.* **2002**, *160*, 1583–1587. [CrossRef]
220. Li, M.; Shang, D.S.; Zhao, W.D.; Tian, L.; Li, B.; Fang, W.G.; Zhu, L.; Man, S.M.; Chen, Y.H. Amyloid beta interaction with receptor for advanced glycation end products up-regulates brain endothelial CCR5 expression and promotes T cells crossing the blood-brain barrier. *J. Immunol.* **2009**, *182*, 5778–5788. [CrossRef]
221. Imhof, B.A.; Dunon, D. Leukocyte Migration and Adhesion. In *Advances in Immunology*; Dixon, F.J., Ed.; Academic Press: Cambridge, MA, USA, 1995; Volume 58, pp. 345–416.
222. Bevilacqua, M.P.; Pober, J.S.; Mendrick, D.L.; Cotran, R.S.; Gimbrone, M.A., Jr. Identification of an inducible endothelial-leukocyte adhesion molecule. *Proc. Natl. Acad. Sci. USA* **1987**, *84*, 9238–9242. [CrossRef]
223. McCoy, M.G.; Nascimento, D.W.; Veleeparambil, M.; Murtazina, R.; Gao, D.; Tkachenko, S.; Podrez, E.; Byzova, T.V. Endothelial TLR2 promotes proangiogenic immune cell recruitment and tumor angiogenesis. *Sci. Signal.* **2021**, *14*, eabc5371. [CrossRef] [PubMed]
224. Li, G.; Xiong, K.; Korff, A.; Pan, C.; Quinn, J.F.; Galasko, D.R.; Liu, C.; Montine, T.J.; Peskind, E.R.; Zhang, J. Increased CSF E-Selectin in Clinical Alzheimer's Disease without Altered CSF Abeta42 and Tau. *J. Alzheimer's Dis.* **2015**, *47*, 883–887. [CrossRef]
225. Govindpani, K.; McNamara, L.G.; Smith, N.R.; Vinnakota, C.; Waldvogel, H.J.; Faull, R.L.; Kwakowsky, A. Vascular Dysfunction in Alzheimer's Disease: A Prelude to the Pathological Process or a Consequence of It? *J. Clin. Med.* **2019**, *8*, 651. [CrossRef] [PubMed]
226. Howland, S.W.; Gun, S.Y.; Claser, C.; Poh, C.M.; Renia, L. Measuring antigen presentation in mouse brain endothelial cells ex vivo and in vitro. *Nat. Protoc.* **2015**, *10*, 2016–2026. [CrossRef] [PubMed]
227. Park, L.; Wang, G.; Zhou, P.; Zhou, J.; Pitstick, R.; Previti, M.L.; Younkin, L.; Younkin, S.G.; Van Nostrand, W.E.; Cho, S.; et al. Scavenger receptor CD36 is essential for the cerebrovascular oxidative stress and neurovascular dysfunction induced by amyloid-beta. *Proc. Natl. Acad. Sci. USA* **2011**, *108*, 5063–5068. [CrossRef]
228. Amersfoort, J.; Eelen, G.; Carmeliet, P. Immunomodulation by endothelial cells-partnering up with the immune system? *Nat. Rev. Immunol.* **2022**, *22*, 576–588. [CrossRef]
229. Winkler, E.A.; Nishida, Y.; Sagare, A.P.; Rege, S.V.; Bell, R.D.; Perlmutter, D.; Sengillo, J.D.; Hillman, S.; Kong, P.; Nelson, A.R.; et al. GLUT1 reductions exacerbate Alzheimer's disease vasculo-neuronal dysfunction and degeneration. *Nat. Neurosci.* **2015**, *18*, 521–530. [CrossRef]
230. VanGuilder, H.D.; Farley, J.A.; Yan, H.; Van Kirk, C.A.; Mitschelen, M.; Sonntag, W.E.; Freeman, W.M. Hippocampal dysregulation of synaptic plasticity-associated proteins with age-related cognitive decline. *Neurobiol. Dis.* **2011**, *43*, 201–212. [CrossRef]
231. Yang, J.; Yao, Y.; Wang, L.; Yang, C.; Wang, F.; Guo, J.; Wang, Z.; Yang, Z.; Ming, D. Gastrin-releasing peptide facilitates glutamatergic transmission in the hippocampus and effectively prevents vascular dementia induced cognitive and synaptic plasticity deficits. *Exp. Neurol.* **2017**, *287 Pt 1*, 75–83. [CrossRef]
232. Wang, F.; Cao, Y.; Ma, L.; Pei, H.; Rausch, W.D.; Li, H. Dysfunction of Cerebrovascular Endothelial Cells: Prelude to Vascular Dementia. *Front. Aging Neurosci.* **2018**, *10*, 376. [CrossRef]
233. Demyanenko, G.P.; Mohan, V.; Zhang, X.; Brennaman, L.H.; Dharbal, K.E.S.; Tran, T.S.; Manis, P.B.; Maness, P.F. Neural Cell Adhesion Molecule NrCAM Regulates Semaphorin 3F-Induced Dendritic Spine Remodeling. *J. Neurosci.* **2014**, *34*, 11274. [CrossRef] [PubMed]

234. Li, W.; Mandeville, E.T.; Duran-Laforet, V.; Fukuda, N.; Yu, Z.; Zheng, Y.; Held, A.; Park, J.H.; Nakano, T.; Tanaka, M.; et al. Endothelial cells regulate astrocyte to neural progenitor cell trans-differentiation in a mouse model of stroke. *Nat. Commun.* **2022**, *13*, 7812. [CrossRef] [PubMed]
235. Park, L.; Anrather, J.; Girouard, H.; Zhou, P.; Iadecola, C. Nox2-Derived Reactive Oxygen Species Mediate Neurovascular Dysregulation in the Aging Mouse Brain. *J. Cereb. Blood Flow Metab.* **2007**, *27*, 1908–1918. [CrossRef] [PubMed]
236. Toth, P.; Tarantini, S.; Tucsek, Z.; Ashpole, N.M.; Sosnowska, D.; Gautam, T.; Ballabh, P.; Koller, A.; Sonntag, W.E.; Csiszar, A.; et al. Resveratrol treatment rescues neurovascular coupling in aged mice: Role of improved cerebromicrovascular endothelial function and downregulation of NADPH oxidase. *Am. J. Physiol. Heart Circ. Physiol.* **2014**, *306*, H299–H308. [CrossRef]
237. Csiszar, A.; Ungvari, Z.; Edwards, J.G.; Kaminski, P.; Wolin, M.S.; Koller, A.; Kaley, G. Aging-induced phenotypic changes and oxidative stress impair coronary arteriolar function. *Circ. Res.* **2002**, *90*, 1159–1166. [CrossRef]
238. Toth, P.; Tarantini, S.; Ashpole, N.M.; Tucsek, Z.; Milne, G.L.; Valcarcel-Ares, N.M.; Menyhart, A.; Farkas, E.; Sonntag, W.E.; Csiszar, A.; et al. IGF-1 deficiency impairs neurovascular coupling in mice: Implications for cerebromicrovascular aging. *Aging Cell* **2015**, *14*, 1034–1044. [CrossRef]
239. Szklarczyk, D.; Gable, A.L.; Nastou, K.C.; Lyon, D.; Kirsch, R.; Pyysalo, S.; Doncheva, N.T.; Legeay, M.; Fang, T.; Bork, P.; et al. The STRING database in 2021: Customizable protein-protein networks, and functional characterization of user-uploaded gene/measurement sets. *Nucleic Acids Res.* **2021**, *49*, D605–D612. [CrossRef]
240. Szklarczyk, D.; Gable, A.L.; Lyon, D.; Junge, A.; Wyder, S.; Huerta-Cepas, J.; Simonovic, M.; Doncheva, N.T.; Morris, J.H.; Bork, P.; et al. STRING v11: Protein–protein association networks with increased coverage, supporting functional discovery in genome-wide experimental datasets. *Nucleic Acids Res.* **2018**, *47*, D607–D613. [CrossRef]
241. Szklarczyk, D.; Morris, J.H.; Cook, H.; Kuhn, M.; Wyder, S.; Simonovic, M.; Santos, A.; Doncheva, N.T.; Roth, A.; Bork, P.; et al. The STRING database in 2017: Quality-controlled protein-protein association networks, made broadly accessible. *Nucleic Acids Res.* **2017**, *45*, D362–D368. [CrossRef]
242. Marie, C.; Pedard, M.; Quirie, A.; Tessier, A.; Garnier, P.; Totoson, P.; Demougeot, C. Brain-derived neurotrophic factor secreted by the cerebral endothelium: A new actor of brain function? *J. Cereb. Blood Flow Metab.* **2018**, *38*, 935–949. [CrossRef]
243. Amenta, F.; Di Tullio, M.A.; Tomassoni, D. Arterial Hypertension and Brain Damage—Evidence from Animal Models (Review). *Clin. Exp. Hypertens.* **2003**, *25*, 359–380. [CrossRef] [PubMed]
244. Tomiga, Y.; Higaki, Y.; Anzai, K.; Takahashi, H. Behavioral defects and downregulation of hippocampal BDNF and nNOS expression in db/db mice did not improved by chronic TGF-β2 treatment. *Front. Physiol.* **2022**, *13*, 969480. [CrossRef] [PubMed]
245. Guo, S.; Kim, W.J.; Lok, J.; Lee, S.R.; Besancon, E.; Luo, B.H.; Stins, M.F.; Wang, X.; Dedhar, S.; Lo, E.H. Neuroprotection via matrix-trophic coupling between cerebral endothelial cells and neurons. *Proc. Natl. Acad. Sci. USA* **2008**, *105*, 7582–7587. [CrossRef]
246. Navaratna, D.; Guo, S.Z.; Hayakawa, K.; Wang, X.; Gerhardinger, C.; Lo, E.H. Decreased cerebrovascular brain-derived neurotrophic factor-mediated neuroprotection in the diabetic brain. *Diabetes* **2011**, *60*, 1789–1796. [CrossRef] [PubMed]
247. Numakawa, T.; Suzuki, S.; Kumamaru, E.; Adachi, N.; Richards, M.; Kunugi, H. BDNF function and intracellular signaling in neurons. *Histol. Histopathol.* **2010**, *25*, 237–258. [CrossRef] [PubMed]
248. Meuchel, L.W.; Thompson, M.A.; Cassivi, S.D.; Pabelick, C.M.; Prakash, Y.S. Neurotrophins induce nitric oxide generation in human pulmonary artery endothelial cells. *Cardiovasc. Res.* **2011**, *91*, 668–676. [CrossRef] [PubMed]
249. Rudge, J.S.; Li, Y.; Pasnikowski, E.M.; Mattsson, K.; Pan, L.; Yancopoulos, G.D.; Wiegand, S.J.; Lindsay, R.M.; Ip, N.Y. Neurotrophic factor receptors and their signal transduction capabilities in rat astrocytes. *Eur. J. Neurosci.* **1994**, *6*, 693–705. [CrossRef]
250. Vignoli, B.; Battistini, G.; Melani, R.; Blum, R.; Santi, S.; Berardi, N.; Canossa, M. Peri-Synaptic Glia Recycles Brain-Derived Neurotrophic Factor for LTP Stabilization and Memory Retention. *Neuron* **2016**, *92*, 873–887. [CrossRef]
251. Grammas, P.; Moore, P.; Weigel, P.H. Microvessels from Alzheimer's disease brains kill neurons in vitro. *Am. J. Pathol.* **1999**, *154*, 337–342. [CrossRef]
252. Yoder, M.C. Human endothelial progenitor cells. *Cold Spring Harb. Perspect. Med.* **2012**, 2, a006692. [CrossRef]
253. Lee, S.T.; Chu, K.; Jung, K.H.; Park, H.K.; Kim, D.H.; Bahn, J.J.; Kim, J.H.; Oh, M.J.; Lee, S.K.; Kim, M.; et al. Reduced circulating angiogenic cells in Alzheimer disease. *Neurology* **2009**, *72*, 1858–1863. [CrossRef] [PubMed]
254. Custodia, A.; Ouro, A.; Romaus-Sanjurjo, D.; Pias-Peleteiro, J.M.; de Vries, H.E.; Castillo, J.; Sobrino, T. Endothelial Progenitor Cells and Vascular Alterations in Alzheimer's Disease. *Front. Aging Neurosci.* **2021**, *13*, 811210. [CrossRef] [PubMed]
255. Zhang, S.; Zhi, Y.; Li, F.; Huang, S.; Gao, H.; Han, Z.; Ge, X.; Li, D.; Chen, F.; Kong, X.; et al. Transplantation of in vitro cultured endothelial progenitor cells repairs the blood-brain barrier and improves cognitive function of APP/PS1 transgenic AD mice. *J. Neurol. Sci.* **2018**, *387*, 6–15. [CrossRef] [PubMed]

MDPI AG
Grosspeteranlage 5
4052 Basel
Switzerland
Tel.: +41 61 683 77 34

Biomolecules Editorial Office
E-mail: biomolecules@mdpi.com
www.mdpi.com/journal/biomolecules

www.ingramcontent.com/pod-product-compliance
Lightning Source LLC
Chambersburg PA
CBHW062041220926
51850CB00010B/2633
* 9 7 8 3 7 2 5 8 6 1 0 5 7 *